Mediterranean Foods

Composition and Processing

Mediterranean Foods

Composition and Processing

Editors

Rui M.S. Cruz & Margarida C. Vieira

MeditBio and Department of Food Engineering, ISE,
University of Algarve, Portugal

CRC Press
Taylor & Francis Group
Boca Raton London New York

CRC Press is an imprint of the
Taylor & Francis Group, an **informa** business

A SCIENCE PUBLISHERS BOOK

CRC Press
Taylor & Francis Group
6000 Broken Sound Parkway NW, Suite 300
Boca Raton, FL 33487-2742

First issued in paperback 2021

© 2017 by Taylor & Francis Group, LLC
CRC Press is an imprint of Taylor & Francis Group, an Informa business

No claim to original U.S. Government works

ISBN-13: 978-0-367-78277-1 (pbk)
ISBN-13: 978-1-4987-4089-0 (hbk)

Library of Congress Cataloging-in-Publication Data

Names: Cruz, Rui M. S., editor. | Vieira, Maria Margarida Cortez, editor.
Title: Mediterranean foods : composition and processing / editors, Rui M.S. Cruz & Margarida C. Vieira.
Description: Boca Raton, FL : Taylor & Francis, [2016] | "A Science Publishers book." | Includes bibliographical references and index.
Identifiers: LCCN 2016030678| ISBN 9781498740890 (hardback) | ISBN 9781498740906 (e-book)
Subjects: LCSH: Food--Mediterranean Region--Composition. | Food--Mediterranean Region--Analysis. | Food industry and trade--Mediterranean Region.
Classification: LCC TX531 .M43 2016 | DDC 664/.07--dc23
LC record available at https://lccn.loc.gov/2016030678

Visit the Taylor & Francis Web site at
http://www.taylorandfrancis.com

and the CRC Press Web site at
http://www.crcpress.com

Preface

The Mediterranean region is well known for its rich culinary history. While the literature available at present tends to focus mainly on the nutritional, gastronomic and/or health aspects of Mediterranean cuisine, this book aims at a more scientific approach.

Each of the eleven chapters that follow present and discuss the composition and characteristics of a specific food product from the Mediterranean basin along with the specific processing methodology applied to produce and preserve it in this part of the world.

Combining the efforts of approximately thirty experts of varying nationalities and diverse backgrounds, this book is an ideal reference source for government, industry, academia and professionals working in the areas of nutrition, food science and technology.

The editors would like to acknowledge all contributors for their invaluable work towards accomplishing this project.

Rui M.S. Cruz
Margarida C. Vieira

Contents

Preface   v

List of Contributors   xiii

1. **Characteristics of Some Important Italian Cheeses:**   1
   Parmigiano Reggiano, Grana Padano, Mozzarella,
   Mascarpone and Ricotta
   Germano Mucchetti, Alessandro Pugliese and Maria Paciulli

   1.1 Introduction   1
   1.2 Parmigiano Reggiano and Grana Padano   2
        1.2.1 Characteristics   2
        1.2.2 Chemical composition   4
        1.2.3 Microbiological characteristics   6
        1.2.4 Technology   7
   1.3 Mozzarella   11
        1.3.1 Characteristics   11
        1.3.2 Chemical composition   14
        1.3.3 Microbiological characteristics   16
        1.3.4 Technology   17
   1.4 Mascarpone   18
        1.4.1 Characteristics   18
        1.4.2 Chemical composition   19
        1.4.3 Microbiological characteristics   20
        1.4.4 Technology   20
   1.5 Ricotta   22
        1.5.1 Characteristics   22
        1.5.2 Chemical composition   24
        1.5.3 Microbiological characteristics   24
        1.5.4 Technology   26
   References   29

2. **Croatian Meat Products with Mediterranean Influence:**   35
   Dry-cured Hams and Other Products
   Helga Medić and Nives Marušić Radovčić

   2.1 Introduction   35

2.2	Types of Croatian dry-cured hams	35
	2.2.1 Dalmatinski pršut	35
	2.2.2 Drniški pršut	39
	2.2.3 Istarski pršut	41
	2.2.4 Krčki pršut	45
	2.2.5 Quality parameters	48
2.3	Other dry-cured products	56
	2.3.1 Dalmatinski šokol	56
	2.3.2 Dalmatinska pečenica	56
	2.3.3 Dalmatinska panceta	56
2.4	Recent and future strategies	57
	2.4.1 Strategies based on production	57
	References	61

3. Fish Products from South Portugal: Dried *Litão*, Tuna *Muxama*, and Canned Mackerel — 65
Eduardo Esteves

3.1	Introduction	65
3.2	Dried *litão*	68
	3.2.1 Processing	70
	3.2.2 Composition	71
3.3	Tuna *muxama*	75
	3.3.1 Processing	76
	3.3.2 Composition	79
3.4	Canned mackerel	81
	3.4.1 Processing	83
	3.4.2 Composition	86
3.5	Conclusion	95
	References	96

4. Egyptian Legumes and Cereal Foods: Traditional and New Methods for Processing — 102
Mohammed Hefni and *Cornelia M. Witthöft*

4.1	Introduction	102
4.2	Egyptian food consumption data	104
4.3	Processing methods for legumes and cereals	104
	4.3.1 Soaking	105
	4.3.2 Germination	105
	4.3.3 Fermentation	107
	4.3.4 Canning	108
	4.3.5 Freezing	109
4.4	Traditional Egyptian foods	109
	4.4.1 Bean stew foul	110
	4.4.2 Falafel	110

4.4.3 Nabet soup   110
4.4.4 Koshari   110
4.4.5 Baladi bread   112
4.5 Chemical composition   113
4.6 Development of novel functional cereal and   114
     legume ingredients
     4.6.1 Cookies with increased folate content   116
     4.6.2 Baladi bread with increased folate content   116
     4.6.3 Canned germinated faba beans—foul with increased   117
            folate content
References   117

5. **Slovenian Dairy Products**   **121**
    Andreja Čanžek Majhenič and *Petra Mohar Lorbeg*

5.1 Introduction   121
5.2 Mohant cheese (PDO)   122
     5.2.1 Processing   122
     5.2.2 Characteristics   123
5.3 Bohinj cheese   123
     5.3.1 Processing   124
     5.3.2 Characteristics   125
5.4 Tolminc cheese (PDO)   125
     5.4.1 Processing   126
     5.4.2 Characteristics   126
5.5 Nanos cheese (PDO)   127
     5.5.1 Processing   127
     5.5.2 Characteristics   128
5.6 Bovec cheese   129
     5.6.1 Processing   129
     5.6.2 Characteristics   130
5.7 Karst ewe cheese   131
     5.7.1 Processing   131
     5.7.2 Characteristics   132
5.8 Dolenjski ewe cheese   132
     5.8.1 Processing   132
     5.8.2 Characteristics   133
5.9 Trnič cheese   133
     5.9.1 Processing   135
     5.9.2 Characteristics   135
5.10 Other dairy products   136
     5.10.1 Pregreta smetana, škrlupec (overheated cream)   136
     5.10.2 Skuta   137
     5.10.3 Kislo mleko (sour milk)   139
References   140

6. **Spanish Dry-cured Hams and Fermented Sausages**   **141**
   Jorge Ruiz, María J. Pagán, Purificación García-Segovía and
   Javier Martínez-Monzó

   6.1   Introduction   141
   6.2   Dry-cured ham   141
         6.2.1 Types of Spanish dry-cured hams   142
         6.2.2 Quality characteristics in fresh meat   143
   6.3   Fermented sausages   155
         6.3.1 Processing   157
         6.3.2 Quality characteristics   164
   References   165

7. **Slovenian Honey and Honey Based Products**   **171**
   Mojca Korošec, Rajko Vidrih and *Jasna Bertoncelj*

   7.1   Introduction   171
   7.2   Beekeeping technology   172
   7.3   Slovenian honey   174
         7.3.1 Types of honey   174
         7.3.2 Composition   176
         7.3.3 Antibacterial activity   183
         7.3.4 Protected honeys   185
         7.3.5 Traditional products with honey   187
   References   191

8. **Portuguese Sheep Cheeses**   **196**
   Gil Fraqueza

   8.1   Introduction   196
   8.2   Characteristics of sheep milk   196
         8.2.1 Factors affecting the production and composition of   201
                sheep milk
   8.3   Types of Cheese   201
         8.3.1 Composition   202
         8.3.2 Traditional Portuguese sheep cheeses   202
         8.3.3 Processing   208
         8.3.4 Milk composition and cheese yield   218
   References   220

9. **Portuguese Galega Kale**   **226**
   Teresa R.S. Brandão, Fátima A. Miller, Elisabete M.C. Alexandre,
   Cristina L.M. Silva and *Sara M. Oliveira*

   9.1   Introduction   226
   9.2   Composition   227
   9.3   Processing   229
         9.3.1 Fresh cut   230

9.3.2 Low temperature preservation   230
9.3.3 MAP   233
9.3.4 Drying   233
9.3.5 Cooking   235
9.3.6 Processing impact on quality   236
9.4 Final remarks   237
Acknowledgements   237
References   237

10. Turkish Meat Products   240
Ismet Ozturk, Salih Karasu, Osman Sagdic and *Hasan Yetim*

10.1 Introduction   240
10.2 Sucuk   241
     10.2.1 Production   241
     10.2.2 Quality characteristics   245
     10.2.3 Microbiological properties   247
10.3 Pastirma   249
     10.3.1 Production   249
     10.3.2 Quality characteristics   251
     10.3.3 Microbiological properties   252
10.4 Kavurma   253
     10.4.1 Production   254
     10.4.2 Quality characteristics   255
     10.4.3 Microbiological properties   256
10.5 Kofte   256
     10.5.1 Production   257
     10.5.2 Quality characteristics   259
     10.5.3 Microbiological properties   260
References   263

11. Greek Dairy Products   267
Ekaterini Moschopoulou and *Golfo Moatsou*

11.1 Introduction   267
11.2 Cheese varieties   268
     11.2.1 Brined cheeses   268
     11.2.2 Hard and very hard type cheeses   276
     11.2.3 Gruyere type cheeses   284
     11.2.4 Pasta Filata and other semi hard cheese varieties 289
     11.2.5 Soft-type cheeses with spreadable texture   296
     11.2.6 Whey cheeses   300
11.3 Greek fermented milk products   304
     11.3.1 Traditional yoghurt   304
     11.3.2 Strained yoghurt   306
     11.3.3 Set type yoghurt from goat or sheep milk   309

11.3.4 Mixed fermented milk/cereal products   310
11.4 Miscellaneous dairy products   312
References   315

Index   **321**

List of Contributors

Alessandro Pugliese
Department of Food Science, University of Parma, Parco Area delle Scienze, 59/A 43124 Parma, Italy.
E-mail: alessandropugliese82@gmail.com

Andreja Čanžek Majhenič
Chair of Dairy Science, Biotechnical Faculty, University of Ljubljana, Groblje 3, Slovenia.
E-mail: andreja.canzek@bf.uni-lj.si

Cornelia M. Witthöft
Department of Chemistry and Biomedical Sciences, Faculty of Health and Life Sciences, Linnaeus University, SE-391 82 Kalmar, Sweden.
E-mail: cornelia.witthoft@lnu.se

Cristina L.M. Silva
CBQF - Centre of Biotechnology and Fine Chemistry, State Associated Laboratory, Faculty of Biotechnology, Catholic University of Portugal, Rua Dr. António Bernardino Almeida, 4200-072 Porto, Portugal.
E-mail: clsilva@porto.ucp.pt

Eduardo Esteves
Department of Food Engineering, Institute of Engineering and CCMAR - Centre of Marine Sciences, University of Algarve, 8005-139 Faro, Portugal.
E-mail: eesteves@ualg.pt

Ekaterini Moschopoulou
Laboratory of Dairy Research, Department of Food Science and Human Nutrition, Agricultural University of Athens, 11855 Athens, Greece.
E-mail: catmos@aua.gr

Elisabete M.C. Alexandre
QOPNA - Organic Chemistry, Natural and Agro-Food Products Research Unit, Department of Chemistry, University of Aveiro, 3810-193 Aveiro, Portugal.
E-mail: elisabete.alexandre@ua.pt

Fátima A. Miller
CBQF - Centre of Biotechnology and Fine Chemistry, State Associated Laboratory, Faculty of Biotechnology, Catholic University of Portugal, Rua Dr. António Bernardino Almeida, 4200-072 Porto, Portugal.
E-mail: fmiller@porto.ucp.pt

Germano Mucchetti
Department of Food Science, University of Parma, Parco Area delle Scienze, 59/A, 43124 Parma, Italy.
E-mail: germano.mucchetti@unipr.it

Gil V.C. Fraqueza
Department of Food Engineering, Institute of Engineering and CCMAR - Centre of Marine Sciences, University of Algarve, 8005-139 Faro, Portugal.
E-mail: gfraque@ualg.pt

Golfo Moatsou
Laboratory of Dairy Research, Department of Food Science and Human Nutrition, Agricultural University of Athens, 11855 Athens, Greece.
E-mail: mg@aua.gr

Hasan Yetim
Department of Food Engineering, Faculty of Engineering, University of Erciyes, 38039 Kayseri, Turkey.
E-mail: hyetim@erciyes.edu.tr

Helga Medić
Faculty of Food Technology and Biotechnology, University of Zagreb, Pierottijeva 6, 10000 Zagreb, Croatia.
E-mail: hmedic@pbf.hr

Ismet Ozturk
Department of Food Engineering, Faculty of Engineering, University of Erciyes, 38039 Kayseri, Turkey.
E-mail: ismet@erciyes.edu.tr

Jasna Bertoncelj
Department of Food Science and Technology, Biotechnical Faculty, University of Ljubljana, Groblje 3, Slovenia.
E-mail: jasna.bertoncelj@bf.uni-lj.si

Javier Martínez-Monzó
Department of Food Technology, Polytechnic University of Valencia Camino de Vera, s/n, 46022 Valencia, Spain.
E-mail: xmartine@tal.upv.es

Jorge Ruiz
Department of Food Science, University of Copenhagen Rolighedsvej 30, 1958 Frederiksberg C, Denmark.
E-mail: jorgeruiz@food.ku.dk

María J. Pagán
Department of Food Technology, Polytechnic University of Valencia Camino de Vera, s/n, 46022 Valencia, Spain.
E-mail: jpagan@tal.upv.es

Maria Paciulli
Department of Food Science, University of Parma, Parco Area delle Scienze, 59/A 43124 Parma, Italy.
E-mail: paciulli.maria@gmail.com

Mohammed Hefni
Food Industries Department, Faculty of Agriculture, Mansoura University 35516, P.O. Box 46, Mansoura, Egypt.
E-mail: mohammed.hefni@mans.edu.eg

Mojca Korošec
Department of Food Science and Technology, Biotechnical Faculty, University of Ljubljana Groblje 3, Slovenia.
E-mail: mojca.korosec@bf.uni-lj.si

Nives Marušić Radovčić
Faculty of Food Technology and Biotechnology, University of Zagreb, Pierottijeva 6, 10000 Zagreb, Croatia.
E-mail: nmarusic@pbf.hr

Osman Sagdic
Department of Food Engineering, Faculty of Chemical and Metallurgical Engineering, Yildiz Technical University, 34210 Istanbul, Turkey.
E-mail: osagdic@yildiz.edu.tr

Petra Mohar Lorbeg
Institute of Dairy Science and Probiotics, Biotechnical Faculty, University of Ljubljana Groblje 3, Slovenia.
E-mail: petra.mohar@bf.uni-lj.si

Purificación García-Segovía
Department of Food Technology, Polytechnic University of Valencia Camino de Vera, s/n, 46022 Valencia, Spain.
E-mail: pugarse@tal.upv.es

Rajko Vidrih
Department of Food Science and Technology, Biotechnical Faculty, University of Ljubljana, Groblje 3, Slovenia.
E-mail: rajko.vidrih@bf.uni-lj.si

Salih Karasu
Department of Food Engineering, Faculty of Chemical and Metallurgical Engineering,Yildiz Technical University, 34210 Istanbul, Turkey.
E-mail: skarasu@yildiz.edu.tr

Sara M. Oliveira
CBQF - Centre of Biotechnology and Fine Chemistry, State Associated Laboratory, Faculty of Biotechnology, Catholic University of Portugal, Rua Dr. António Bernardino Almeida, 4200-072 Porto, Portugal.
E-mail: soliveira@porto.ucp.pt

Teresa R.S. Brandão
CBQF - Centre of Biotechnology and Fine Chemistry, State Associated Laboratory, Faculty of Biotechnology, Catholic University of Portugal, Rua Dr. António Bernardino Almeida, 4200-072 Porto, Portugal.
E-mail: tbrandao@porto.ucp.pt

CHAPTER 1

Characteristics of Some Important Italian Cheeses

Parmigiano Reggiano, Grana Padano, Mozzarella, Mascarpone and Ricotta

Germano Mucchetti,* Alessandro Pugliese and *Maria Paciulli*

1.1 Introduction

More than 73% out of 13 million tonnes of milk processed in Italy in 2013 was transformed into cheeses (Table 1.1) with 56.8% of cheese milk used to produce cheeses with a Protected Designation of Origin (PDO), according to CE Regulation 1151/2012 (EU 2012). PDO cheeses make up 45.5% of the total Italian cheese production. There are 50 Italian PDO cheeses (DOOR 2015).

Yet despite the abundance of cheese varieties reflecting the heritage of the Italian dairy culture, only some of them are recognized and widely exported worldwide. Within the limits of a book chapter, it is impossible to illustrate all of the most famous Italian cheeses.

The characteristics and processes of Parmigiano Reggiano, Grana Padano, Mozzarella, Mascarpone and Ricotta will be discussed. The first three cheeses cover about 60% of Italian cheese production (Table 1.2) and the other two are representative of the traditional Italian way of adding value to cream and whey. Furthermore, these five products combined represent more than 70% of the Italian cheese export (Assolatte 2014) and so deserve to be better acknowledged.

Department of Food Science, University of Parma, Parma, Italy.
* Corresponding author: germano.mucchetti@unipr.it

Table 1.1 Milk processing in Italy in 2013 (Assolatte 2014).

Product variety	Milk Products	Milk
Cheeses obtained from	ton	ton
cow milk	891982	8878146
sheep milk	54800	319680
goat milk	6144	58366
buffalo milk	49602	193299
a mixture of milk from different species	58419	436897
Total cheeses	1060947	9886388
PDO cheeses	*483224*	*5612806*
ratio PDO cheeses/total cheeses (%)	*45.5*	*56.8*
Fluid milk (pasteurized, UHT, ESL)	2664500	2664500
Other milk products	333808	905423
Total (cheeses, fluid milk, other)	4059255	13456311

Table 1.2 Actual cheese yield (kg cheese/100 kg milk * 100) (Assolatte 2014).

Cheese variety	Cheese (ton)	Milk (ton)	Yield
Mozzarella	220030	1760240	12.5
Mozzarella di bufala campana DOP	37302	144099	25.9
Mozzarella di latte di bufala non DOP	12300	49200	25.0
Grana Padano	173917	2418090	7.2
Parmigiano Reggiano	132189	1905522	6.9
Other hard cheeses	13750	192500	7.1
Mascarpone	50000	137500[a]	36.4

[a] = cream instead of milk.

1.2 Parmigiano Reggiano and Grana Padano

1.2.1 Characteristics

Parmigiano Reggiano (PR) and Grana Padano (GP) PDO are hard cooked cheeses produced with raw, partially skimmed milk added to natural whey starter, and ripened for a long time.

PR, formerly called Grana Parmigiano Reggiano until 1951, and GP derived the name from their typical granular and flaky structure, as Grana in Italian means "grain". The granular structure is given by the interactions among the effects of casein coagulation, coagulum cutting and curd particles cooking. It is a common belief that some of the best final structural cheese characteristics are largely dependent on the fresh cheese curd properties. Flaky or brittle structure can be instrumentally measured. Data from the stress-strain curves of PR cheeses with different ripening time showed two main properties: i) the strain at fracture decreases with age, as proteolysis makes the cheese more brittle; ii) the difference between the strain at the linear limit and the fracture strain decreases with age, suggesting that cracks within the cheese body are faster in older than in younger cheese (Noel et al. 1996).

The description of properties and processing methods of PR and GP cheeses made by the official standards (EU 2009a, EU 2011a, MIPAAF 2014) outlines the key factors of their quality.

According to point 3.2 of PR standard (EU 2009a), "*PR shall have the following characteristics: cylindrical in form with a slightly convex or virtually straight heel, with flat faces with a slightly raised edge; dimensions: diameter of the flat faces 35 to 45 cm, heel height 20 to 26 cm; minimum weight of each wheel of cheese: 30 kg; external appearance: crust of a natural straw colour; thickness of the crust: approximately 6 mm; colour of the body of the cheese: between light straw-coloured and straw-coloured; characteristic aroma and taste: fragrant, delicate, flavoursome but not pungent; texture of the body of the cheese: fine, grained, flaky; fat content per dry matter: 32% minimum*". The point 5.2 of the standard highlights: "*The specific characteristics of PR cheese are the structure of the body of the cheese, fine grained and flaky, the fragrant aroma and delicate taste, which is flavoursome without being pungent, and its high level of solubility and digestibility. These characteristics stem from the specific features of and selection criteria applied to the milk used raw on a daily basis in copper vats, coagulated with calf rennet with a high chymosin content, from the curing in saturated brine and prolonged maturation period*".

Grana Padano standard (EU 2011a) is similar for many aspects and "*The specificity of 'Grana Padano PDO' may be ascribed to the following elements: size and weight of the cheese; particular morphology of the paste, linked to the production technique, characterised by a granular texture which gives rise to its typical flakiness; white or straw colour, with a delicate flavour and fragrant aroma, due essentially to the widespread use of waxy corn in the fodder fed to the cattle; water and fat content largely similar to the protein content, high level of natural breakdown of the proteins in peptones, peptides and free amino acids; resistance to lengthy ripening, even beyond 20 months*".

The standards (MIPAAF 2014) establish the criterium according to which the origin of the cheese is ascertained on the basis of the specific free amino acid (FAA) content (both for PR and GP) and/or on the basis of specific isotopic ratios (only GP). Thresholds for FAA content of PR and GP cheeses (Table 1.3) are reported in technical literature.

The lack of the identification mark, printed on the rind of PR and GP cheeses at the origin, when marketed as grated cheese or as rindless pieces, makes the risk of falsification high. The GP standard forces dairies to immediately place the grated cheese in packs bearing the designation of

Table 1.3 Free amino acids (FAA) suitable for quality assessment of Parmigiano Reggiano and Grana Padano PDO cheeses.

Amino acid	Threshold value (%)[a]	
	PR	GP
Total content of FAA	≥15.0	
Glutamic acid		≥16.0
Lysine		≥10.8
Serine	≥4.0	≥3.4
Glutamine	≤2.5	≤2.3
Ornithine	≤2.5	≤2.0
γ-Aminobutyric acid	≤0.6	≤0.6
Source	Hogenboom and Pellegrino 2007	Cattaneo et al. 2008

[a] = All data expressed as % of total FAA, except total content of FAA (% of total cheese protein).

origin because *"is more likely to ensure the authenticity of the grated product, which by nature is more difficult to identify than a whole cheese"* (EU 2011a). Furthermore, a non immediate packing would increase the risk of damage to the quality of the grated cheese (oxidation, drying, loss of volatile molecules, etc.) because of its highest surface to volume ratio.

The need to prevent fraud introduced supplementary duties for grated PR or GP cheese (MIPAAF 2014), as the maximum percent of rind allowed (18%), the range of moisture content from 25 to 35%, the mesh of the grated cheese (a maximum 25% is able to pass through an opening with a 0.5 mm diameter).

1.2.2 Chemical composition

The accurate definition of the chemical and physical composition of mature PR and GP cheeses is difficult because of the large range of ripening time beyond the shortest period (9 and 12 months, respectively) foreseen by the PDO regulations. PR cheese ripening may be prolonged up to 36 months, while GP often overcomes 20 months. Within this extra ripening time, cheese moisture continues to decrease, while products resulting from proteolysis, lypolysis and other biochemical pathways increase (Sforza et al. 2012, Malacarne et al. 2009).

PR and GP cheeses have many common characteristics (Table 1.4) and some discriminant properties; the latter mostly given by the methods of milk collection, milk management before coagulation and finally by the ripening conditions (Gatti et al. 2014).

The moisture content ranges from 35 to 28%, according to the ripening time, while the ratio of protein to fat in the cheeses is in the order of

Table 1.4 Chemical composition of Grana Padano and Parmigiano Reggiano cheese (g/100 g) (Mucchetti and Neviani 2006).

Cheese component	GP	PR
Moisture	31.3 ± 0.9	30.8 ± 0.9
Fat	29.7 ± 1.1	28.4 ± 1.4
Protein	32.9 ± 1.3	33.0 ± 0.9
Lactose	0	0
Galactose	0	0
Lactic acid	1.3 ± 0.0	1.6 ± 0.1
Ash	4.5 ± 0.3	4.6 ± 0.3
NaCl	1.6 ± 0.2	1.4 ± 0.2
Calcium	1.2 ± 0.1	1.2 ± 0.1
Phosphorous	0.7 ± 0.1	0.7 ± 0.1
Copper (ppm)	4.9 ± 1.6	8.3 ± 1.8
Proteolysis Index[a]	33.2 ± 3.1	33.8 ± 2.8
pH	5.7 ± 0.1	5.3

[a] = nitrogen water soluble/total nitrogen x 100.

1.10 ± 0.04, according to the cheese milk characteristics and the amount of fat losses into the whey. Both PR and GP cheeses are "naturally" free of lactose and galactose, as sugars are completely fermented by starter bacteria. The amount of lactic acid (about 1.5%) is nearly constant during ripening in regular cheeses. Both stereo isomers of lactic acid are present, with an L to D ratio of approximately 1.1 (De Dea Lindner et al. 2008). The average citric acid content is about 50 mg/100 g (Gatti et al. 2014).

Bacci et al. (2002) proposed a threshold of 2 mg propionic acid/100 g cheese as a marker for the evidence of a propionic blowing defect in PR cheese. The threshold value of butyric acid, produced by the lactate fermentation achieved by clostridia or by lipolysis, is usually lower than 50 mg/100 g (Tosi et al. 2008).

The ash content of PR and GP cheeses is about 4.5%, and the main contributors are sodium chloride (NaCl), calcium and phosphorous. NaCl content is about 1.5% with a value of about 4.5 to 5.5% when expressed as salt in moisture, and it contributes to lower the water activity (a_w).

A survey on a_w values of GP cheeses ripened from 5 to more than 24 months showed values from 0.935 to 0.875 (Pellegrino, personal communication). Monfredini et al. (2012) showed a difference of a_w between the under rind zone of Trentino Grana aged 9 months and the inner zone, measuring values from 0.908 to 0.936.

The synergistic actions of several enzymes (rennet, plasmin, cathepsins, bacterial proteases and peptidases) within the long ripening time is responsible for the casein proteolysis, and it is fundamental to obtain the characteristics of both the cheeses, as indicated by the proteolysis index values (Table 1.4). The contribution of rennet to casein proteolysis during cooked cheese ripening is controversial and likely depends on rennet inactivation by the heat load applied during cheese-making (Mucchetti and Neviani 2006, Bansal et al. 2009).

The traditional proteolysis index shows that more than one third of cheese proteins are hydrolyzed to an extent that makes peptides soluble in water. β-casein is largely hydrolyzed by plasmin both in PR and in GP cheese: Pizzano et al. (2000) and Gaiaschi et al. (2001) measured 2 g of residual β-casein/100 g of cheese within 20 and 15 months of ripening of PR and GP cheeses, respectively. Considering an initial amount of about 14 g of β-casein, the hydrolysis went up by 85%.

αS-casein is more resistant to proteolysis, remaining at an amount of about 4.5 and 3.1 g/100 g after 15 and 22 months of GP ripening, respectively (Gaiaschi et al. 2000).

A large part of peptides hydrolyzed by proteases are further hydrolyzed by Lactic Acid Bacteria (LAB) peptidases to FAA. Among FAA, the non-protein amino acid (AA) pyroglutamic acid (pGlu), derived from the cyclization of glutamic acid, is present up to 0.6 g/100 g cheese, and it was proposed as a possible marker of age ripening (Mucchetti et al. 2000, Mucchetti et al. 2002a, Masotti et al. 2010). On the contrary, some peptides are resistant to further proteolysis (Sforza et al. 2012). Interactions between some amino acids and lactate led to the accumulation of aminoacyl derivatives, believed to be positively involved in the flavour of aged cheese (Sforza et al. 2009).

Free fatty acids (FFA) are relatively abundant (up to 3.5 g/100 g) in PR and GP ripened cheeses, occurring mainly in the outer part of the cheese wheel (Malacarne et al. 2009). Short chain fatty acids (SCFA) (up to C 8:0) represent less than 7.1% of total FFA (Malacarne et al. 2009, Prandini et al. 2009), a fraction lower than the original in milk fat. This proportion shows that lipases active in cheese during ripening are not responsible for a selective hydrolysis of piquant FFA and the low ratio of SCFA to total FFA explains how Italian hard cooked cheeses are not usually recognized as lypolyzed cheeses (Gatti et al. 2014).

1.2.3 Microbiological characteristics

Despite the long time of ripening, both PR and GP cheeses harbour a viable LAB microbiota, ranging from 3.0 to 7.0 log cfu/g, inversely to cheese age (from nine to 24 months) and mainly composed of non starter LAB (NSLAB) originating from the raw milk (Gatti et al. 2014).

An epifluorescence microscopy-based count of viable cells present in PR cheese samples aged 12 to 20 months showed values approximately ten times lower than the total count (De Dea Lindner et al. 2008), suggesting that a large number of dead cells are not lysed/autolysed during the time of ripening, even though it is well known that lysis of LAB starter cells starts within a few days of cheese-making.

Microbial counts in cheese differ according to the spatial coordinates of the samples taken from the cheese wheel, but results are conflicting. Reason for this discrepancy are not clear, as the cheesemaking technology and the consequent stress are often not fully described. Viable microbial counts of ripened cheeses are the result of the ability of the microbes to survive the severe heat and acid stress during the first steps of cheesemaking, and to grow using energy sources other than lactose during the ripening, in a condition of progressive decrease of a_w.

The LAB species most frequently found in Italian hard cooked cheeses are *Lactobacillus rhamnosus*, *Lactobacillus casei*, *Lactobacillus paracasei* and *Pediococcus acidilactici* (Gala et al. 2008, Gatti et al. 2008, Solieri et al. 2012). Thermophilic homo and heterofermentative LAB, most of them belonging to whey starter microbiota, were occasionally found (Gala et al. 2008, Monfredini et al. 2012, Solieri et al. 2012).

Non LAB microbiota of ripened hard cooked cheeses occurs in low quantities and usually is associated with alterative genera or species. One of the most relevant defects of PR and GP is the late blowing mainly caused by *Clostridium tyrobutirycum*, contaminating milk via silage, forage and bedding. Studying the prevalence of different species of clostridia in milk and GP curd Feligni et al. (2014), showed that beside *C. tyrobutirycum*, other presences of *Clostridium butyricum*, *Clostridium sporogenes*, *Clostridium beijerinckii*, *Clostridium tertium* and *Clostridium perfringens* were usually seen. To contrast spore germination and vegetative cell growth, GP producers add lysozyme to milk (Carini et al. 1985), while PR producers do not add any additive to milk.

Despite PR and GP are raw milk cheeses, the usual processing conditions make the cheese safe even when pathogenic bacteria such as *Listeria monocytogenes*, *Salmonella* spp., *Staphylococcus aureus* and *Escherichia coli* O157:H7 are experimentally added to raw milk during challenge tests (Panari et al. 2001, Mucchetti 2005). *In vitro* studies (Ercolini et al. 2004) showed that the heat stress caused by curd cooking is not the only responsible factor for early bacterial death.

1.2.4 Technology

The PR and GP cheesemaking process is summarized in Fig. 1.1. While GP milk is frequently collected only once a day and all the mass of milk is partially skimmed by spontaneous creaming at a temperature between 8 and 20°C, PR milk is collected twice a day at a temperature ≥18°C, and

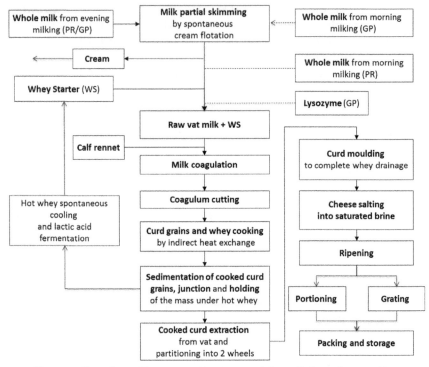

Figure 1.1 Flow sheet of Parmigiano Reggiano and Grana Padano cheesemaking.

only the milk from the evening milking is skimmed at a temperature of approximately 16 to 18°C.

Milk is usually moved from the equipments (open flat basins or closed tanks) used for cream flotation to the coagulation vats by gravity, using open channels, or by pumps. A growing number of dairies opted for mixing all the partially skimmed milk separated from each basin before moving it into the vat, thus aiming to reduce the variability of vat milk composition.

Whey starter (WS) is added to cheese milk in a percentage of approximately 2.7 to 3.5%, increasing milk acidity from 3.2 to about 4.0 °SH/50 ml, and reducing milk pH from 6.7 to about 6.4 (Gatti et al. 2014).

The operations performed into the cheese vat are crucial to obtain perfect PR and GP cheeses, because they determine the grainy structure, the heat stress and the amount and rate of whey separation from the curd, the key factor determining the availability of nutrients for the microorganisms.

To manage the whey syneresis, both the moment of coagulum cutting and the cheese curd grains size have to be controlled (Iezzi et al. 2012).

Coagulum is cut about 2 minutes after the gelation time, when it is still very soft (Table 1.5).

Traditionally, both in PR and GP dairies, cutting is manually performed using a tool called "spino" made by a stick that ends with a network of elliptical shaped blades. To overcome the physical stress of the cheesemaker (given the repetitiveness of this operation) and to reduce variability, many dairies opted for a mixed manual-mechanical cutting

Table 1.5 Process parameters of Parmigiano Reggiano cheesemaking (Mucchetti et al. 2014).

skimmed milk from evening milking (kg)	515.48 ± 56.13
temperature of milk after creaming (°C)	16.90 ± 2.21
milk pH	6.71 ± 0.03
whole milk from morning milking (kg)	535.62 ± 42.65
temperature of milk arriving at the dairy (°C)	20.51 ± 2.78
milk pH	6.70 ± 0.03
total amount of milk into the vat (kg)	1051.11 ± 52.86
milk temperature into the vat (°C)	18.99 ± 1.93
milk pH	6.71 ± 0.03
whey starter (WS) acidity (°SH/50 ml)	30.45 ± 1.55
milk acidity increase after WS addition (°SH)	0.80 ± 0.06
whey starter quantity (kg/100 kg milk)	32.27 ± 3.51
milk pH after whey starter addition	6.39 ± 0.04
normalized amount of rennet (g/100 kg milk)	4.01 ± 1.01
milk temperature at renneting (°C)	33.75 ± 0.40
gelation time (GT) (s)	615.11 ± 97.83
time of coagulum hardening (CH) (s)	105.54 ± 35.17
ratio HT to GT	0.17 ± 0.05
duration of cutting (s)	151.04 ± 47.04
ratio cutting time to coagulation time (GT+HT)	0.22 ± 0.08
curd temperature at the end of cutting (°C)	34.28 ± 1.07
curd temperature at the end of cooking (°C)	55.08 ± 0.28
duration of curd cooking (s)	314.46 ± 84.42
rate of curd cooking (°C/min)	4.14 ± 1.27
overall time from rennet addition to cooking end (TT) (s)	1181.04 ± 166.58
ratio TT to GT	1.93 ± 0.12
time of cooked curd lying under hot whey (min)	64.84 ± 9.35

technology. The first manual step is followed by a mechanical cutting, using a similar shaped tool, adapted to be connected to a motor (Iezzi et al. 2012). To date, however, a complete mechanical cutting is not used, because it is believed to be responsible for higher fat losses into whey.

Analyzing curd size by 2D image analysis, mean and median projected areas of fresh curd grains particles were 2.72 and 1.53 mm^2, respectively.

GP and PR curd grains are cooked to about 53 and 55°C within 3.5 to 6 minutes respectively. The curd grain shrinkage, caused by the cooking process, reduces the area values of cooked grain particles by a factor of 1.76 ± 0.4 (Iezzi et al. 2012).

The cooked grains sediment on the round bottom of the vat, where they lie under hot whey for 60 minutes, fuse themselves and continue to drain off whey. The total solids content of cooked curd may be estimated at about 55.5%, immediately after the extraction from the vat.

The sedimented curd is extracted from the vat by means of a shovel and a cloth, and it is cut into two twins wheels. The curd, wrapped in the cloth, is hung over the vat full of hot whey for 10 to 15 minutes. Then the curd is put into a plastic mould with its cloth and starts to cool. The large size of the cheese curd, together with its low heat conductivity, is responsible for the temperature gradient within the molded curd, which increases by up to 15°C within seven hours from molding (Giraffa et al. 1998). Starter LAB (SLAB), other bacteria and enzymes undergo a different heat stress according to their spatial location in the cheese. Lactic acid fermentation is faster in the external zones, while galactose fermentation occurs more slowly in the inner zones, subjected to the highest heat stress; however, within 48 hours of molding, all the lactose and galactose are fermented. Cheeses with a residual content of galactose are considered defective.

Molding lasts for approximately three days, and on the second day, the curd without cloth is put into a finely pierced steel mold to obtain its convex final shape.

PR and GP cheeses are salted into saturated brines at room temperature for a period of 18 to 25 days, proportionate to the size of the cheese and the brining procedure. Methods of brining are variable according to the size of the dairy, including the traditional static brining with floating wheels, the static brining by full immersion of the wheels placed on cages with five to seven shelves or the dynamic brining in pools. The diffusion process of NaCl from the rind towards the inner zones is very slow, and NaCl distribution across the PR wheel is quite homogeneous only after the first year of ripening (Malacarne et al. 2009).

Typically, both PR and GP cheeses (Fig. 1.2) are ripened in large warehouses with temperature and relative humidity (RH) control to drive moisture losses and create a difference between the air RH and the a_w of the cheese surface.

Figure 1.2 (a) Parmigiano Reggiano and (b) Grana Padano cheeses.

The standard of production for GP cheeses established a temperature range between 15 and 22°C. Typically, the temperature of PR warehouses varies from 10 to 20°C, with the most common temperatures being 14 to 16.5°C; meanwhile, RH values range from 70 to 84% (Guidetti et al. 1995).

1.3 Mozzarella

1.3.1 Characteristics

The name Mozzarella (MOZ) includes several types of pasta filata cheeses, sometimes sharing only the stretching of the molten kneaded cheese curd. MOZ cheese worldwide diffusion has been probably determined in the last two centuries by the migration of many cheesemakers out of Italy that continued their activity in the hosting countries, from Oceania to America and by the diffusion of pizza consumption. The melting pot given by the mixture of different food habits contributed to the deep changes of MOZ cheese characteristics, originating new varieties as MOZ cheese for pizza, low moisture MOZ cheese and new processes, as the pizza cheeses obtained without starter (by means of milk direct acidification with acetic, lactic or hydrochloric acids). Direct MOZ cheese was ideated in US (Breene et al. 1964) and tailored in Italy both as the traditional raw milk MOZ cheese from Apulia (Italy) and the more diffused high moisture cheese variety produced by industrial dairies, using pasteurized milk and fully automated equipments (Mucchetti and Neviani 2006).

According to Codex Alimentarius, MOZ may be defined as an unripened, rindless, smooth and elastic cheese with a long-stranded parallel-orientated fibrous protein structure without evidence of curd granules. The cheese may be formed into various shapes and sizes and has a near white colour. High moisture MOZ is a soft cheese with overlying layers that may form pockets containing liquid of milky appearance. It is usually packed with a keeping liquid, including water and sometimes NaCl and organic acids, even if the largest sizes may be packed without liquid. Low moisture MOZ is a firm/semi-hard homogeneous cheese without holes, suitable for shredding, and usually packed without liquid. MOZ cheese is made by pasta filata processing, which consists of heating curd of a suitable pH value, kneading and stretching until the curd is smooth and free from lumps. Still warm, the curd is cut and molded, then firmed by cooling (Codex Alimentarius 2011a, b, c).

However, the pH value of the curd is only an indirect measure of the suitability of the curd to be plasticized and stretched, as the real condition to obtain a good product is a suitable dissociation of casein-bound calcium from the casein matrix to the whey fraction of cheese curd (Kindsted et al. 2004). The amount of calcium ions solubilized into the whey is pH dependent, however the pH values allowing a comparable degree of casein-bound calcium dissociation are variable, ranging from 4.8 to 5.9. This large range depends on different causes: milk casein content variability (higher for buffalo MOZ cheese), timing of lactic acid production and whey syneresis (starter MOZ cheese) or acid addition (direct MOZ cheese), and also the type of added acid (lactic acid, mono carboxylic acids or citric acid, as is used mostly in Italy).

The structure of MOZ cheese also depends on the method of curd stretching (manual with a stick, or mechanical with the use of screws, hooks, diving arms) (Mucchetti and Neviani 2006). When screws are used, the aggregated para-casein matrix of curd before stretching is disrupted by the shearing forces of the rotating screws, resulting in a reorganization of the matrix into roughly parallel-aligned para-casein fibers (McMahon et al. 1999). When diving arms are used, the asymmetric movement of the arms elongates the casein matrix creating a network of long filamentous fibres vertically and horizontally oriented, which are able to entrap a large amount of water (Mucchetti and Neviani 2006).

Both the varieties of Italian MOZ cheese, with starter or direct, are characterised by a milky fresh taste and a high moisture content, very often higher than 60%, corresponding to the upper limit for US whole milk MOZ cheese (CFR 2015), with little or no proteolysis.

In Italy several types of MOZ cheese are produced and may be classified as follows:

i. "Mozzarella di Bufala Campana" (MBC) PDO (EU 2007), from water buffalo milk produced in a specified area of Southern Italy, with the use of a natural whey culture;
ii. MOZ with buffalo milk, not PDO, often produced with selected starter in a larger production area;
iii. Traditional MOZ (UNI 1995), later MOZ Traditional Speciality Guaranteed (TSG) (EC 1998), obtained using a LAB starter according to the standard, mainly composed by *Streptococcus thermophilus* with the presence of *Enterococci* and other non spore forming unidentified heat resistant bacteria;

iv. MOZ cheese in light brine (UNI 2002a), obtained by natural or selected LAB starter or by direct acidification with lactic acid, citric acid or glucono-δ-lactone;

v. Pizza cheese (UNI 2000), a MOZ cheese mainly used as food ingredient for pizza;

vi. MOZ cheese obtained with a blend of cow and buffalo milk.

The layout of MOZ cheese production can be split into two steps: the cheese curd production and the stretching of the curd. The split of the process phases made possible the using of cheese curd (produced in the same dairy or elsewhere, in Italy or in Europe) as an additional ingredient of MOZ cheese.

UNI standards document the use of refrigerated or frozen cheese curds among the potential ingredients of the cheese. However, according to CE Regulation on food products labelling (EU 2011b), the use of curds as an ingredient is not declared on cheese labels, creating a hidden conflict between producers.

To avoid the use of cheese curds as an ingredient, the standards of MBC PDO cheese and MOZ TSG cheese require that cheese production must to be carried out within 60 hours from milking or in a continuous cycle in the same plant.

To date, as in Europe, a common legal denomination of cheese is lacking and European cheese producers follow their national rules. As a consequence, in Italy, the use of powder or concentrated milk to produce cheese is forbidden (RI 1974), because Italy is a net milk importing country and the Italian government tends to protect Italian milk producers. This prohibition does not exist in many other European countries. So, non Italian cheese producers can use concentrated or dry milk as a commodity, increasing their competitiveness. To limit the use of imported frozen curds and to prevent the use of imported dry milk, the Italian Ministry of Agriculture issued a decree establishing a limit of furosine in pasta filata cheeses (MIPAF 2000), with a clause excluding the application of the limit of furosine for the cheeses imported from Europe, to avoid infringement procedures by the European Commission.

Italian MOZ cheese varieties can be differentiated among themselves according to many variables. The most influencing are:

i. use of raw, thermized or pasteurized milk;

ii. duration of shelf life (from one to about 30 days);

iii. composition of the packing liquid used for its keeping (with or without NaCl and/or lactic/citric acid added);

iv. pH of milk at renneting, which affects the degree of solubilisation of the calcium bound to casein into the whey fraction;

v. pH of curd at stretching (direct method or starter or a combination).

The expected duration of shelf life and pH at renneting are the two variables most influencing the amount of total moisture of cheese, while the mode of stretching the curd may be responsible for the partitioning of moisture into the cheese.

Water in MOZ cheese is considered partitioned among: i) expressible water (H_2O molecules remote from the casein network whose motion is unaffected by the presence of the protein); ii) entrapped water (H_2O molecules inside the meshes of the gel network made by casein molecules), given by the difference between the freezable water and the expressible water; iii) junction water (H_2O molecules trapped within casein junction zones) (Gianferri et al. 2007, McMahon et al. 1999). In MBC PDO cheese the expressible water was measured as 11% of total water and fat content (Gianferri et al. 2007).

MBC PDO standard provides that the cheese when cut *"exudes fatty, whitish whey-like droplets with the aroma of lactic ferments"*, with MOZ TSG *"releasing a milky liquid when cut or squeezed lightly"*. Release of expressible water caused by cheese cutting is a characteristic trait of short shelf life MOZ, because during shelf life the expressible water interacts with casein and some packing liquid can be absorbed by the cheese (Kuo et al. 2001) determining an increase of its weight (Paonessa 2004, Giangiacomo et al. 1991).

MBC PDO cheese has an economic value higher than cow MOZ cheese and for this reason it is the cheese with the highest number of attempts of fraud, mainly substituting part of buffalo milk with cow milk. For the same reason MBC PDO cheese is the most controlled Italian cheese. The presence of cow milk can be verified by many methods, two of them with legal value in Europe or in Italy. The European method (EC 2001) is based on detection of cows' milk and caseinate in cheeses made from buffalo milk by isoelectric focusing of γ-caseins after plasminolysis, while the Italian method is based on the characterization of whey proteins by RP-HPLC (MIRAAF 1996). PCR and PCR real time methods, based on the detection of mitochondrial DNA from somatic cells, have been proposed (Feligini et al. 2005).

1.3.2 Chemical composition

The gross chemical composition of different varieties of Italian MOZ cheese (Table 1.6) shows that the main component is water, almost always over 56% except for MOZ cheese for pizza. As a fresh cheese, proteolysis and lypolysis should be absent or negligible. Within shelf life, heat stable plasmin and coagulant enzymes, if not denatured by the stretching process, are able to start primary proteolysis, changing the original structure and taste of MOZ cheese.

Table 1.6 Mozzarella cheese composition (g/100 g) according to nutritional labels of Italian products.

	Moisture	Fat	Protein	Carbohydrates	NaCl	Ca	Ratio Fat to Protein
Mozzarella with starter							
Mean	62.28	17.02	17.80	0.84	0.70	0.44	0.96
CV%	2.32	5.95	6.69	52.97	16.50	31.49	7.51
Direct Mozzarella							
Mean	60.50	18.53	17.50	1.53	0.43	0.31	1.06
CV%	2.54	4.36	4.95	32.83	35.25	7.53	0.61
Mozzarella as Pizza cheese							
Mean	54.14	20.72	22.13	1.30	0.27	0.43	0.94
CV%	4.09	7.63	10.28	34.40	62.85	18.30	12.92
Mozzarella Bufala Campana PDO cheese							
Mean	60.8	23.4	14.3	0.21	0.21	0.23	1.64
CV%	4.77	9.83	6.29	44.32	28.50	0.01	9.30

To differentiate MOZ cheese produced with or without starter, the carbohydrates fraction should be split between sugars and organic acids. *Streptococcus thermophilus*, the most important bacterium among the strains composing natural or selected starters, produces L-lactic acid and does not use galactose that residues into the cheese. L-lactic acid and galactose are components characterizing MOZ cheeses produced with starter. The carbohydrate pattern of MBC is more complex, because a natural whey culture is used as starter, and among the rich microbiota, the presence of LAB can produce D-lactic acid and can partially utilize galactose. On the contrary, direct MOZ is characterized by a slight higher presence of citric acid, when it is used as milk acidulant, and by the absence of galactose (Ghiglietti et al. 2004) (Table 1.7).

The amount of soluble component (sugars, organic acids, minerals, non protein nitrogen, etc.) in the serum fraction of MOZ cheese is largely

Table 1.7 Sugars and organic acids (g/100 g) of different Italian Mozzarella cheese varieties ready for consumption (Ghiglietti et al. 2004).

Cheese	Moisture	Lactose	Galactose	Lactic acid	Citric acid	pH
MBC	59.15	0.3	0.25	0.44	0.13	5.25
MBC	60.19	0.42	0.40	0.57	0.13	5.13
Mozzarella with starter	59.56	0.21	0.54	0.57	0.12	5.33
Mozzarella without starter	61.20	1.92	0.00	0.00	0.21	5.98

dependent on the matter exchanges that occur during the stretching of the curd.

Furthermore, during the shelf life, the soluble component interacts with the packing liquid and changes with an extent depending on the composition of the latter. Characteristics of "virgin" packing liquid are very variable as the aim of each cheesemaker is to find an equilibrium between the serum phase of MOZ cheese and the packing liquid, slowing the matter exchanges. However, as touched on, the composition of the serum phase is largely dependent on the stretching procedure.

1.3.3 Microbiological characteristics

All MOZ cheese varieties are characterized by the stretching of the acid curd. This is an operation which induces a severe heat stress on microorganisms, reducing their viability according to their thermo tolerance. At the opposite, cheese hardening by dipping into fresh water and brining are two steps responsible for potential post-contamination of the surface of MOZ cheese, with a probability inversely correlated to the ability of managing good manufacturing practices (GMP).

The equivalence of the effects of curd stretching and those given by milk pasteurization on the survival of non spore forming pathogenic bacteria is a discussed item, as different results are reported in literature as a function of the strain tested (*Salmonella, Listeria monocytogenes, Escherichia coli* O157:H7). Survivors were sometimes found in laboratory stretched cheeses obtained with artificially contaminated milk, but rarely were the stretching conditions fully described and the laboratory experimental conditions were often less severe than those applied in industrial cheese-making (Mucchetti et al. 1997, Villani et al. 1996, Spano et al. 2003, Cortesi et al. 2007).

Surveys on artisanal dairies from Southern Italy did not show *Salmonella, L. monocytogenes, E. coli* O157:H7 and *Brucella* in MOZ cheese, while presence of *E. coli* or *Staphylococcus aureus* was sometimes observed (Di Giannatale et al. 2009, Casalinuovo et al. 2014).

The event of blue discoloration of MOZ which occurred in Germany and in Italy in 2010, demonstrated the role of tap water as a way to disseminate *Pseudomonas* spp. (RASFF 2010). *Pseudomonas fluorescens* was the strain able to produce the blue discoloration (Caputo et al. 2015). Among the non LAB, *Pseudomonas, Acinetobacter* spp. and psychrotrophic bacteria were recovered. *P. fluorescens* and *Pseudomonas lundensis* were indicated as potential responsible for wrinkling and exfoliation of the outer surface of MOZ cheese, because of their hydrolytic activity on α_{S1}, β and γ casein (Baruzzi et al. 2012).

LAB counts and species are variable according to the MOZ cheese variety, and usually *S. thermophilus* is recovered, together with other mesophilic (*Lactococcus lactis*) or thermophilic LAB. Biodiversity, as expected, is larger for cheeses produced with raw milk and natural starters according to traditional practices (Ercolini et al. 2004).

1.3.4 Technology

The key points of MOZ cheesemaking are summarized in Fig. 1.3. First of all, the cheesemaker has to regulate the fat to protein ratio and the cheese moisture content. This is different in other cheese varieties, where the first target is readily obtained standardizing the milk composition, taking into account the fat losses into the whey (specific for each cheese) and the recovery of milk casein to curd (about 94% according to Emmons et al. 2010). MOZ cheesemakers also have to consider the stretching step (Fig. 1.4), where an additional amount of molten fat is lost in the stretching water (Francolino et al. 2010) together with soluble peptides originated by proteolysis during the acidification of the curd before stretching.

MOZ moisture management is rather more complex. The cheesemaker has to manage the pH of the curd at stretching, because pH drop determines the solubilization degree of calcium bound to casein and the

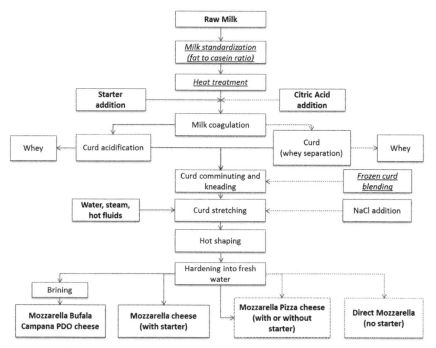

Figure 1.3 Flow sheet of Mozzarella cheesemaking (the dotted lines represent the path of direct Mozzarella varieties; the continuous lines the starter Mozzarella; the underlined operations in italic are optional).

Figure 1.4 PDO Buffalo Mozzarella hand shaping.

ability of casein to be stretched, but also influences the ability of casein to bound water.

Finally, as most Italian MOZ cheeses are salted during stretching, with the exception of MBC PDO cheese, the cheesemaker has to consider the supplementary roles of sodium when added in this moment, as the enhancement of the release of calcium bound to casein, and the ability of casein to swell bounding water in presence of low NaCl concentration (Locci et al. 2012).

For these reasons each MOZ cheese curd variety has a specific narrow range of pH suitable for a good stretching practice. Outside these borders, MOZ cheese cannot have a correct structure, despite the attempts of the cheesemaker to modulate the negative effects managing the given energy (temperature and duration of the mechanic stress) (Mucchetti and Neviani 2006).

This narrow range of pH is time dependent for starter MOZ with an increasing lack of predictability when natural starters are used and curd temperature is not finely tuned. On the contrary, the range of pH is time independent for direct MOZ cheese, making easier automation and/or continuous processes leading to a cost effective product. The cheese final moisture content during stretching may be controlled on line by means of near infrared (NIR) instruments equipped with optic fibers.

1.4 Mascarpone

1.4.1 Characteristics

Mascarpone is a rindless, soft, spreadable and unripened dairy product, obtained by the acid heat coagulation of the milk cream, to which is sometimes added milk. It is usually assigned to the sub-group of "heat heated cream cheeses", belonging to the group of cream cheeses, but it has the specificity to be obtained without any bacterial fermentation.

Among the ingredients (lactic, tartaric, acetic acids, fermented whey and citrus juice) used to lower the pH of the cream and listed by UNI (1998), today citric acid is preferred and industrially used, as documented by the labels of some of the main Italian Mascarpone producers.

Mascarpone is typically white or, when cows are fed with grass, light pale in colour. The colour of Mascarpone is little more yellow than the corresponding cream, because the content of β carotene in Mascarpone is higher than in the corresponding cream (Battelli et al. 1995).

Mascarpone has a mild sweet taste of cream and a delicate, slightly cooked, nutty and fresh butter-like flavour. Bitterness, salt and piquancy are not perceivable. The cooked flavour may be related to the interaction between Mascarpone processing conditions and the severity of the heat treatment of the cream. Consistency is smooth and spreadable, also at low temperature (5°C).

Mascarpone can be eaten as it is, but more often it is used as an ingredient of other foods. It is mainly used in the dessert such as Tiramisù (literally "pull me up"), an Italian cake made with biscuits dipped in coffee and layered with a whipped mixture of Mascarpone cheese, egg yolk, liquor and cocoa. It is also used as a filling or topping of the Italian traditional sweet leavened soft cakes (e.g., Panettone and Pandoro).

A typical industrial dairy use of Mascarpone is in its layering with slices of Gorgonzola PDO cheese, pressing together alternate layers of Gorgonzola and Mascarpone in a ratio of about 1 to 1.

1.4.2 Chemical composition

The main components of Mascarpone are milk fat (usually higher than 40%) and moisture. Today, milk proteins (casein and whey protein) are usually less than 6% (Table 1.8), lower than in the past, when Mascarpone had a stronger consistency and was sold wrapped in waxed paper and not packed into plastic cups (Mucchetti and Neviani 2006).

Table 1.8 Approximate composition (g/100 g) of Mascarpone.

Moisture	Fat	Protein	Lactose	Ratio fat to protein	
48.5	42.9	5.2	2.8	8.4	Mean
8.1	9.8	16.6	32.0	25.4	CV%

Source: Nutritional facts from WEB sites of the largest Italian Mascarpone producers (accessed on January 2, 2015). Moisture content was calculated as the difference between 100 and the sum of component plus 1%, an estimated content of ash plus citric acid.

46.7	47.1	4.0	2.2	12.1	Mean
3.4	3.7	21.8	8.1	21.3	CV%

Source: Resmini et al. 1984.

In recent decades, the trend to the reduction of animal fats in the human diet pulled the producers to slightly reduce the usual Mascarpone fat content. When Mascarpone light varieties are produced (with fat less than 35%), the amount of protein (casein and whey proteins) required for keeping a correct structure are increased.

The introduction of new technologies, as the membrane separation processes (Mucchetti 2013, Hinrichs 2004, Battelli et al. 1995, Resmini et al. 1984), made possible a larger variability of the fat to protein ratio and of the Mascarpone structure.

The physical data characterizing Mascarpone are few and not always consistent. Franciosa et al. (1999) analyzing 1017 Mascarpone samples collected at retail markets, found pH values ranging from 5.40 to 6.60, a_w values from 0.945 to 0.988 and positive redox potential (Eh) values from +78 to +285 mV.

The temperature, a_w and pH values do not guarantee the destruction or the inhibition of the germination of *Clostridium botulinum* spores, so Mascarpone has to be kept refrigerated (4°C) prior to consumption.

1.4.3 Microbiological characteristics

The microbiota of Mascarpone is composed by heat resistant and post-contamination microorganisms. The number of the latter depends on the processing conditions, e.g., the application of a second heat treatment to Mascarpone before the hot filling of the cups.

Mesophilic aerobic and anaerobic spore counts of Mascarpone is up to 10^4 and to 3.6×10^3 cfu/g, respectively (Franciosa et al. 1999). Among them *C. Botulinum* and *Bacillus cereus* can be found.

The outbreak of *C. Botulinum* poisoning caused by "Tiramisù" ingestion in 1996 (Aureli et al. 1996) was associated to Mascarpone contamination and to the suspension of a correct cold storage during the supply chain. The experimental inoculation of samples of Mascarpone with *C. botulinum* type A spores have shown that the production of toxin is favoured by the abuse in the storage temperature, and confirmed that refrigeration alone can be sufficient for controlling *C. botulinum* growth in Mascarpone (Franciosa et al. 1999). An episode of post-contamination of Mascarpone by *Salmonella enteritidis* was signaled by Panico et al. (2000).

1.4.4 Technology

Mascarpone was traditionally obtained draining off whey by filtration with a linen cloth from the acid heat clotted cream (Fig. 1.5). The origin of the cream (obtained by centrifugal separation or by spontaneous flotation), its temperature and pH of coagulation and, above all, the conditions of whey drainage make it difficult to standardize the composition of Mascarpone.

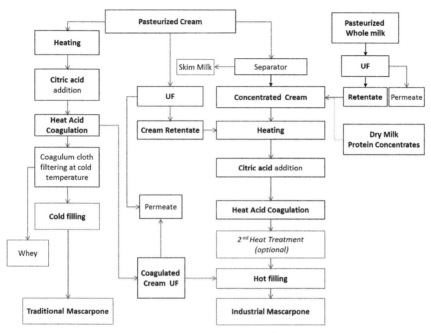

Figure 1.5 Flow sheet of Mascarpone processing (the dotted lines represent an alternative option, while the operations in italic are optional).

Whey drainage is a time-consuming process (more than 12 hours) that enhances the risk of microbial spoilage, as the temperature gradient originated by the spontaneous cooling of the hot coagulum maintains the product in conditions favourable to the microbial growth for some hours.

Today the large part of the cream used for producing Mascarpone comes from fluid milk processing, and whey or permeate is drained off by centrifugal separation or by ultrafiltration (UF) membrane process.

As in the Quark process, the system of whey separation impacts product composition, changing the amount of whey proteins retained into the cheese. Using cream UF, all the whey proteins are retained, while the proportion of whey proteins retained by centrifugal separation process depends on the degree of aggregation to casein by means of sulphur bridges, caused by the heat treatment of the raw cream and by the following acid heat coagulation of the standardized cream.

In addition to the traditional and the centrifugal separation processes, other technologies are used, all excluding the whey drainage from the process (Fig. 1.3). The simplest technology is the batch or continuous coagulation of the pasteurized cream concentrated by UF, without any further whey drainage. Cream (20 to 25% fat; 2.5 to 2.8% protein) is pasteurized and concentrated by spiral wound UF membranes to a Volume Concentration Factor (VCF) x 2, to obtain the target Mascarpone composition. The UF concentrate is then acid heat coagulated at about 90–95°C for some minutes. To lower cream viscosity, membrane concentration is usually carried out a temperature from 50 to 55°C (Mucchetti 2013, Hinrichs 2004, Franciosa et al. 1999).

Alternatively, the suitable amount of acid whey is separated by UF from heat acid coagulated cream (19 to 22% fat) until the Mascarpone reaches the target composition (55% total solids, 45% fat). To eliminate spores of *C. Botulinum* the cream can be Ultra High Temperature (UHT) treated before acidification and UF.

The third option is the mixing of cream with a high fat content (50 to 55%) with the UF retentate of whole milk (VCF x 5 times), followed by the acid heat coagulation of the mixture and hot filling (Resmini et al. 1984). The ratio of cream to milk UF retentate can be varied according to the protein content of Mascarpone cheese, ranging from 4.5 to 6%.

The fourth option is the blending of high fat cream (45 to 54%) with dry milk protein concentrates added in the amount proportional to the Mascarpone target protein content, followed by acid heat coagulation and hot packing.

The smoothness of Mascarpone can be improved by the introduction of a low pressure homogenization stage or by the passage through a smoothing equipment. Safety and shelf life can be prolonged introducing a second heat treatment (105°C) by means of a scraped heat exchanger.

After coagulation, Mascarpone is usually hot filled in packs suitable for its specific final use (consumers, HoReCa users or the food industry) and then cooled down to 4°C, obtaining its final structure. Shelf life is usually up to 60 days. Some small dairies continue to wrap Mascarpone in waxed paper, foreseeing a shorter shelf life.

1.5 Ricotta

1.5.1 Characteristics

Ricotta production in Italy continues to be the most widespread way to add value to the main by-product of cheesemaking, the whey, and it is one the reasons for the poor development of the Italian whey derivatives industry.

Ricotta is a dairy product belonging to the group of whey cheeses obtained by the coagulation of whey proteins by heat with or without the addition of lactic or citric acid, as in Codex Stan 284 (Codex Alimentarius 2011d) and in Standard UNI 10978 (UNI 2002b), or calcium and/or magnesium salts modifying the ionic strength (Mucchetti et al. 2002b).

The term Ricotta was used in the Eastern US states for naming cheeses obtained by heat coagulation of whole or part skim milk (Maubois and

Kosikowski 1978) and this linguistic disorder sometimes continues to be made.

Whole milk, skim milk, milk cream or whey cream, and sometimes NaCl, may be added to the whey. The use of conservative agents, e.g., potassium sorbate or nisin to extend the shelf life, is disappearing in Italy.

Ricotta may be obtained using whey and other milk ingredients obtained from cow, sheep, buffalo and goat milk or their blends (Mucchetti et al. 2002b, UNI 2002b).

In Italy the denomination of two products, Ricotta Romana and Ricotta di Bufala Campana, is protected by the PDO system. However, the largest amount of Italian Ricotta production is not composed by PDO Ricotta products obtained with whey from cow milk and it is marketed with a shelf life up to 20 or 50 days, with the longest duration for Ricotta submitted to a second heat treatment and hot packing (Extended Shelf Life or ESL Ricotta).

Ricotta may be defined as a rindless soft fresh dairy product, with a grainy appearance but smooth to the taste, and spreadable. It is usually sold at retail packed in plastic cups of 100 to 1500 g or in bags up to 20 kg for use as a food ingredient.

According to the model proposed by Hinrichs (2004), the structure of Ricotta is given by a protein network (whey protein alone or with casein) made mainly by intermolecular covalent disulphide bonds contributing to gel stabilization, with the network entrapping the fat globules and the abundant moisture fraction. All the factors interacting with the gel formation (process temperature, time, ionic strength, type of cations, and so on) influence the structure of fresh Ricotta and its stability within the shelf life (Mucchetti et al. 2002b).

Its colour is white, with different tones according to the animal species of origin of the raw materials, characterized by a different amount of carotenes, and the applied technology (Piazza et al. 2004, Pizzillo et al. 2005). The addition of different amounts of milk or cream to whey changes the colour and structure of Ricotta.

Ricotta has a mild sweet taste, because of its lactose content, and the flavour of slightly cooked milk. Bitterness and piquancy are not perceivable. The cooked flavour may be more pronounced in ESL Ricotta.

There are more than 25 varieties of traditional Ricotta produced in Italy (Mucchetti et al. 2002b) and some of them can be semisoft or even hard, as a consequence of salting, ripening and/or smoking processes. Among them, one of the more widespread is the oven baked Ricotta, typical from Sicily, characterized by a prolonged oven cooking and responsible for the high degree of surface caramelization.

Ricotta can be eaten as it is, but often it is used as an ingredient of other foods, sweet (e.g., Sicilian "cannoli") or salty (e.g., "ravioli"). Ricotta used as a food ingredient may be different from fresh Ricotta. When used for "cannoli", ovine Ricotta contains 10% or more added sugar and, as ovine milk availability varies with the season, it is often stored frozen. Ricotta for "ravioli" is coagulated at a higher temperature (95°C) to improve the degree of denaturation of α-Lactalbumin (α-LA) and the holding water ability.

1.5.2 Chemical composition

The composition of different fresh Ricotta varieties, according to the nutritional labels of several Italian dairies and to literature, are reported in Tables 1.9 and 1.10 respectively. Ricotta has a large variability of composition because of the different characteristics and proportion of the dairy ingredients used.

The UNI standard for Ricotta (UNI 2002b) foresees that the whey component should be at least 90% of the blend. The PDO standard of Ricotta di Bufala Campana allows for a maximum of 6% of milk and/or cream (EU 2009b), while PDO Ricotta Romana allows up to 15% of milk addition to whey (EU 2010). Larger additions of milk are not unusual for other Ricotta products. Casein could represent up to 40% of Ricotta proteins, while fat from milk could be up to 50% of total Ricotta fat (Mucchetti et al. 2002b).

1.5.3 Microbiological characteristics

The microbiota of Ricotta is composed by heat resistant and by post-contamination microorganisms. The number of the latter is influenced by the presence of the second heat treatment of Ricotta after the drainage of the deproteinized whey (called "scotta"), and/or by the packing method (hot or cold filling).

Counts of Ricotta samples collected at retail markets showed that the safety and hygienic quality of Ricotta were good (Mucchetti et al. 2002b), but at the same time showed a large variability of microbiological characteristics (from <10^0 cfu/g to more than 10^6 cfu/g). LAB and enterococci were the most represented microbiota, in that order. Aerobic spore formers were not detected in seven out 17 samples, while the highest count was less than 10^3 cfu/g. Cosseddu et al. (1997) found an episodic contamination by *B. cereus*.

Ovine Ricotta, on the contrary, is characterized by higher microbial counts, including also heat labile microbiota as coliform bacteria (Cherchi et al. 1999) and moderately heat resistant spores of *B. cereus*. A prevalence of *B. cereus* in about 50% of the samples collected in Sardinia (Cosentino et al. 1997, Fadda et al. 2012) and a decreasing prevalence of *E. coli* from 22 to 0% of the samples in the years from 2009 to 2012 was observed (Fadda et al. 2014).

Table 1.9 Approximate composition of unripened Ricotta (g/100 g).

Moisture	Fat	Protein	Lactose	NaCl	Ratio fat to protein	
Ricotta from whey of cow milk						
75.3	12.2	8.6	3.4	0.3	1.5	Mean
3.8	25.2	16.3	21.9	35.2	31.7	CV%
Ricotta from whey of sheep milk						
70.1	13.6	10.6	6.8		1.3	Mean
7.6	34.3	15.1	57.6		43.2	CV%
Ricotta from whey of buffalo milk						
71.9	16.4	7.9	2.9	0.6	2.1	Mean
5.8	35.4	15.5	93.1		45.6	CV%
Ricotta from whey of goat milk						
67.8	18.7	10.5	2.1	0.3	1.8	Mean
13.5	31.1	41.0	40.3	99.2	21.7	CV%

Source: Nutritional facts from 23 web sites or labels of Italian producers (accessed on January 5, 2015). Moisture content was calculated as the difference between 100 minus the sum of component plus 1%, as estimated content of ash plus organic acids.

Table 1.10 Composition of Italian Ricotta (g/100 g) according to the species of origin of the raw material.

Origin of raw material	Cow[A]	Sheep[A]	Sheep[B]	Sheep[C]	Girgentana goat[D]	Buffalo[E]
Moisture	75.72	70.03	75.24	70.59	67.87	67.40
Fat	9.73	18.21	11.67	16.43	20.74	20.78
Protein	9.12	7.82	8.64	8.84	6.67	6.65
Fat to protein ratio	1.06	2.38	1.35	1.86	3.05	3.12
Lactose	3.35	3.74			3.71	2.79
Galactose	0.04					
Lactic acid	0.16					0.14
Citric acid	0.50					
Ash	1.16	0.53			1.00	1.81
NaCl			0.90	0.31		
pH	6.26	6.50	6.54		6.27	
Redox (Eh)	183.00					
a_w			0.97			

[A] Mucchetti et al. 2002b, [B] Mancuso et al. 2014, [C] Giancolini et al. 2009, [D] Pizzillo et al. 2005, [E] Mucchetti and Neviani 2006.

Since the pH (6.5) and a_w (0.97) of artisanal ovine Ricotta (Mancuso et al. 2014) do not limit the microbial growth, good manufacturing practices are mandatory to warrant safety requirements.

Heat treatment applied to ovine Ricotta processing (about 80°C) is generally lower than that applied to bovine (cow and buffalo) Ricotta (90 to 95°C), and the lower temperature is responsible for a higher degree of microbial survival. Survivors and post-contaminant strains can easily grow during the step of "scotta" drainage. The higher microbial load of Ricotta surface compared to the interior (Carminati et al. 2002) is a proof of the incidence of post-contamination.

1.5.4 Technology

The milestones of Ricotta production are in Fig. 1.6. The main ingredient is the whey and its freshness is fundamental, as the content of organic acids and the fermented taste and flavour of acidified whey contrast the typical properties of Ricotta.

Usually all the ingredients of the blend are mixed together before the heating starts. In some industrial productions for ESL Ricotta, to reduce

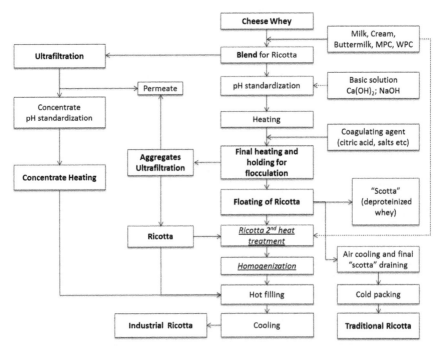

Figure 1.6 Flow sheet of Ricotta processing (the dotted lines represent an alternative option, while the underlined operations in italics are optional).

the taste of "cooked", milk or cream are added directly to the drained Ricotta before the second heat treatment (Toppino et al. 2004).

The properties of the coagulum and the yield of aggregation are the result of the interactions among the protein content and quality of the raw material, pH, ionic strength and process temperature. The higher the protein content, the easier the intermolecular disulphide bonds, the better the aggregation and the greater the capacity to entrap fat and bound moisture.

The origin of whey (sheep > buffalo > cow > goat) and the amount of milk added, influence the protein content of the blend, as well as the addition of whey powders, whey protein concentrates (WPC) or milk protein concentrates (MPC). The amount of proteins may be also increased concentrating the whey by UF or evaporation.

The pH values used in Ricotta production range from 6.5 to 5.6. Ovine whey proteins are easily heat coagulated also at the natural pH of milk, while cow and buffalo whey has to be slightly acidified to allow protein aggregation.

Acid whey (pH below 5.5) does not coagulate easily, as it is well known that acidification of whey proteins under their isoelectric point is a tool to stabilize them, preventing from denaturation when used in acidic beverages.

The pH of whey at the moment of heat denaturation of the proteins affects Ricotta texture. According to the general rule, the heat coagulation at pH close to neutrality will give a more elastic structure, while at pH tending to acidic values the structure will be more grainy. Furthermore, α-LA undergoes more extensive denaturation at higher pH values (Modler 1988).

The natural whey mineral content affects Ricotta production. Whey proteins may be heat denatured at pH >6.5 in absence of calcium ions, but they cannot aggregate because of electric charge repulsion.

A higher cation concentration, as in the case of NaCl addition to the blend, increases the interactions with negative charged proteins and improves aggregation. The traditional use of magnesium sulphate (Mucchetti et al. 2002b), or other calcium and magnesium salts may integrate or substitute the addition of organic acids to whey. The modification of the mineral-protein interactions leads to coagulate the proteins. The use of calcium or magnesium sulphate is traditional also in the coagulation of legumes proteins, among those, soy to obtain tofu (Cai et al. 2001). The mechanism of coagulation can be hypothesized according to the model of Kawachi et al. (1998). When sweet whey with its low natural protein and mineral content and a pH far from isoelectric point is heated, denaturation occurs without gelling: the addition of bivalent cations as Mg or Ca facilitates the formation of hydrophobic bonds in the soluble denatured proteins, thus forming a network structure and hence a homogeneous gel.

Finally, the process temperature is crucial to obtain the coagulum. The operating temperatures vary from 75 to 90°C depending on the type of whey (e.g., sheep 75°C, cow 85°C, buffalo 90°C). The higher the temperature, the higher the α-LA denaturation and the Ricotta yield, but the resulting Ricotta structure will be more grainy and sandy and less moist and smooth.

The heating rate and the type of heat exchange may be also important for process and yield. The slower the heating rate (e.g., 1.8°C/min as in indirect heat exchange), the higher the yield. The direct heating by steam injection with a deep sparger increases the heating rate (e.g., 3.2°C/min) and facilitates the rising of Ricotta grains, because of the simultaneous rising of the steam bubbles (Mucchetti et al. 2002b), creating a gas assisted flotation process.

When the blend has a temperature of few degrees close to the maximum, acids or salts are added to hot whey to improve protein aggregation. The fat entrapped in the protein network makes the density of Ricotta flakes lower than that of the "scotta", allowing the fast rising of flakes to surface. Holding of Ricotta flakes at the surface at high temperature for some minutes makes the curd network more compact and cohesive and suitable to be separated. The size of protein aggregates is important for the structure of the product, but also for the efficient removal of Ricotta from its "scotta".

Traditionally, collection and removal of Ricotta, floating on the surface of the "scotta" within the tank, are performed manually, with the intervention of a team of workers under uncomfortable environment conditions. Manual removal is done by spooning the Ricotta layer off the top of the vat by means of a perforated spoon and by filling plastic moulds ("fiscella"), from which "scotta" drainage will be completed. At the end of drainage, the moulds containing Ricotta are poured into plastic pots and sealed.

Industrially, to speed up the process, Ricotta is collected and loaded onto dripping cloths which in turn are folded into bags and raised in a suspended condition above the tank to complete the dripping of the "scotta" (Magnanini et al. 1995).

Alternatively the process can be mechanized. A series of paddles shoves the Ricotta curd across the surface of the vat onto a perforated conveyor belt (Modler 1988), building a thin bed allowing a fast drainage of "scotta" and the thickening of Ricotta, that at the end of the belt, is poured into the cups.

The method patented by Sordi (1984) avoids the manual discharging of Ricotta, as protein flocculation is into a conical heated tank and "scotta" and Ricotta are discharged in order from the bottom of the tank. "Scotta" drainage from the bed of Ricotta is completed by means of a suitable dripping belt.

The duration of "scotta" drainage varies from some minutes up to 24 hours, depending on the target moisture of the product, its size and the separation system used. During this step, when the temperature of the Ricotta surface in direct contact with the ambient falls below 60°C, the risk of microbial post contamination and growth starts.

There are two common practices used to prevent or reduce the risk of post-contamination: direct hot packing and a second heat treatment followed by hot packing.

The second heat treatment is the preferred technology to produce ESL Ricotta. Ricotta is heat treated (80 to 110°C) by means of a scraped surface heat exchanger with a heat load proportionally to the expected shelf life. To reduce the risk of a structure that is too grainy because of the highest protein aggregation, Ricotta is then smoothed by homogeneization at low pressure or by shearing given by the rotation of scraped blades or gate/anchor impellers in a smoothing equipment. Finally the homogenized Ricotta is hot packed.

Hot packed Ricotta is cooled in a cold room (4°C), or by immersion into an iced water bath or by passage in a tunnel with air flowing at a temperature below 0°C.

References

Assolatte (Associazione Italiana Lattiero Casearia). 2014. Industria Lattiero Casearia Italiana. Rapporto 2013. Mondo del Latte. Supplement to n° 7.

Aureli, P., G. Franciosa and M. Pourshaban. 1996. Foodborne botulism in Italy. Lancet 348, 1594.

Bacci, C., A. Paris and F. Brindani. 2002. Ruolo di *Clostridium* spp. in alterazioni del Parmigiano Reggiano riconducibili a gonfiore tardivo. Ann. Fac. Medic. Vet. Università di Parma. 22: 221–231.

Bansal, N., P.F. Fox and P.L.H. McSweeney. 2009. Comparison of the level of residual coagulant activity in different cheese varieties. J. Dairy Res. 76: 290–293.

Baruzzi, F., R. Lagonigro, L. Quintieri, M. Morea and L. Caputo. 2012. Occurrence of non lactic acid bacteria populations involved in protein hydrolysis of cold-stored high moisture Mozzarella cheese. Food Microbiol. 30: 37–44.

Battelli, G., L. Pellegrino and G. Ostini. 1995. Alcuni aspetti della qualità della panna e del Mascarpone industriale da essa ottenuto. Latte 20: 1098–1112.

Breene, W.M., W.V. Price and C.A. Ernstrom. 1964. Manufacture of pizza cheese without starter. J. Dairy Sci. 47: 1173–1180.

Cai, R., B. Klamczynska and B.K. Baik. 2001. Preparation of bean curds from protein fractions of six legumes. J. Agric. Food Chem. 49: 3068–3073.

Caputo, L., L. Quintieri, D.M. Bianchi, L. Decastelli, L. Monaci, A. Visconti and F. Baruzzi. 2015. Pepsin-digested bovine lactoferrin prevents Mozzarella cheese blue discoloration caused by *Pseudomonas fluorescens*. Food Microbiol. 46: 15–24.

Carini, S., G. Mucchetti and E. Neviani. 1985. Lysozyme: activity against clostridia and use in cheese production—a review. Microbiologie – Ailments – Nutrition 3: 299–320.

Carminati, D., E. Bellini, A. Perrone, E. Neviani and G. Mucchetti. 2002. Qualità microbiologica e conservabilità della Ricotta vaccina tradizionale. Ind. Alim. 41: 549–555.

Casalinuovo, F., P. Rippa, L. Battaglia and N. Parisi. 2014. Isolation of *Cronobacter* spp. (*Enterobacter sakazakii*) from artisanal Mozzarella. Italian Journal of Food Safety 3: 1526, 29–32.

Cattaneo, S., J.A. Hogenboom, F. Masotti, V. Rosi, L. Pellegrino and P. Resmini. 2008. Grated Grana Padano cheese: new hints on how to control quality and recognize imitations. Dairy Sci. Technol. 88: 595–605.

CFR (Code Federal Regulation). 2015. Mozzarella cheese and Scamorza cheese. 21CFR133.155. http://www.accessdata.fda.gov/scripts/cdrh/cfdocs/cfcfr/ CFRSearch.cfm?fr=133.155 accessed on March 7, 2015.

Cherchi, A., M. Porcu, L. Spanedda and C.I.G. Tuberoso. 1999. Caratteristiche chimico-fisiche, organolettiche e microbiologiche di Ricotta ovina prodotta in Sardegna. Riv. It. Sci. Alimen. 28: 307–314.

Codex Alimentarius. 2011a. Codex General Standard for Cheese. Codex Stan 283-1978. In Codex Alimentarius. Milk and Milk products. 2nd ed. WHO and FAO, Rome.

Codex Alimentarius. 2011b. Group standard for unripened cheese including fresh cheese. Codex Stan 221-2001. In Codex Alimentarius. Milk and Milk products. 2nd ed. WHO and FAO, Rome.

Codex Alimentarius. 2011c. Individual cheese standards: Mozzarella. Codex Stan 262-2006. In Codex Alimentarius. Milk and Milk products. 2nd ed. WHO and FAO, Rome.

Codex Alimentarius. 2011d. Standard for Whey cheeses. Codex Stan 284-1971. In Codex Alimentarius. Milk and Milk products. 2nd ed. WHO and FAO, Rome.

Cortesi, M.L. and N. Murru. 2007. Safety and quality along the buffalo milk and cheese chain. Ital. J. Anim. Sci. 6: 207–216.

Cosentino, S., A.F. Mulargia, B. Pisano, P. Tuveri and F. Palmas. 1997. Incidence and biochemical characteristics of *Bacillus* flora in Sardinian dairy products. Int. J. Food Microbiol. 38: 235–238.

Cosseddu, A.M., E.P.L. De Santis, R. Mazzette, A. Fresi and G. Lai. 1997. Ricotta bovina fresca confezionata: caratteristiche microbiologiche di interesse igienico-sanitario. Latte 22: 76–81.

De Dea Lindner, J., V. Bernini, A. De Lorentiis, A. Pecorari, E. Neviani and M. Gatti. 2008. Parmigiano–Reggiano cheese: evolution of cultivable and total lactic microflora and peptidase activities during manufacture and ripening. Dairy Sci. Technol. 88: 511–523.

Di Giannatale, E., A. Alessiani, V. Prencipe, O. Matteucci, T. Persiani, K. Zilli and G. Migliorati. 2009. Polymerase chain reaction and bacteriological comparative analysis of raw milk samples and buffalo mozzarella produced and marketed in Caserta in the Campania region of Italy. Veterinaria Italiana 45: 437–442.

DOOR. 2015. (http://ec.europa.eu/agriculture/quality/door/registeredName. html?denominationId=271, accessed on February 28, 2015).

Emmons, D.B. and H.W. Modler. 2010. Invited review: a commentary on predictive cheese yield formulas. J. Dairy Sci. 93: 5517–5537.

Ercolini, D., G. Mauriello, G. Blaiotta, G. Moschetti and S. Coppola. 2004. PCR–DGGE fingerprints of microbial succession during a manufacture of traditional water buffalo mozzarella cheese. J. Appl. Microbiol. 96: 263–270.

EC [European Communities]. 1998. COMMISSION REGULATION (EC) No. 2527/98 of November 25, 1998 supplementing the Annex to Regulation (EC) No. 2301/97 on the entry of certain names in the "Register of certificates of specific character" provided for in Council Regulation (EEC) No. 2082/92 on certificates of specific character for agricultural products and foodstuffs. Official Journal of the 99 Communities. L 317 date November 26, 1998.

EC [European Communities]. 2001. COMMISSION REGULATION (EC) No. 213/2001 of 9 January 2001 laying down detailed rules for the application of Council Regulation (EC) No. 1255/1999 as regards methods for the analysis and quality evaluation of milk and milk products and amending Regulations (EC) No. 2771/1999 and (EC) No. 2799/1999. Official Journal of the European Communities. L 37 date February 7, 2001.

EU [European Union]. 2007. Publication of an amendment application pursuant to Article 6(2) of Council Regulation (EC) No. 510/2006 on the protection of geographical indications and designations of origin for agricultural products and foodstuffs. Official Journal of the European Union C 90, 5–9, date April 25, 2007.

EU [European Union]. 2009a. Publication of an amendment application pursuant to Article 6(2) of Council Regulation (EC) No. 510/2006 on the protection of geographical indications and designations of origin for agricultural products and foodstuffs. Official Journal of the European Union C 87, date April 16, 2009.

EU [European Union]. 2009b. Publication of an amendment application pursuant to Article 6(2) of Council Regulation (EC) No. 510/2006 on the protection of geographical indications and designations of origin for agricultural products and foodstuffs. Official Journal of the European Union C 260, date October 30, 2009.

EU [European Union]. 2010. Publication of an amendment application pursuant to Article 6(2) of Council Regulation (EC) No. 510/2006 on the protection of geographical indications and designations of origin for agricultural products and foodstuffs. Official Journal of the European Union C 101, date April 20, 2010.

EU [European Union]. 2011a. COMMISSION IMPLEMENTING REGULATION (EU) No. 584/2011 of June 17, 2011 approving non minor amendments to the specification for a name entered in the register of protected designations of origin and protected geographical indications (Grana Padano (PDO)). Official Journal of the European Union L 160, date June 18, 2011.

EU [European Union]. 2011b. Regulation (EU) No. 1169/2011 of the European Parliament and of the Council of October 25, 2011 on the provision of food information to consumers. Official Journal of the European Union, L 304 date November 22, 2011.

EU [European Union]. 2012. Regulation (EU) No. 1151/2012 of the European Parliament and of the Council of November 21, 2012 on quality schemes for agricultural products and foodstuffs. Official Journal of the European Union L 343 date December 14, 2012.

Fadda, A., A. Delogu, E. Mura, A.C. Noli, G. Porqueddu, M.L. Rossi and G. Terrosu. 2012. Presence of *Bacillus cereus*, *Escherichia coli* and Enterobacteriaceae in fresh and salted Ricotta cheese: official controls in Sardinia during the period 2009–2012. Italian Journal of Food Safety 1: 43–45.

Feligini, M., I. Bonizzi, V.C. Curik, P. Parma, G.F. Greppi and G. Enne. 2005. Detection of adulteration in Mozzarella cheese, Food Technol. Biotechnol. 43: 91–95.

Feligini, M., E. Brambati, S. Panelli, M. Ghitti, R. Sacchi, E. Capelli and C. Bonacina. 2014. One year investigation of *Clostridium* spp. occurrence in raw milk and curd of Grana Padano cheese by the automated ribosomal intergenic spacer analysis. Food Control 42: 71–77.

Franciosa, G., M. Pourshaban, M. Gianfranceschi, A. Gattuso, L. Fenicia, A.M. Ferrini, V. Mannoni, G. De Luca and P. Aureli. 1999. *Clostridium botulinum* spores and toxin in Mascarpone cheese and other milk products. J. Food Prot. 62(8): 867–871.

Francolino, S., F. Locci, R. Ghiglietti, R. Iezzi and G. Mucchetti. 2010. Use of milk protein concentrate to standardize milk composition in italian citric Mozzarella cheesemaking LWT-Food Sci. Technol. 43: 310–314.

Gaiaschi, A., B. Beretta, C. Poiesi, A. Conti, M.G. Giuffrida, C.L. Galli and P. Restani. 2000. Proteolysis of αs-casein as a marker of Grana Padano cheese ripening. J. Dairy Sci. 83: 2733–2739.

Gaiaschi, A., B. Beretta, C. Poiesi, A. Conti, M.G. Giuffrida, C.L. Galli and P. Restani. 2001. Proteolysis of β-casein as a marker of Grana Padano cheese ripening. J. Dairy Sci. 84: 60–65.

Gala, E., S. Landi, L. Solieri, M. Nocetti, A. Pulvirenti and P. Giudici. 2008. Diversity of lactic acid bacteria population in ripened Parmigiano Reggiano cheese. Int. J. Food Microbiol. 125: 347–351.

Gatti, M., B. Bottari, C. Lazzi, E. Neviani and G. Mucchetti. 2014. Invited review: Microbial evolution in raw-milk, long ripened cheeses produced using undefined natural whey starters. J. Dairy Sci. 97: 573–591.

Gatti, M., J. De Dea Lindner, A. De Lorentiis, B. Bottari, M. Santarelli, V. Bernini and E. Neviani. 2008. Dynamics of whole and lysed bacterial cells during Parmigiano Reggiano cheese production and ripening. Appl. Environ. Microbiol. 74: 6161–6167.

Ghiglietti, R., S. Santarelli, S. Francolino, A. Perrone, F. Locci and G. Mucchetti. 2004. Evoluzione del contenuto di zuccheri, acidi lattico e citrico e calcio durante la produzione e la conservazione di "Mozzarella di bufala Campana": confronto con la mozzarella di vacca. Sci. Tec. Latt. Casear. 55: 227–249.

Gianferri, R., V. D'Aiuto, R. Curini, M. Delfini and E. Brosio. 2007. Proton NMR transverse relaxation measurements to study water dynamic states and age-related changes in Mozzarella di Bufala Campana cheese. Food Chem. 105: 720–726.

Giangiacomo, R., L. Piergiovanni, G. Messina and P. Fava. 1991. L'effetto di diversi materiali di confezionamento sulla conservazione della mozzarella prodotta per fermentazione biologica. Latte 16: 1160–1172.

Giraffa, G., L. Rossetti, G. Mucchetti, F. Addeo and E. Neviani. 1998. Influence of the temperature gradient on the growth of thermophilic lactobacilli used as natural starters in Grana cheese. J. Dairy Sci. 81: 31–36.

Guidetti, R., R. Mora and M. Zannoni. 1995. Influenza delle condizioni di stagionatura sul formaggio Parmigiano-Reggiano. Nota II: formaggio conservato in magazzino con o senza climatizzazione. Sci. Tec. Latt. Casear. 46: 178–189.

Hinrichs, J. 2004. Mediterranean milk and milk products. Eur. J. Nutr. 43: I/12–I/17.

Hogenboom, J. and M.L. Pellegrino. 2007. Verifica della validità del quadro degli amminoacidi liberi a 15 anni dalla sua introduzione per la caratterizzazione del Parmigiano Reggiano. Sci. Tec. Latt. Casear. 58: 5–16.

Iezzi, R., F. Locci, R. Ghiglietti, C. Belingheri, S. Francolino and G. Mucchetti. 2012. Parmigiano Reggiano and Grana Padano cheese curd grains size and distribution by image analysis. LWT -Food Sci. Technol. 47: 380–385.

Kawachi, K., M. Takeuchi and T. Nishiya. 1998. Solution containing whey protein, whey protein gel, whey protein powder and processed food product produced by using the same. United States Patent N° 5,217,741.

Kindsted, P., M. Caric and S. Milanovic. 2004. Pasta Filata cheeses. *In*: P.F. Fox, P.L.H. McSweeney, T.M. Cogan and T.P. Guinee (eds.). Cheese. Chemistry, Physics and Microbiology. Vol. 2 Major Cheese Groups. Elsevier Academic Press.

Kuo, M.I., S. Gunasekaran, M. Johnson and C. Chen. 2001. Nuclear magnetic resonance study of water mobility in pasta filata and non-pasta filata Mozzarella. J. Dairy Sci. 84: 1950–1958.

Locci, F., R. Ghiglietti, S. Francolino, R. Iezzi and G. Mucchetti. 2012. Effect of stretching with brine on the composition and yield of high moisture Mozzarella cheese. Milchwissenschaft 67: 81–85.

Magnanini, F., E. Tabaglio and D. Bonizzoni. 1995. Apparatus for the Production of Ricotta. European Patent Application N° EP 0 636 311 A1.

Malacarne, M., A. Summer, P. Franceschi, P. Formaggioni, M. Pecorari, G. Panari and P. Mariani. 2009. Free fatty acid profile of Parmigiano Reggiano cheese throughout ripening: comparison between the inner and outer regions of the wheel. Int. Dairy J. 19: 637–641.

Mancuso, I., C. Cardamone, G. Fiorenza, G. Macaluso, L. Arcuri, V. Miraglia and M.L. Scatassa. 2014. Sensory and microbiological evaluation of traditional ovine Ricotta cheese in modified atmosphere packaging. Italian Journal of Food Safety 3: 1725, 122–124.

Masotti, F., J.A. Hogenboom, V. Rosi, I. De Noni and L. Pellegrino. 2010. Proteolysis indices related to cheese ripening and typicalness in PDO Grana Padano cheese. Int. Dairy J. 20: 352–359.

Maubois, J.L. and F. Kosikowski. 1978. Making Ricotta cheese by ultrafiltration. J. Dairy Sci. 61: 881–884.

McMahon, D.J., R.L. Fife and C.J. Oberg. 1999. Water partitioning in Mozzarella cheese and its relationship to cheese meltability. J. Dairy Sci. 82: 1361–1369.

MIPAF [Ministero Politiche Agricole e Forestali]. 2000. Fissazione dei valori massimi di furosina nei formaggi freschi a pasta filata e nel latte (crudo e pastorizzato perossidasi-positivo). Decreto Ministeriale 15.12.2000 in Gazzetta Ufficiale Repubblica Italiana 31, date February 7, 2001.

MIPAAF [Ministero Politiche Agricole Alimentari e Forestali]. 2014. https://www.politicheagricole.it/flex/cm/pages/ServeBLOB.php/L/IT/IDPagina/3340 (accessed on December 30, 2014).

MIRAAF [Ministero Risorse Agricole Alimentari e Forestali]. 1996. Approvazione del metodo ufficiale di analisi reIativo al "Riconoscimento e dosaggio del siero di latte vaccino nel latte di bufala e nei formaggi prodotti con l'lmpiego totale o parziale di latte di bufala mediante RP-HPLC delle sieroproteine specifiche." Decreto 10.4.1996. in Gazzetta Ufficiale Repubblica Italiana 135, date June 11, 1996.

Modler, H.W. 1988. Development of a continuous process for the production of Ricotta cheese. J. Dairy Sci. 71: 2003–2009.

Monfredini, L., L. Settanni, E. Poznanski, A. Cavazza and E. Franciosi. 2012. The spatial distribution of bacteria in Grana cheese during ripening. Syst. Appl. Microbiol. 35: 54–63.

Mucchetti, G. 2005. Relazione finale del Coordinatore del Progetto CNR – MIUR "Sicurezza Formaggi Tipici Italiani -SiForTi" CNR-MIUR. http://www.pnragrobio.unina.it/SIFORTI.html#Risultati_Finali1. Accessed on March 12, 2015.

Mucchetti, G. 2013. Integrated membrane and conventional processes applied to milk processing. *In*: A. Cassano and E. Drioli (eds.). Integrated Membrane Operations in the Food Production. De Gruyter, Berlin, Germany.

Mucchetti, G., D. Carminati and F. Addeo. 1997. Tradition and innovation in the manufacture of the water buffalo Mozzarella cheese produced in Campania. In Proceed. 5th World Buffalo Congress, Caserta, Italy.

Mucchetti, G., F. Locci, P. Massara, R. Vitale and E. Neviani. 2002a. Production of pyroglutamic acid by thermophilic lactic acid bacteria in hard-cooked mini-cheeses. J. Dairy Sci. 85: 2489–2496.

Mucchetti, G., D. Carminati and A. Pirisi. 2002b. Ricotta fresca vaccina ed ovina: osservazioni sulle tecniche di produzione e sul prodotto. Latte 27: 154–166.

Mucchetti, G., F. Locci, M. Gatti, E. Neviani, F. Addeo, A. Dossena and R. Marchelli. 2000. Pyroglutamic acid in cheese: presence, origin and correlation with the ripening time of Grana Padano cheese. J. Dairy Sci. 83: 659–665.

Mucchetti, G., M. Gatti, M. Nocetti, P. Reverberi, A. Bianchi, F. Galati and A. Petroni. 2014. Segmentation of Parmigiano Reggiano dairies according to cheesemaking technology and relationships with the aspect of the cheese curd surface at the moment of its extraction from the cheese vat. J. Dairy Sci. 97: 1202–1209.

Mucchetti, G. and E. Neviani. 2006. Microbiologia e tecnologia lattiero casearia. Qualità e Sicurezza. Tecniche Nuove Editore, Milano, Italy.

Noel, Y., M. Zannoni and E.A. Hunter. 1996. Texture of Parmigiano Reggiano cheese: statistical relationship between rheology and sensory varieties. Lait. 76: 243–254.

Panari, G., S. Perini, R. Guidetti, M. Pecorari, G. Merialdi and A. Albertini. 2001. Indagine sul comportamento di ceppi potenzialmente patogeni nella tecnologia del Parmigiano-Reggiano. Sci. Tec. Latt.-Cas. 52: 13–22.

Panico, M.G., F. Primiano, F. Nappi and F. Attena. 1999. Epidemia di tossinfezione alimentare da *Salmonella enteritidis* causata da un formaggio fresco commerciale. Euro Surveill. 4(4):pii=74. Available online: http://www.eurosurveillance.org/ViewArticle.aspx?ArticleId=74.

Paonessa, A. 2004. Influence of the preservation liquid of Mozzarella di Bufala Campana DOP on some aspects of its preservation. Bubalus Bubalus IV: 30–35.

Piazza, L., M. Bartoccini and S. Barzaghi. 2004. Instrumental texture determination of Ricotta cheese during storage. 1° Convegno Nazionale GSICA "Shelf-life degli alimenti confezionati", Milano 11–13/06/2003. In Special Issue It. J. Food Sci. 41–52.

Pizzano, R., M.A. Nicolai, P. Padovano, P. Ferranti, F. Barone and F. Addeo. 2000. Immunochemical evaluation of bovine ß-Casein and its 1–28 phosphopeptide in cheese during ripening. J. Agric. Food Chem. 48: 4555–4560.

Pizzillo, M., S. Claps, G.F. Cifuni, V. Fedele and R. Rubino. 2005. Effect of goat breed on the sensory chemical and nutritional characteristics of Ricotta. Lives Prod. Sci. 94: 33–40.

Prandini, A., S. Sigolo and G. Piva. 2009. Conjugated linoleic acid (CLA) and fatty acid composition of milk, curd and Grana Padano cheese in conventional and organic farming systems. J. Dairy Res. 76: 278–282.

RASFF—Rapid Alert System for Food and Feed. 2010. Blue Mozzarella. In: Rapid Alert System for Food and Feed, Annual Report 2010. Office for Official Publications of the European Communities, Luxembourg. http://ec.europa.eu/food/safety/rasff/docs/rasff_annual_report_2010_en.pdf, accessed on January 31, 2015.

RI—Repubblica Italiana. 1974. Nuove norme concernenti il divieto di ricostituzione del latte in polvere per l'alimentazione umana. Legge 138/1974 in Gazzetta Ufficiale Repubblica Italiana, 117 date May 7, 1974.

Resmini, P., M.A. Pagani and F. Prati. 1984. L'ultrafiltrazione del latte nella tecnologia del Mascarpone. Sci. Tecn. Latt. Casear. 35: 213–230.

Sforza, S., V. Cavatorta, F. Lambertini, G. Galaverna, A. Dossena and R. Marchelli. 2012. Cheese peptidomics: A detailed study on the evolution of the oligopeptide fraction in Parmigiano Reggiano cheese from curd to 24 months of aging. J. Dairy Sci. 95: 3514–3526.

Sforza, S., V. Cavatorta, G. Galaverna, A. Dossena and R. Marchelli. 2009. Accumulation of non proteolytic aminoacyl derivatives in Parmigiano Reggiano cheese during ripening. Int. Dairy J. 19: 582–587.

Solieri, L., A. Bianchi and P. Giudici. 2012. Inventory of non starter lactic acid bacteria from ripened Parmigiano Reggiano cheese as assessed by a culture dependent multiphasic approach. Syst. Appl. Microbiol. 35: 270–277.

Sordi, M. 1984. Method of producing soft cheese of the type of Italian cottage cheese, and system for the implementation thereof. European Patent Application N° EP 0 094 542.

Spano, G., E. Goffredo, L. Beneduce, D. Tarantino, A. Dupuy and S. Massa. 2003. Fate of *Esherichia coli* O157:H7 during the manufacture of Mozzarella Cheese. Lett. Appl. Microbiol. 36: 73–76.

Toppino, P.M., L. Campagnol, D. Carminati, G. Mucchetti, M. Povolo, S. Benedetti and M. Riva. 2004. Shelf-life study of packed industrial Ricotta cheese. 1° Convegno Nazionale GSICA "Shelf-life degli alimenti confezionati", Milano 11–13/06/2003. In Special Issue It. J. Food Sci. 252–266.

Tosi, F., S. Sandri, G. Tedeschi, M. Malacarne and E. Fossa. 2008. Variazioni di composizione e proprietà fisico-chimiche del Parmigiano-Reggiano durante la maturazione e in differenti zone della forma. Sci. Tecn. Latt.-Cas. 59: 507–528.

UNI—Ente Nazionale Italiano di Unificazione. 1995. Formaggio Mozzarella tradizionale. Definizione, composizione, caratteristiche e confezionamento. Norma Italiana 10537.

UNI—Ente Nazionale Italiano di Unificazione. 1998. Formaggio Mascarpone. Definizione di specificità, composizione, caratteristiche. Norma Italiana 10710.

UNI—Ente Nazionale Italiano di Unificazione. 2000. Formaggio Mozzarella per pizza. Definizione, composizione, caratteristiche e confezionamento. Norma Italiana 10848.

UNI—Ente Nazionale Italiano di Unificazione. 2002a. Formaggio Mozzarella in liquido di governo. Definizione, composizione, caratteristiche e confezionamento. Norma Italiana 10979.

UNI—Ente Nazionale Italiano di Unificazione. 2002b. Ricotta Fresca. Definizione, composizione, caratteristiche. Norma Italiana 10978.

Villani, F., O. Pepe, G. Mauriello, G. Moschetti, L. Sannino and S. Coppola. 1996. Behaviour of *Listeria monocytogenes* during the traditional manufacture of water-buffalo Mozzarella cheese. Lett. Applied Microbiol. 22: 357–360.

CHAPTER 2

Croatian Meat Products with Mediterranean Influence

Dry-cured Hams and Other Products

Helga Medić* and Nives Marušić Radovčić

2.1 Introduction

Dry-cured ham production is related to Mediterranean countries, the origin of a large number of different kinds of dry-cured hams. The production process basically includes the dry salting of a dressed pork ham, drying and long ripening. The characteristics of dry-cured hams depend on many factors such as different pig breeds, the breeding and feeding of animals, age and body weight, the quality of raw hams, applied processing and others. There is a long tradition of production of dry-cured meat products in Croatia and dry-cured ham is one of its most representative products. Croatia has recognized the importance of its autochthonous meat products and four types of dry-cured hams have been protected (Marušić et al. 2014). Istarski pršut has a Protected Designation of Origin (PDO) while Dalmatinski, Drniški and Krčki pršut have Protected Geographical Indication (PGI). Croatian dry-cured hams have some specific characteristics compared to other Mediterranean dry-cured hams which will be discussed in this chapter.

2.2 Types of Croatian dry-cured hams

2.2.1 Dalmatinski pršut

Dalmatinski pršut is a dry-cured meat product made of pork leg with bone, rind and subcutaneous fat tissue.

Faculty of Food Technology and Biotechnology, University of Zagreb, Zagreb, Croatia.
* Corresponding author: hmedic@pbf.hr

2.2.1.1 Raw material

Dalmatinski pršut must be produced from fresh hams, obtained from pigs of commercial meat breeds, crossbreeds or breeding lines, or crossbreeds of any combination thereof.

Raw ham is separated from the pig carcass between the last lumbar vertebra (*v. lumbales*) and the first sacral vertebra (*v. sacrales*). The pelvic bone, i.e., the ilium bone (*os ilium*), the ischial bone (*os ishii*) and the pubic bone (*os pubis*), and the sacrum (*os sacrum*) and also caudal vertebrae (*v. caudales*) must be removed. Ham must be separated from the pelvis in the hip joint (*articulatio coxae*) that connects the femoral head (*caput femoris*) and pelvic socket (*acetabulum*) in the pelvic bone. In the ham remains only a part of the ischial bones with cartilage (*tuber ishii*). Ham must have semicircular shape so that the proximal edge of the ham is approximately 8 to 10 cm away from the head femur (*caput femoris*). Feet are separated in the tarsal joint (*articulus tarsi*). In connection with the tibia and fibula there must remain only heel bumps (*tuber calcanei*) above which the ham is hung or bound during drying. On medial and lateral sides, the ham has the rind and subcutaneous adipose tissue and the mass of the dressed leg is at least 11 kg.

Fresh ham (Fig. 2.1) should not have visible signs of any defects. Meat must have reddish pink colour, compact structure and dry surfaces (RFN). The use of pale, soft and exudative meat (PSE) or dark, firm and dry meat (DFD), or reddish pink colour, but soft and exudative meat (RSE) and not exudative, firm but pale coloured meat (PFN) is forbidden. At the time of its delivery to the production site, the pH of a ham, as measured in the area of the semimembranosus muscle, should be between 5.5 and 6.1. Fat thickness with the skin on the outside of the fresh trimmed ham, measured vertically below the head of the femur, should be at least 15 mm, and preferably fat thickness should be 20 to 25 mm. On the edge of the entire leg, fat tissue must prevent separation of the skin and muscles that are below it.

Fresh hams must not be subjected to any preservation process except chilling. This means that during storage and transport of the hams, temperature must be kept in the range of 1 to 4°C, while freezing of hams

Figure 2.1 Dalmatinski pršut raw material (Association of Dalmatinski pršut).

is not allowed. Time between slaughtering and salting of fresh hams should not be less than 24 or longer than 96 hours (Kos et al. 2015).

2.2.1.2 Description of the finished product

Dalmatinski pršut (Fig. 2.2) does not contain any additives (nitrites, nitrates, potassium sorbate, ascorbic and propionic acids) and spices except sea salt. It has the following sensory characteristics: the exterior of the ham should not have cracks, cuts and hanging parts of muscle and rind, and no large skin folds. At the cross-section of the ham, subcutaneous fat tissue must be white to pinkish white in colour and muscle tissue uniform red to bright red. The smell of the finished product should be of the ripened, salted, dried and smoked pork, with no foreign odours (tar, oil, fresh meat, wet or dry grass) and the smell of smoke must be expressed. Flavour

Figure 2.2 Dalmatinski pršut.

should be slightly brackish to saline and chewing consistency should be soft. Physico-chemical characteristics of Dalmatinski pršut are presented in Table 2.1. The mass of Dalmatinski pršut at the time of placing on the market is at least 6.5 kg, and its age, counting from the first date of processing, is at least 12 months.

2.2.1.3 Processing

Salting. This phase is most critical in the production of a ham. Low temperature should be maintained over the entire phase of salting and pressing to avoid spoilage of raw hams. Salting is therefore performed at 2 to 6°C and at relative humidity higher than 80%. Mandatory massage is performed before salting to squeeze out residual blood from the whole ham, particularly of the femoral artery.

Fast and even penetration of salt in the ham has outstanding importance for the quality of the finished product. This is one reason why it is very

Table 2.1 Physico-chemical characteristics of Croatian dry-cured hams.

	Water content (%)	NaCl content (%)	a_w
Dalmatinski pršut	40–55	4.5–7.5	<0.930
Drniški pršut	≤40	≤7.0	<0.900
Istarski pršut	<55	<8	<0.930
Krčki pršut	40–60	4–8	<0.930

important that hams have the same temperature (1 to 4°C), because very cold hams absorb less salt, and underchilled hams tend to spoilage.

The whole surface of the ham is rubbed with a dry salt and left with the medial side up. After seven to 10 days (depending on weight), it is necessary to re-rub hams with salt and let them rest for the next 10 days with the medial side down.

Pressing. After salting, the hams are pressed. The main goal of this additional phase is the proper formation of ham, which is especially important when the ham is placed on the market in one piece, with the bone. Hams are pressed under the constant pressure of approximately 0.1 kgcm^{-2}. Pressing lasts up to 10 days, after which the hams are washed with clean water and drained, then readied for smoking, drying and ripening.

If the pressing phase is omitted, salted hams, after 14 to 20 days of salting, rest for seven to 10 days without rearrangement, followed by washing with clean water and draining. As in the phase of salting, temperature during pressing should be 2 to 6°C and relative humidity must be higher than 80%.

Smoking and drying. Salted hams are transferred to another chamber in order to level the temperature and to allow salt to diffuse to the inner area of the ham before smoking. The chamber must have air vents protected by mesh to prevent entry of insects.

Smoking is carried out using cold smoke derived from burning hardwood sawdust of beech (*Fagus* sp.), oak (*Quercus* sp.) and hornbeam (*Carpinus* sp.). If smoking is done in the classical way with open fire, it is necessary to take special care of the temperature in the chamber, which must not exceed 22°C. Higher temperatures cause a protein denaturation in the surface layer of the ham. This can create undesirable barriers for the free exit of water from the internal parts of the ham, and thus the deterioration of the ham. Smoking and drying lasts upto 45 days.

Ripening. After smoking and drying, hams are moved to the chamber with stable microclimate which have openings for air exchange (windows) for proper ripening. All openings must be protected by a dense mesh that prevents free entry of insects, rodents and other pests. Preferably, the ripening room temperature should not exceed 20°C and the relative humidity is below 90%. Such conditions allow hams to evenly lose moisture and properly mature.

During the ripening of hams, the cracks on the medial side are allowed to be filled with a mixture made of chopped pork fat, wheat or rice flour and salt. Ripening takes place in a dark area with a moderate air circulation. After one year from the date of salting, the ham is ripened and ready for consumption.

Packaging and manner of placing on the market. The product with the geographical indication Dalmatinski pršut must be placed on the market only after the completion of the last phase of production and after the certification body finds the conformity of products to their specification. The product can be placed on the market as a whole ham, in pieces or sliced.

2.2.2 Drniški pršut

Drniški pršut is salted with coarse sea salt, pressed, cold smoked and dried. Trimmed hams are without pelvic bones and feet and the duration of production is at least 12 months in a limited geographical area.

2.2.2.1 Raw material

Drniški pršut (Fig. 2.3) is produced only from fresh hams; they must not have been frozen to ensure the quality of finished product. Meat of a poor quality, pale, soft, and exudative (PSE) meat or dark, firm and dry (DFD) meat cannot be used. At the time of salting, hams are aged from two to four days from the date of slaughter and the minimum weight of the dressed leg for salting is 11 kg.

2.2.2.2 Description of the finished product

Upon completion of the production, ham (Fig. 2.4) has a distinctive look and a correctly rounded edge. It should be without pelvic bones and feet, and part of the rind and fat tissue on the inner side of the leg. Externally,

Figure 2.3 Drniški pršut raw material (Association of Producers of Drniški pršut).

the ham must not have visible defects and the surface may contain the remains of a thin layer of mold which usually grows on the hams during ripening.

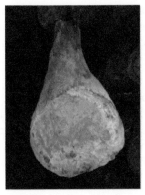

Figure 2.4 Drniški pršut (Association of Producers of Drniški pršut).

The finished product presents the following sensory characteristics:

a) a uniform intense ruby red colour of cold meat, except in the area of white adipose tissue;
b) an intense aroma of ripened, slightly smoked dried pork, free of foreign odours;
c) a good interconnection of the muscles in cross-section, good cutting performance and an unhardened edge;
d) a typical dryness but also good chewiness, ready to swallow due to favourable juiciness and smoothness;
e) a full mild sweet flavour, moderate salinity; without acidity, bitterness or rancidity.

The mass of the finished product is at least 6.5 kg (Karolyi and Guarina 2015).

2.2.2.3 Processing

Drniški pršut is processed traditionally. Raw hams are salted with sea salt, pressed, cold smoked, dried and ripened in a specific climate. All stages in the production of Drniški pršut are performed manually.

Raw material. Fresh ham for the production of Drniški pršut is processed without feet, pelvic and sacral bones and caudal vertebrae. In the ham remains the femur, tibia and fibula with the patella, ischial bumps and, depending on the amount of cut that removes feet, the tarsal bones. Dressed ham is without the rind and fat tissue from the medial side to the height of the knee joint, and without the muscles removed. Dressed ham should be without visible defects and with a minimal weight of 11 kg.

Dry curing. Hams are salted manually with coarse sea salt after the removal of excess blood from the leg. The amount of salt is adjusted to the weight of the ham. Initially salted hams rest horizontally and are subsequently salted after a week. Salting is performed at the temperature of 0 to 5°C and the length of salting depends on the mass of the ham.

Pressing. After salting, hams are pressed for about 10 days at low temperature with a rearrangement at the half time of the pressing period (top to bottom and vice versa).

Washing. Upon completion of the pressing, hams are rinsed of excess salt with cold running tap water, tied with a rope in the area of the hock and hung to drain.

Smoking. Smoking is done in smokehouses with the hams placed perpendicular to the direction of blowing dominant winds. Smoking lasts for 30 to 45 days. The traditional production of smoke is applied, where smoke is produced in a metal firebox using logs of beech (*Fagus sylvatica*) and hornbeam (*Carpiniusbetulus* L.). According to custom, the local vegetation is added, such as deadwood spruce (*Juniperus communis*), wood and shells of almonds (*Amygdalus communis*) and dry immortelle (*Helichrysum arenarium*) for better aroma of the smoke. The regime of smoking depends on the weather and cold smoke (<25°C). In the middle of the smoking process, hams are once again pressed for about five days, in order to achieve the final shape.

Air drying. Upon completion of smoking, smoked hams remain in the same premises exposed to natural air circulation. Frequent winds at a certain time of the year (November to March), mainly Bura – cold and dry north-easterly winds, enable the continuous process of drying. With the arrival of warmer days in April and early May, smoked hams are moved to basement chambers for ripening.

Ripening. Ripening takes place in dark stable microclimate premises with an air temperature between 12 and 18°C and relative humidity between 60 and 75%. Ham achieves maturity in 12 to 18 months from salting. The average weight loss of production is estimated to be in the interval of 40.7 to 41.7% (Karolyi and Đikić 2013).

Labelling and packaging. Drniški pršut can be marketed whole, cut into pieces of various shapes or sliced. Pieces and slices of ham are placed on the market in vacuum packaging of various shapes, sizes and weights.

2.2.3 Istarski pršut

Istarski pršut is a dry-cured meat product, produced from pork ham (hind leg), without rind and subcutaneous fat tissue, and with pelvic bones.

2.2.3.1 Raw material

The fresh hams (Fig. 2.5) are firstly dry salted with sea salt and spices are added, then air dried with a long maturation period. If fresh ham weighs up to 16 kg, the production process will last for at least 12 months, and if it weighs over 16 kg, it will last for at least 15 months.

2.2.3.2 Description of the finished product

Istarski pršut contains the pelvic bones and has elongated shape. The foot is detached in the tarsal joint, and the rind is also removed, except

Figure 2.5 Istarski pršut raw material (Association of Producers of Istarski pršut).

for the part under the tarsal joint (10 to 15 cm). The surfaces of the ham are clean or with a thin layer of mold, which influences longer shelf life. The cross section of the muscle tissue has a uniform pinkish red colour without discolouration, and fat tissue must be white. Istarski pršut has the distinctive characteristic smell of the dried ripened pork meat and herbs. It has a typical intense flavour and moderate salinity.

The final product (Fig. 2.6) must weigh at least seven kg with the physico-chemical properties shown in Table 2.1.

2.2.3.3 Processing

Raw material. Istarski pršut is produced from pork fresh ham: the offspring of pure breed Swedish Landrace, German Landrace and Large White from domesticated breeding; two breed crosses between these breeds; backcross between these breeds; three breed crosses between F1 generation (Landrace x Large White) x Duroc breed.

During the fattening phase, the pigs used for Istarski pršut production are fed with compound feed. All ingredients, except minerals, vitamins and other

Figure 2.6 Istarski pršut (Association of Producers of Istarski pršut).

additives, must originate from the area allowed for pig raising. All ingredients have to be traceable and verified by the authorized certification body.

In the last fattening phase (above 110 kg), it is permissible to replace up to 50% of corn with barley. Apart from that, wheat or wheat bran, fresh clover and alfalfa, pumpkins, cabbages, fodder beet, sugar beet pulp, whey and cooked potatoes can be used for feeding the pigs.

For the production of Istarski pršut, the pigs must be at least nine months old with the average body mass at slaughtering in the range of 180 kg ± 10%. Fresh hams are trimmed in a way that the pelvic bones are left while the sacrum bone and caudal vertebrae are removed. The rind and the subcutaneous fat tissue are removed from the ham upto 10 to 15 cm from the tarsal joint and fresh ham must weigh at least 13 kg. Fresh hams can only be chilled in the temperature range of –1 to +4°C. Freezing of fresh hams is not allowed. Time between slaughtering and start of production should not be less than 24 or longer than 96 hours.

Processing. All production phases of Istarski pršut, from salting and pressing, through drying and ripening to the finished product, must take place within the administrative borders of Istria County, with the exception of the islands that belong to the Istria County, where production is not allowed.

The specificities of the raw material are the result of controlled breeding of certain genotypes of pigs which are fed according to special requirements in extended fattening in order to get the body mass of 180 kg. Heavy pigs are needed at slaughter to get large weights of fresh ham, necessary due the loss of considerable amount of water in the drying and ripening stages.

The traditional technology requires the processing of ham in the way that leaves the pelvic bones attached (*os ilium, os pubis* and *os ischii*), which is not a standard procedure in the production of cured ham in other regions. Specifically, after the ham is processed and rounded off, the femur head (*caput femoris*) is hidden, while in other types of cured ham, it can be seen from the inside, the medial side, and is one of their main visual characteristics. In Istarski pršut this head is hidden because it is inside the pelvic bone (*acetabulum*) (European Commission 2013).

A completely unusual technique used in processing the ham is also the removal of rind and subcutaneous fat tissue from the entire surface. The processed hams are sea salt dried with the addition of pepper (*Piper nigru*), garlic (*Allium sativum*), laurel (*Laurus nobilis*) and rosemary (*Rosmarinus officinalis*). The addition of so many spices during salting represents another distinguishing feature in the production of Istarski pršut. The lack of the rind and fat tissue not only causes slightly more drying than is the case with other hams, but also one more distinguishing feature: the growth of molds on external surfaces of the ham. The presence of surface molds during the drying and ripening phases represents another distinctive external characteristic of Istarski pršut.

The specific sensory properties of Istarski pršut are evident in its external appearance since it lacks rind and subcutaneous fat tissue. It also contains the pelvic bones, which give it an unusually elongated shape, and has often on the surface, big or small clusters of mold.

The final product is characterized by a particular flavour, a mild, slightly salty taste, uniform pinkish red colour and the desirably consistency of muscle tissue. It contains no additives because it is produced in a traditional way.

Because of the peculiar processing of ham, Istarski pršut is therefore always slightly bigger and heavier than the cured hams produced in other regions, although the latter have the rind and subcutaneous fat tissue.

The drying of the ham without rind and subcutaneous fat tissue, which causes direct exposure to the air of the whole surface of the muscle tissue and covering of the surface with mold during the ripening phase, certainly results in a slightly different development of the ripening process and later in the distinctive sensory properties of Istarski pršut.

The specificity of the aroma and taste of Istarski pršut is influenced by different added spices. The influence of spices on the aroma of Istarski pršut is very important, because—unlike other types of cured ham—the entire surface of the ham is in contact with the spices, so their flavour can penetrate more easily and more deeply into all parts of the ham.

Salting and pressing. Prior to salting the hams are massaged to extract residual blood. After, salt is applied by hand and hams are placed on racks for at least seven days at a temperature of 0 to 6°C. Salting can be done only from mid October till the end of March for obtaining a high quality product, because of favourable natural conditions to which are hams exposed during the drying phase. This depends primarily on the occurrence of cold and dry winds that blow in the winter.

After the salting phase, hams are pressed for at least a week under the same microclimate conditions like salting.

Drying and ripening. Drying and ripening of Istarski pršut is carried out in appropriate premises with controlled temperature and humidity. Whenever natural conditions allow, appropriate microclimate is created in the drying chamber by a natural flow of air which is controlled by opening or closing the window. In the production of Istarski pršut, smoking is not allowed.

The chambers for drying and ripening of Istarski pršut must be dark with a stable microclimate and air temperature that must not exceed 19°C throughout the year.

During the ripening phase it is necessary to maintain a microclimate that will enable growth of desirable molds on the ham surface to give Istarski pršut a distinctive look. In many European countries dry sausages with a mold coating are very popular, but dried ham coated with molds is specific for Istria (Comi et al. 2004).

Specific rules concerning slicing and packaging. Istarski pršut can be marketed in integral form with bones or in the form of packages that may contain deboned ham, larger or smaller chunks of cured ham or sliced cured ham. These types of products must be vacuum packed or packed in a modified atmosphere.

By limiting the portioning and packaging of cured ham to the area of its production, the process control is greatly simplified, the application of traceability is facilitated and the possibility of fraud and the abuse of labels is minimized.

2.2.4 Krčki pršut

Krčki pršut is a dry-cured meat product made from pork ham, excluding pelvic bones, dry-cured with sea salt and spices, air dried without smoking and then dried and ripened for a minimum of one year. Production of Krčki pršut is restricted exclusively to the area of the island of Krk.

2.2.4.1 Raw material

Krčki pršut (Fig. 2.7) is produced only from fresh meat, obtained from pigs from commercial meat breeds, cross-breeds or breeding lines, or cross-breeds of any combination thereof.

The fresh ham must be free of the pelvic bones and of the tail vertebrae (*v. caudales*). The ham must be separated from the hip at the hip joint (*articulatio coxae*) where the femur head (*caput femoris*) connects with the acetabulum on the pelvic bone. Only the cartilaginous part of the ischium (*tuber ischii*) must be left in the ham. The muscles of the ham must be

Figure 2.7 Krčki pršut raw material (Mesnica Market, "Žužić").

nicely rounded in a semi circle so that the proximal edge of a trimmed ham is some 10 to 15 cm from the femur head. The feet, together with the proximal tarsals, are removed from the ham at the tarsal joint (*articulatio tarsi*). Only the point of the hock (*tuber calcanei*), above which the ham is tied and hung to dry, may remain attached to the tibia and the fibula. The ham is covered with rind and subcutaneous fat on the medial and lateral sides. The distal part of the rind, with the subcutaneous fat, must be rounded with no loosely hanging parts of muscle on the open medial side (European Commission 2014). A fresh ham must display no visible signs of trauma whatsoever. The meat must be reddish pink, firm in texture and free of surface exudate (RFN). The pH of a ham at the beginning of the processing, as measured in the area of the semimembranosus muscle, should be between 5.5 and 6.0. The thickness of the fat should be approximately 25 mm measured vertically below the femur head and the layer of fat, including the skin, must be at least 15 mm thick.

The hams must not undergo any preservation process with the exception of chilling and the time between slaughtering and salting of the hams must be between 24 and 120 hours. The minimum weight of a dressed ham is 12 kg.

The production of Krčki pršut has some specificities and differs from hams produced in neighbouring regions (e.g., Istria and Dalmatia).

2.2.4.2 Description of the finished product

Krčki pršut (Fig. 2.8) must be spherical in shape, with properly rounded edges, without the distal part (foot) and with limited visible muscle tissue below the femur heads no more than 12 cm in length. Cracks that can occur during maturation may be coated with a mixture made up of pork fat, wheat or rice flour, sea salt and ground pepper. It has sweet and moderately salty flavour, the mild characteristic aroma of ripened dried pork meat. At the cross section, meat has a uniform pink to red colour, marbled with pieces of white fat. It has soft consistency, which makes it easier to slice correctly, and with no hard and dark outer edge. The mass of the final product is more than 6.5 kg (European Commission 2014). Physico-chemical parameters of the finished product are given in Table 2.1.

Figure 2.8 Krčki pršut (Mesnica Market, "Žužić").

2.2.4.3 Processing

Salting. Just before salting, the hams are squeezed to extract residual blood. Hams are salted with a mixture of sea salt and ground black pepper (*Piper nigru*). Salting is done manually so that the dry cure is firmly rubbed in the ham. After salting, the hams with the medial side up are placed along the shelves and remain at least a week. Hams can also be stacked in multiple rows.

During the salting, addition of laurel leaves (*Laurus nobilis*) and branches of rosemary (*Rosmarinus officinalis*) are allowed. The use of preservatives, such as sodium nitrite and sodium nitrate is not permitted. In the salting chamber, temperature should be from 0 to 6°C with a relative humidity greater than 75%. Given that the salting is one of the most critical stages in the production of ham, in addition to the temperature, it is necessary to take into account the process hygiene. After seven days, the hams are again rubbed with the cure of the same composition and redeposited on the shelves or stacks. This second phase of salting takes at least ten days.

Pressing. Upon completion of the second phase of salting, the pressing of hams begins and lasts for at least a week. Pressing is usually done in the same room in which the phase of salting took place and in the same conditions of temperature and moisture. After the lapse of a given period of pressing and salting of 24 days, hams are washed with clean water and drained.

Drying. The drying chamber must, in addition to the system of controlling temperature and humidity, have openings (windows) that are exposed to prevailing winds and arranged to allow natural airflow. When natural conditions allow, it is recommended to create natural air flow by opening or closing openings. When this is not possible, one may control the temperature and humidity. Smoking of hams is not allowed. The drying phase lasts a minimum of 90 days during which the temperature should not exceed 10°C, and the relative humidity should be in the range of 65 to 75%.

Ripening. Ripening is the last stage in the production process and begins after the drying phase. Maturation takes place in dark rooms at a temperature of 9 to 18°C, relative humidity between 60 and 80% and with a slight air exchange. The production of Krčki pršut, since the beginning of salting until the end of the ripening phase, lasts at least 12 months.

Specific rules concerning slicing and packaging. Krčki pršut may be placed on the market in one piece or in packages that may contain the whole ham without the bone, larger or smaller pieces of ham or sliced ham. These forms of product must be packaged in vacuum or in a protective atmosphere.

2.2.5 Quality parameters

Physico-chemical properties in different types of dry-cured hams are shown in Table 2.2. There is a lack of data for Krčki and Drniški pršut in literature, so Istarski and Dalmatinski pršut will be compared to other Mediterranean dry-cured hams.

The difference in water content of Istarski pršut compared to others is mainly because of the different trimming of raw hams which are dressed without rind and fat tissue. During the processing, a higher amount of water is removed from the hams. Fat content is one of the most important quality parameters of dry-cured ham which influences its acceptability (Marušić et al. 2013) and San Daniele dry-cured ham has the highest amount of fat. Croatian dry-cured hams have similar fat content to Iberian and Parma dry-cured hams but are higher in fat content than Serrano dry-cured ham.

Table 2.2 Physico-chemical characteristics in the *biceps femoris* muscle in different types of dry-cured hams.

	Istarski	Dalmatinski	Iberian	Serrano	Parma	San Daniele
Water (g/100 g)	37.9	43.2*	49.0	48.5	54.1–61.8	54.7–60.4
Protein (g/100 g)	43.1	32.6*	17.9–30.6	27.9–30.6	27.3–30.8	27.3–30.8
Fat (g/100 g)	17.0	14.2*	19.2	12.0	18.4	23.0
%NaCl	6.3	7.45*	4.0–5.9	5.0–6.0	4.5–6.9	4.5–6.9
mg MDA/kg	0.4		0.4–0.5	–	0.3–0.5	–
a_w	0.89	0.83*	0.85	0.85	0.94	0.93
L*	34.7	37.3*	38.8	34.8	37.9–38.0	37.9–38.0
a*	9.7	8.4*	18.9	15.6	17.7–15.9	17.7–15.9
b*	6.1	8.4*	7.6	10.5	5.9–6.1	5.9–6.1
Literature	a		a		A	a

*Author's unpublished data; a - Marušić et al. 2014.

2.2.5.1 Fatty acid composition

The fatty acids profile of different dry-cured hams is shown in Table 2.3. Dietary fat content plays a significant role in the prevention and treatment of a number of chronic disorders, particularly coronary heart disease. Dietary fat intake should account for 15 to 30% of total diet energy, saturated fatty acids (SFA) no more than 10%, polyunsaturated fatty acids (PUFA) is 6 to 10% (n-6: 5 to 8%; n-3: 1 to 2%), around 10 to 15% from monounsaturated fatty acids (MUFA), and less than 1% from trans fatty acids. It is also recommended to limit cholesterol intake to 300 mg/day (WHO 2003).

Table 2.3 Fatty acids profile of *biceps femoris* of different dry-cured hams (Marušić 2013).

Fatty acids (%)	Iberian	Serrano	Parma	Nero Siciliano	Bayonne	Istarski	Dalmatinski
C12:0	0.07	0.07	–	0.08	–	0.10	0.11
C14:0	1.27	1.37	1.18	1.09	1.08	1.42	1.47
C16:0	22.92	24.48	21.65	22.55	22.91	25.05	25.85
C18:0	7.45	10.98	12.67	11.08	12.53	12.17	13.51
C20:0	0.22	–	0.14	0.14	–	0.19	0.22
SFA	31.93	37.00	35.99	34.94	36.52	39.21	41.42
C16:1	3.39	3.41	3.05	2.94	3.32	3.51	3.10
C18:1	54.51	47.99	49.99	39.53	43.60	48.26	46.51
C20:1	–	0.97	0.86	0.80	0.57	0.76	0.76
MUFA	57.90	53.37	54.04	43.29	47.49	52.96	50.65
C18:2	9.41	9.62	7.77	16.75	11.70	6.80	6.87
C18:3	0.65	0.53	0.21	0.93	0.50	0.46	0.46
C20:4	0.11	0.97	0.61	2.35	3.10	0.16	0.17
PUFA	10.17	11.01	8.59	20.03	15.30	7.84	7.93

The fatty acid composition of lipids of dry-cured hams from white pigs includes 35 to 40% of SFA, 45 to 50% of MUFA and 10 to 15% of PUFA (Jiménez-Colmenero et al. 2010). Istarski and Dalmatinski pršut have a similar ratio of fatty acids. Istarski pršut contains 39% of SFA, MUFA, 53 to 54% and 7 to 8% of PUFA, and Dalmatinski pršut 41% of SFA, 51% of MUFA and 8% of PUFA. Iberian hams contain a higher proportion of MUFA (54 to 58%), and a lower percentage of SFA (30 to 35%) and PUFA (8 to 12%), which is explained by the high proportion of oleic acid in the acorns eaten by the pigs during fattening (Isabel et al. 2003). The most abundant saturated fatty acids in dry-cured hams are palmitic acid (25%), stearic acid (12%) and myristic acid (1.5%) (Fernández et al. 2007). The same trend is seen in Istarski and Dalmatinski pršut.

Among the various factors affecting the sensory and technological ham quality, fatty acid composition is recorded as one of the most important factors (Bosi et al. 2000). Fatty acid composition depends on the pig breed, and the proportion of saturated, monounsaturated and polyunsaturated fatty acids. Share of SFA in Parma ham is 38%, MUFA 51% and PUFA 11%, which is in line with the results of fatty acid in other commercial Italian, French and Spanish hams (Bosi et al. 2000, Gandemer 2002). Compared to the fatty acid composition of other types of dry-cured hams (e.g., Iberian, Serrano, Parma, Nero Siciliano, Bayonne) Istarski and Dalmatinski pršut contain higher content of SFA and lower content of PUFA. MUFA content in Istarski pršut is 52.96% and 50.65% in Dalmatinski which is lower than in Iberian (57.90%), Serrano (53.37%) and Parma (54.04%) ham. Italian Nero Siciliano (43.29%) and French Bayonne (47.49%) have a slightly lower proportion of MUFA than Croatian dry-cured hams.

The proportion of PUFA in lipids in Istarski and Dalmatinski pršut is 7 to 8%. Iberian hams have similar values (6 to 8%) while Seranno dry-cured hams have 11 to 15% of PUFA (Jiménez-Colmenero et al. 2010). Nutritionists currently tend to focus more on the PUFA/SFA balance and the n-6/n-3 PUFA ratio rather than the absolute content or individual levels of fatty acids. PUFA-rich diets reduce LDL–cholesterol levels in the blood whereas SFA exerts the opposite effects, and so a PUFA/SFA ratio above 0.4 is recommended for healthy foods and diets (DoH 1994). The PUFA/SFA ratios in Istarski and Dalmatinski pršut are 0.20 (Marušić 2013). General range in dry-cured hams range is 0.17 to 0.35, the highest levels occurring in hams from white pigs, such as Serrano or Parma hams (Jiménez-Colmenero et al. 2010).

Excessive amounts of n-6 PUFA and very high n-6/n-3 PUFA ratios promotes cardiovascular disease, cancer and inflammatory and autoimmune diseases, whereas increased levels of n-3 PUFA (and low n-6/n-3 PUFA ratios) exert suppressive effects (Simopoulos 2002). Dry-cured hams contain higher n-6/n-3 ratios than recommended; 15 to 20 (Simopoulos 2002). The n-6/n-3 ratio in dry-cured hams is generally near the upper limits of the recommended ratio. Genetic and feeding strategies have proven to be effective in producing dry-cured hams with PUFA/SFA and n-6/n-3 ratios characteristic of healthy fats.

2.2.5.2 Aroma

The aroma of meat is one of the most important quality parameters and it is dependent on the raw material as well as the production process. Flavour is the overall perception of taste and aroma. Taste is usually associated with non volatile compounds, such as free amino acids and small peptides that are formed at the end of the manufacturing process, while aroma is associated with the formation of volatile compounds with important aromatic characteristics.

Proteolysis and lipolysis are the two main biochemical reactions responsible for the generation of flavour precursors (Harkouss et al. 2015,

Toldrá 2006). Muscle proteins undergo an intense proteolysis resulting in a great number of small peptides and high amounts of free amino acids. The enzymes responsible for these changes are proteinases (cathepsins, calpains, peptidases and aminopeptidases). The combination of small peptides and free amino acids contribute to the characteristic flavour of dry-cured ham (Aristoy and Toldrá 1995). Non volatile components, amino acids and peptides are the active ingredients of flavours that have a major impact on the final flavour of the ham.

Proteolysis results by increasing the amount of glutamic acid, aspartic acid, methionine, isoleucine, leucine and lysine. These free amino acids contribute to the flavour of ham through their mutual interaction. In addition, the amino acids contribute directly to the taste (Toldrá 1998) and have the role of precursors of the volatile aroma components through Strecker degradation of amino acids by forming the methyl-branched alcohols, aldehydes and sulfides as well as through the Maillard reaction by forming pyrazines (Flores et al. 1998).

Peptides and free amino acids are generated in large amounts from the progressive enzymatic degradation of major sarcoplasmic and myofibrillar proteins (Toldrá 2007). More specifically, lysine and tyrosine have been correlated with aged taste, and glutamic acid, aspartic acid, methionine, phenylalanine, tryptophan, lysine, leucine, and isoleucine have been correlated with the length of the drying and the fully ripened ham taste. Bitter tastes are found in hams with excessive amounts of tryptophan, tyrosine, and phenylalanine (Toldrá 2002).

Muscle and adipose tissue lipids are subject to intense lipolysis generating free fatty acids by the action of lipases (lysosomal acid lipase, acid phospholipase and adipose tissue lipase) that are transformed to volatiles as a result of oxidation. Most of the volatile compounds of the dry-cured hams are formed by oxidation of fatty acids through a process of auto oxidation and beta oxidation (Gandemer 2002).

The volatiles contribute individually and with mutual interactions to exceptional flavour characteristics of dry-cured ham. Ketones, esters, aromatic hydrocarbons and pyrazines are basic volatiles responsible for a pleasant and desirable ham flavour (Flores et al. 1998). The overall aroma of the ham depends on the balance among the volatile components derived from the oxidation of fatty acids and those resulting from the degradation of amino acids. There is evidence of independence between the reaction of lipid oxidation and degradation of amino acids that form the volatile compounds. Free fatty acids are formed by enzymatic degradation of triglycerides and phospholipids (Motilva et al. 1992, Buscailhon et al. 1994b). Lipolytic enzymes are found in muscle and adipose tissue, but also show very good stability (Motilva et al. 1993, Toldrá 1998, 2007).

Approximately 200 volatiles have been identified in dry-cured ham. These volatile compounds are representative of most classes of organic compounds, such as aldehydes, alcohols, hydrocarbons, pyrazines, ketones, esters, lactones, furans, sulfur and chloride compounds and carboxylic acid (Buscailhon et al. 1994a, Flores et al. 1998). Some volatile compounds such as pyrazines, sulfur compounds and branched aldehydes have significant effects on flavour. They can be formed by amino acids degradation, or their formation may depend on the processing conditions (Flores et al. 1998). The final flavour of dry-cured ham depends on the specific aroma for each particular volatile compound.

The dry-cured hams that have a longer production process have more intense aroma due to a higher concentration of all types of volatile compounds generated by proteolysis and lipolysis (Ruiz et al. 1999). During production, ham aroma changes from fat aroma, pork aroma, fresh aroma, dry-cured aroma and aroma of aged meat (Dirinck et al. 1997, Flores et al. 1998, Ruiz et al. 1999). Except volatile compounds derived from lipolysis and proteolysis some volatile compounds can be formed from spices added in the production process like garlic (Ansorena et al. 2000) or pepper (Sabio et al. 1998).

2.2.5.2.1 Aroma compounds

Aldehydes. Aldehydes constitute the most important family of volatile compounds from a quantitative point of view. These compounds play an important role in the aroma of the dry-cured ham because they have a low perception threshold. Aldehydes are the main secondary products of lipid oxidation. Straight chain aldehydes are typical products of lipid oxidation and play an important role in the flavour of dry-cured ham (Muriel et al. 2004, Ramirez and Cava 2007). Linear aldehydes, such as hexanal, heptanal, octanal and nonanal, come mainly from an oxidative degradation of unsaturated fatty acids like oleic, linoleic, linolenic, and arachidonic (Pastorelli et al. 2003). On the other hand, the major pathway of the branched chain aldehydes (like 3-methylbutanal) seems to be the oxidative determination decarboxylation, probably via Strecker degradation. Hexanal is described in literature as the major oxidation product in other dry-cured meat products (Ramirez and Cava 2007, Ruiz et al. 1999). The aroma of hexanal can be described as green, grassy and fatty (García-Gonzalez et al. 2008).

Aldehydes are the most abundant group of compounds in Dalmatinski (36.12%) and Istarski pršut (51.4%) (Marušić 2013). This result is comparable to the results of other European dry-cured hams where aldehydes were also the most abundant group of compounds like in San Daniele (31.53%) (Gasparado et al. 2008). Aldehydes have a low odour threshold value and present, in small amounts, contribute significantly to the flavour of dry-cured ham. Hexanal, which is derived from oxidation of n-6 fatty acids like linoleic and arachidonic acid, is one of the most abundant compounds in Istarski pršut (Marušić et al. 2014). The most potent odorants found in Iberian ham are hexanal (*green*), (Z)-3-hexenal (*acornlike*), 3-methylbutanal (*malty, nutty, toasted*), 1-octen-3-one (*mushroom*), 1-octen-3-ol (*mushroom, rustlike*), hydrogen sulfide (*boiled or rotten eggs, sewagelike*), methanethiol (*rotten eggs or meat, sewagelike*) and 2-methyl-3-furanthiol (*nutty, dry-cured hamlike, toasted*) (Carrapiso and Garcia 2004). Also found in Istarski pršut are 3-methylbutanal, 1-octen-3-one and 1-octen-3-ol. The most abundant aldehyde in Dalmatinski pršut is nonanal. In Istarski pršut, nonanal is also present in higher amounts. Nonanal comes from the oxidation of oleic acid that is the most abundant unsaturated fatty acid in hams (Pham et al. 2008). Nonanal contributes to the flavour with its sweet and fruity aroma (Nunes et al. 2008). The most abundant saturated aldehydes in Iberian dry-cured ham are octanal, nonanal and hexanal. Benzaldehyde is one of the most abundant aldehydes in Dalmatinski pršut. This compound has been found in Iberian hams at a very high concentration (García-González et al. 2013). Branched aldehydes such as 2-methylbutanal and 3-methylbutanal are the largest contributors to the flavour of Spanish and Italian dry-cured hams (Marušić et al. 2014).

Alcohols. The origin of alcohols may be a chemical degradation or in part, microbial activity may be involved. Alcohols have a higher threshold value than aldehydes and their impact on the flavour is small; exceptions are alcohols such as 1-octene-3-ol and 1-penten-3-ol, which have a low threshold value (Sabio et al. 1998). The 3-methyl-1-butanol, present in a large amount in French and Spanish dry-cured ham, may be formed by microbial activity. Alcohols contribute to the flavor of dry-cured ham with herbaceous, woody and fatty aroma notes (García and Timon 2001). Generally, a higher content of alcohols occurs if there is a higher degree of lipid oxidation, which can be confirmed with a high content of hexanal.

Linear and branched alcohols are among the main lipid oxidation products. The methyl branched alcohols can also be derived from the Strecker degradation of amino acids. It is known that branched alcohols originate from microbial degradation of the respective branched aldehydes. Thus, the formation and release of branched alcohols is affected by the salting conditions due to the antimicrobial activity of NaCl. Thus, a higher production of branched alcohols is observed when NaCl is partially replaced by other formulations. Alcohols contribute to ham flavour with herbaceous, woody and fatty notes. The most abundant alcohols in Istarski pršut are 1-octene-3-ol, phenylethyl alcohol, octanol and benzyl alcohol while in Dalmatinski pršut, they are 1-octene-3-ol, benzyl alcohol, phenylethyl alcohol and 1-octenol (Marušić 2013); 3-methyl-1-butanol is by far the most abundant alcohol in Iberian dry-cured ham compared to other breeds. The high concentration of 3-methyl-1-butanol can be due to the activity of the microorganisms present in the ham. Another odour compound whose concentration is higher in Iberian hams is hexanol, which contributes to a fruity green odour perception. Hexanol is also found in Dalmatinski pršut. Unsaturated alcohols (1-octene-3-ol, 1-penten-3-ol) and pentanol are the most abundant compounds in Corsican dry-cured hams with the branched alcohols (2-methylpropanol, 2- and 3-methylbutanol) and ethanol in French Bayonne and Spanish Serrano dry-cured hams. Italian San Daniele dry-cured ham is characterized with ethanol, isobutanol, 1-propanol and 1-penten-3-ol (Gasparado et al. 2008).

Ketones. Ketones are formed by lipid oxidation (Flores et al. 1998), but may also be formed by microbial activity. The formation of methyl ketones has been well documented because these compounds are responsible for the aroma of many blue cheeses. Methyl ketones are formed by chemical reactions in the presence of a large number of microorganisms, but high concentration of ketones is a symptom of bad quality dry-cured ham (Pastorelli et al. 2003). In dry-cured ham the number of microorganisms in relatively low, so it is assumed that these compounds can be formed by other chemical reactions (Sabio et al. 1998). Only if the concentration of ketones is extremely high, is it possible to conclude that the microorganisms are involved in the formation of these compounds.

Methyl ketones are produced by lipid oxidation, by means of auto oxidation or beta oxidation of fatty acids. These compounds contribute to the dry-cured ham aroma and they are considered responsible for fatty aromas associated with cooked meat (García-González et al. 2013). Two methyl ketones (2-heptanone and 2-nonanone) are also present in Dalmatinski and Istarski pršut; 2-propanone is the most abundant ketone in French and Spanish dry-cured hams. Dirinck et al. (1997) reported that 2-propanone has the highest concentration among the ketones identified in Iberian dry-cured ham. In general, the concentration of methyl ketones is higher in Iberian hams. Octen-3-one is a remarkable ketone since its very low odour threshold allows contributing to ham aroma with floral/fresh sensory notes, and also distinguishes Iberian from non Iberian hams, the latter having higher concentrations. The 2,3-octadienone is also present in Istarski and Dalmatinski pršut.

Esters. Esters are formed by esterification of carboxylic acids and alcohols. These compounds have fruity notes, mainly those formed from short chain acids. Esters formed from long chain acids have fatty odour. Esters are responsible for the aroma of Italian (Sabio et al. 1998) and Spanish dry-cured hams and are found in slightly higher concentrations.

Esters are described in dry-cured hams at the end of the maturation process and it seems that NaCl concentration affects the ester production through the activation of esterases (Armenteros et al. 2012). Esters can be formed from the interaction of free fatty acids and alcohols by lipid oxidation in the intramuscular tissue so that the higher the content of alcohols, the higher the concentration of esters. In Dalmatinski pršut only three esters, hexyl hexanoate, isohexyl hexanoate and dodecenyl acetate were found. The low portion of esters is probably related to the antimicrobial activity of sodium chloride to the long curing period (Gasparado et al. 2008). Ester, nonanyl acetate found in Istarski pršut can be one compound distinguishing Istarski pršut from Iberian dry-cured ham.

Sulphur compounds. Sulphur compounds come from the degradation of sulphur amino acids (methionine, cysteine and cystine) and have a strong odour (Sabio et al. 1998). Sulphides, for example dimethyl sulphide, which is formed from the oxidation of methanethiol, gives the ham the aroma of dirty socks. During the long curing process, the concentration of dimethyl disulfide is reduced by 50% (Flores et al. 1998).

Terpenes. Terpenes are generally associated with the addition of spices, in particular pepper (Hinrichsen and Pedersen 1995, Marušić et al. 2011) and some of them have been found in meat as a consequence of their presence in animal feedstuffs (Buscailhon et al. 1993, Sabio et al. 1998). Terpenes such as α-terpinene, terpinolene, limonene, α- and β-pinene, α-thujene, sabinene, α- and β-selinene, β-caryophyllene, α-copaene and linalool are derived from the added pepper, laurel and rosemary during the salting phase of the production process (Maarse and Visscher 1989).

The most abundant terpenes in Dalmatinski pršut are: d-elemene, d-limonene, β-myrcene, sabinene and α-guaiene. Some terpenes were found in Bayonne and Corsican hams, due to the black pepper treatment on the surface of the ham during processing, because these compounds constitute 90% of pepper essential oil (Sabio et al. 1998). Terpenes are found in large content in Istarski pršut and are derived from the added spices. Terpenes with high concentration in Istarski pršut are: α-pinene, β-pinene, sabiene, d-limonene, linalool, β-caryophyllene and p-cymene (Marušić et al. 2014).

Phenols. Phenols are the most classical smoke components but also can originate from added spices like eugenol (4-allyl-2-methoxyphenol) (Maarse and Visscher 1989). The phenolic derivatives volatiles in dry-cured products should be related to smoking during manufacturing. Phenols such as phenol, o- and m-cresol; 2,6- and 2,5-xylenol are responsible for smoked ham aroma. But some phenols such as p-cresol, guaiacol and methyl guaiacol may be of microbial origin (Stahnke 2002).

Phenols are the second most abundant group of compounds in Dalmatinski pršut due to the smoking process of the ham (Marušić 2013). The smoke provides typical colour and flavour, specific components have antioxidative properties and smoke inhibits the growth of surface microbiota. The most abundant phenols in Dalmatinski pršut are: 4-methylphenol, 3-methylphenol, 2-methoxy-4-methylphenol, 2-methylphenol, 2,6-dimethoxyphenol, 4-ethyl-2-methoxyphenol and 3,4-dimethylphenol. Smoked meat flavour is mainly due to the phenols and methoxyphenols present in the wood smoke (Jerković et al. 2007). Methoxyphenols are components of great importance for smoke flavour and for their preserving and antioxidant effect. Methoxyphenols and phenols have pungent, cresolic, heavy, burnt and smoky notes (Guillén and Manzanos 2002). Phenols have low threshold value so their impact on the flavour is significant.

2.3 Other dry-cured products

2.3.1 Dalmatinski šokol

Dalmatian smoked shoulder butt is made from pork neck processed in a special way. The first stage is salting, which takes two to three days. Then it is soaked in wine and rolled in a special blend of spices. It is filled in natural casing and sockinette and then smoked for 30 days, after which it ripens in the wind for two months.

2.3.2 Dalmatinska pečenica

This product is a cured meat made from quality pork loin, produced in a traditional local style; a procedure similar to dry-cured ham but shorter in salting, and smoking in a natural way. This is a deli product cured with aromatic spices and dried on dry wind under controlled technical requirements and modern hygiene knowledge. The production cycle lasts 45 days.

2.3.3 Dalmatinska panceta

This is a dry salted meat product from lean deboned pork belly. The production cycle includes preparation of raw materials, salting, curing, cold smoking, drying and ripening, and lasts for 45 days.

Istarska panceta is produced similarly to Dalmatinska panceta but without smoking, and ripening lasts up to 90 days. Deboned pork belly is also dry salted and black pepper and laurel are added.

There is also a variety of dry fermented sausages which are produced from pork minced meat and fat with addition of different additives and spices. The main production steps are grinding meat ingredients, adding non meat ingredients, blending, stuffing, smoking (optional), drying and ripening.

In Istria there is also a variety of dry-cured meat products like in Dalmatia but their production processes exclude smoking and more spices and herbs are used by the producers.

2.4 Recent and future strategies

Dry curing of hams is a very long and slow process, and for this reason there have been many proposals to accelerate it (Marriott et al. 1987, 1992). Initial strategies were based on boning and skinning of hams for better penetration and diffusion of salt into them (Montgomery et al. 1976, Kemp et al. 1980, Marriott et al. 1983); in other cases, better diffusion was achieved by tumbling of hams in rotating drums, even though some physical damage did occur (Leak et al. 1984). Freezing and thawing of hams produces a membrane disruption that can also facilitate salt diffusion (Kemp et al. 1982) and an accelerated protein and lipid hydrolysis during the initial months has been reported (Motilva et al. 1994). Other recent proposals have been based on the simultaneous vacuum brine impregnation method that can be applied with fresh hams or while frozen hams are thawed. This process gives a substantial reduction in the time needed for thawing and salting, without affecting the biochemical reactions taking place during the processing (Barat et al. 2006) or the sensory quality of the final product (Flores et al. 2006). The control of proteolysis in dry-cured ham is an important development that has been recently proposed (Toldrá 2006). This uses process parameters (pH, salt content, water activity, etc.) to control the muscle endoproteases, mainly involved in texture degradation, and exoproteases, directly involved in the generation of small peptides and free amino acids related to flavour. Another relevant trend is salt reduction in dry-cured ham, especially since high salt content raises blood pressure-sensitive hypertensive consumers (Toldrá and Aristoy 2010).

Dry-cured ham is a product with a high protein content and moderate caloric value, and is a good source of vitamins and minerals. Nevertheless, there are still aspects of its composition that could be improved. These improvements are related to genetic modification of animals and the changes in the production process: aiming to reduce the concentrations of compounds with negative impact on human health. Of course any optimization must aim to produce a healthier product without affecting its quality.

2.4.1 Strategies based on production

2.4.1.1 Genetic strategies

One of the main goals for dry-cured ham production will be to find pig genes that serve to enhance the IMF content of meat pieces intended for curing, with moderate fattening of carcasses. However, numerous studies indicate that loin and ham muscles from Iberian pure-breed pigs contain significantly higher amounts of IMF, haem pigments and iron than those of crossed Iberian x Duroc pigs, and could be more suitable for the production of dry-cured products (Ventanas et al. 2007, Jiménez-Colmenero et al. 2010). Genetic strategies also present interesting opportunities to improve the fatty acid profile (increased n-3 content and reduced n-6/n-3 ratio) of pork and pork products (Lai et al. 2006).

2.4.1.2 Feeding strategy

Feeding practices, aimed at achieving healthier meat products, are used to produce smaller proportions of SFAs and larger proportions of MUFA or PUFA, better n-6/n-3 PUFA and PUFA/SFA ratios and higher antioxidant activity. The fatty acid composition in pig tissues depends on the proportion of fatty acids supplied by the feed and those produced endogenously. Because of the traditional system, whereby Iberian pigs are reared on freely available acorn and grass, Iberian hams contain a higher proportion of MUFA (55.8 to 57.4%) than Serrano or Teruel hams (46.9 to 48.7%) and significant amounts of long chain PUFA (Fernández et al. 2007). This is a result of the high fat content (>6%) and high proportion (>60%) of oleic acid in acorns and the high proportion of linolenic acid in grass. Such a successful approach in terms of nutritional and sensory quality has also been employed to produce hams from Iberian pigs fed in confinement with a mixed diet containing high oleic sunflower oil and α-tocopherol. This significantly increases the levels of oleic acid and antioxidants with respect to Iberian hams from non supplemented control animals and improves some sensory attributes (appearance, texture and odour) and overall organoleptic quality (Ventanas et al. 2007). Isabel et al. (2003) reported positive effects of feeding with MUFA-enriched diets on levels of oleic acid in muscles and dry-cured hams from white genotypes. High oleic supplementation at 6% in concentrate feed could be used to turn the fat softer, causing technological problems for the production of white-pig hams like Serrano or Parma, restricting then, the added amount to 2% (Bosi et al. 2000). Although changes in MUFA content present health benefits, they do not affect n-6/n-3 ratios, which are generally >10. These can be improved by using feeds enriched with polyunsaturated fats, especially of the n-3 family and preferably long-chain, from sources like linseed or marine meals and oils (Santos et al. 2008). In order to prevent cholesterol oxidation and excessive lipid oxidation, several researchers have suggested increasing α-tocopherol content that acts as antioxidant (Jiménez-Colmenero et al. 2010).

2.4.1.3 Processing strategies

Salt reduction. Salt has major advantages for ham such as good microbial stability through the reduction of water activity, a pleasant salty taste and partial solubilization and cohesiveness of myofibrillar proteins. Sodium chloride must therefore be reduced without altering the curing process since it is an important inhibitor of most muscle proteases (Armenteros et al. 2008). A reduction in the total amount of salt has been found to cause excessive proteolysis (Pugliese et al. 2015) and considerable softening due to the intense action of muscle endopeptidase enzymes (Virgili et al. 1995). Furthermore, hams cured with 6% salt have been found to be significantly drier, harder and more fibrous than hams cured with 3% of salt, and a better, saltier taste (Andrés et al. 2004). There are therefore numerous strategies for partial replacement of NaCl by other salts like KCl, $CaCl_2$ and $MgCl_2$, but these may entail changes in processing techniques because of their different diffusion rates. In addition, alternative salts must not exceed certain levels because they can affect the sensory quality.

Bioactive peptides. Several bioactive peptides (antihypertensive, antioxidant or prebiotic peptides) derived from meat proteins have been found in postmortem meat. Some may also be generated during processing of dry-cured ham and remain in the final product. Some of the most important bioactive peptides are the angiotensin I-converting enzyme (ACE) inhibitory peptides. ACE plays an important role in the regulation of blood pressure and has been reported in extracts of dry-cured ham (Arhiara and Ohata 2008). Furthermore, several small peptides have been identified in dry-cured ham (Sentandreu et al. 2003); their origin was attributed to the action of muscle dipeptidylpeptidases, enzymes that are quite active and stable during the curing process (Sentandreu and Toldrá 2001). Some of these dipeptides (Arg-Ser, Gly-Phe, Arg-Phe and Met-Ala) have been found to inhibit over 50% of ACE activity, and Val-Tyr in particular, inhibited more than 90% of such enzyme activity (Sentandreu and Toldrá 2007). Also, some bioactive peptides that produce improvements in immunological responses or satiety effects may be generated as a result of the intense proteolysis (Ventanas 2006).

2.4.1.4 Nutritional compounds in dry-cured ham

Current trends investigate bioactive compounds in foods. In clinical studies, where Iberian ham was included in the normal nutrition of the elderly, it was found that there has been a reduction in total plasma cholesterol, triglycerides and LDL-cholesterol (García Rebollo et al. 1998). Mayoral et al. (2003) concluded that Iberian ham in the diet increases the concentration of antioxidants and reduces lipid oxidation and blood pressure. Martínez-Gonzalez (2009) investigated the incidence of cardiovascular disease and increased body weight over six years on 13,293 healthy participants who consumed dry-cured ham. The research results showed no correlation between dry-cured ham consumption and a high risk of cardiovascular disease or increased body mass.

Some substances naturally present in meat are of nutritional interest because when consumed they exert antihypertensive, antioxidant or antimicrobial activity among others, with health benefits to the consumer. The high concentration of free amino acids in dry-cured ham is the result of intense proteolysis. Taurine is an essential amino acid of great importance during lactation and at times of immune challenge (it may protect human body from oxidative stress). Because of the high concentration of these amino acids, dry-cured ham can be considered as a source of amino acids necessary for normal body functioning. Both carnosine and anserine are antioxidative histidyl dipeptides which are the most abundant antioxidative compounds in meats. These dipeptides help to control oxidation through the prevention of lipid oxidation by inactivating catalysts and/or free radicals in the cytosol. The function they perform is to reduce rancid taste and improve colour stability (Jiménez-Colmenero et al. 2010). They are also considered as anti-aging compounds. L-carnitine is a compound also present in meat which assists the human body in producing energy and in lowering the levels of cholesterol. It is a vitamin-like nutrient essential for energy production and lipid metabolism in many organs and tissues such as skeletal muscle and in the heart. It is also known that it helps the body to absorb calcium to improve skeletal strength and chromium picolinate to help build lean muscle mass. Coenzyme Q10 is a component present in dry-cured hams in high amounts. Coenzyme Q10 is a lipid soluble, endogenous hydroxybenzoquinone compound found in the majority of aerobic organisms. It is a key component of the mitochondrial respiratory chain and is mainly known for its role in oxidative phosphorylation (Small et al. 2012); its presence was then demonstrated in other subcellular fractions and in plasma lipoproteins, where it is endowed with antioxidant properties. Cardiovascular effects of CoQ10 can be ascribed to its bioenergetic role, to its capability of antagonizing oxidation of plasma low density lipoprotein, and to its effect in ameliorating endothelial function (Belardinelli et al. 2006). For these reasons Coenzyme Q10 can be considered as a bioactive compound. Creatine and its phosphorylated derivative phosphocreatine are constituents of muscle tissue implicated in energy delivery. Creatine plays an important role in the energy metabolism of skeletal muscle, providing the necessary energy for vigorous muscle contraction. There is also extensive evidence that, under some circumstances, creatine supplements can enhance muscle performance (Demant and Rhodes 1999). Because of the high concentration of essential amino acids and other nutritionally important compounds present in dry-cured ham, this product can be considered a valuable nutritive food even though its high salt content makes it unviable for hypertensive consumers (Marušić et al. 2013).

References

Andrés, A.I., R. Cava, J. Ventanas, V. Thovar and J. Ruiz. 2004. Sensory characteristics of Iberian ham: Influence of salt content and processing conditions. Meat Sci. 68: 45–51.

Ansorena, D., O. Gimeno, I. Astiasaran and J. Bello. 2001. Analysis of volatile compounds by GC-MS of a dry fermented sausage: Chorizo De Pamplona. Food Res. Int. 34: 67–75.

Ansorena, D., I. Astiasarán and J. Bello. 2000. Changes in volatile compounds during ripening of chorizo de Pamplona elaborated with *Lactobacillus plantarum* and *Staphylococcus carnosus*. Food Sci. Technol. Int. 6: 439–447.

Arhiara, K. and M. Ohata. 2008. Bioactive compounds in meat. *In*: F. Toldrá (ed.). Meat Biotechnology. Springer, New York, USA.

Aristoy, M.C. and F. Toldrá. 1995. Isolation of flavour peptides from raw pork meat and dry-cured ham. Developments in Food Science 37: 1323–1344.

Armenteros, M., M.C. Aristoy and F. Toldrá. 2008. Effect of sodium, potassium, calcium and magnesium chloride salts on porcine muscle proteases. Eur. Food Res. Technol. 229: 93–98.

Armenteros, M., F. Toldrá, M.C. Aristoy, J. Ventanas and M. Estevez. 2012. Effect of the partial replacement of sodium chloride by other salts on the formation of volatile compounds during ripening of dry-cured ham. J. Agric. Food Chem. 60: 7607–7615.

Barat, J.M., R. Grau, J.B. Ibáñez, M.J. Pagán, M. Flores, F. Toldrá and P. Fito. 2006. Accelerated processing of dry-cured ham. Part I. Viability of the use of brine thawing/salting operation. Meat Sci. 72: 757–765.

Belardinelli, R., F. Capestro, A. Misiani, P. Scipione and D. Georgiou. 2006. Moderate exercise training improves functional capacity, quality of life, and endothelium-dependent vasodilation in chronic heart failure patients with implantable cardioverter defibrillators and cardiac resynchronization therapy. Eur. J. Cardiovasc. Prev. Rehabil. 13: 818–825.

Bosi, P., J.A. Cacciavillani, L. Casini, D.P. Lo Fiego, M. Marchetti and S. Mattuzzi. 2000. Effects of dietary high oleic acid sunflower oil, copper and vitamin E levels on the fatty acid composition and the quality of dry-cured Parma ham. Meat Sci. 54: 119–126.

Buscailhon, S., J.L. Berdague and G. Monin. 1993. Time-related changes in volatile compounds of lean tissue during processing of French dry-cured ham. J. Sci. Food and Agric. 63: 69–75.

Buscailhon, S., J.L. Berdagué, J. Bousset, M. Cornet, G. Gandemer, C. Touraille and G. Monin. 1994a. Relations between compositional traits and sensory qualities of French dry-cured ham. Meat Sci. 37: 229–243.

Buscailhon, S., G. Gandemer and G. Monin. 1994b. Time-related changes in intramuscular lipids of French dry-cured ham. Meat Sci. 37: 245–255.

Carrapiso, A.I. and C. Garcia. 2004. Iberian ham headspace: Odorants of intermuscular fat and differences with lean. J. Sci. Food Agric. 84: 2047–2051.

Comi, G., S. Orlic, S. Redzepovic, R. Urso and L. Iacumin. 2004. Molds isolated from Istrian dried ham at the pre-ripening and ripening level. Int. J. Food Micro. 96: 29–34.

Demant, T.W. and E.C. Rhodes. 1999. Effects of creatine supplementation on exercise performance. Sports Med. 28: 49–60.

Dirinck, P., F. Van Opstaele and F. Vandendriessche. 1997. Flavour differences between northern and southern European cured hams. Food Chem. 59: 511–521.

DoH—UK Department of Health. 1994. Nutritional aspects of cardiovascular disease. Report on Health and Social Subject No. 46, London: Her Majesty's Stationery Office, London, UK.

European Commission. Publication of an application pursuant to Article 50(2)(a) of Regulation (EU) No. 1151/2012 of the European Parliament and of the Council on quality schemes for agricultural products and foodstuffs, OJ C 155, June 1, 2013, p. 3–8.

European Commission. Publication of an application pursuant to Article 50(2)(a) of Regulation (EU) No. 1151/2012 of the European Parliament and of the Council on quality schemes for agricultural products and foodstuff, OJ C 412, November 19, 2014, p. 11–14.

Fernández, M., J.A. Ordóñez, I. Cambero, C. Santos, C. Pin and L. de la Hoz. 2007. Fatty acids compositions of selected varieties of Spanish dry ham related to their nutritional implications. Food Chem. 101: 107–112.

Flores, M., J.M. Barat, M.C. Aristoy, M.M. Peris, R. Grau and F. Toldrá. 2006. Accelerated processing of dry-cured ham. Part 2: Influence of brine thawing/salting operation on proteolysis and sensory acceptability. Meat Sci. 72: 766–772.

Flores, M., Y. Sanz, A.M. Spanier, M.C. Aristoy and F. Toldrá. 1998. Contribution of muscle and microbial aminopeptidases to flavour development in dry-cured meat products. Developments in Food Sci. 40: 547–557.

Gandemer, G. 2002. Lipids in muscles and adipose tissues, changes during processing and sensory properties of meat products. Meat Sci. 62: 309–321.

García, C. and M.L. Timón. 2001. Los compuestos responsables del "flavor" del jamón Ibérico. Variaciones en los distintos tipos de jamones. *In*: J. Ventanas (ed.). Tecnología del jamón Ibérico: de los sistemas tradicionales a la explotación racional del sabory elaroma. Ediciones Mundi-Prensa, Madrid, Spain.

García Rebollo, A.J.G., E. Macia Botejera, A. Ortiz, P.J. Morales, M. Martin, A. Falluia, P. Mena and J.E. Campillo. 1998. Effects of consumption of meat product rich in monounsaturated fatty acids (the ham from the Iberian pig) on plasma lipids. Nutri. Res. 18: 743–750.

García-González, D.L., N. Tena, R. Aparicio-Ruiz and M.T. Morales. 2008. Relationship between sensory attributes and volatile compounds qualifying dry-cured hams. Meat Sci. 80: 315–325.

García-González, D.L., R. Aparicio and R. Aparicio-Ruiz. 2013. Volatile and amino acid profiling of dry-cured hams from different swine breeds and processing methods. Molecules 18: 3927–3947.

Gasparado, B., G. Procida, B. Toso and B. Stefanon. 2008. Determination of volatile compounds in San Daniele ham using headspace GC–MS. Meat Sci. 80: 204–209.

Guillén, M.D. and M.J. Manzanos. 2002. Study of the volatile composition of an aqueous oaksmoke preparation. Food Chem. 79: 283–292.

Harkouss, R., T. Astruc, A. Lebert, P. Gatellier, O. Loison, H. Safa, S. Portanguen, E. Parafita and P.S. Mirade. 2015. Quantitative study of the relationships among proteolysis, lipidoxidation, structure and texture throughout the dry-cured ham process. Food Chem. 166: 522–530.

Hinrichsen, L.L. and S.B. Pedersen. 1995. Relationship among flavour, volatile compounds, chemical changes, and microflora in Italian dry-cured ham during processing. J. Agric. Food Chem. 43: 2932–2940.

Isabel, B., C.J. López-Bote, L. de la Hoz, M. Timón, C. Garcia and J. Ruiz. 2003. Effects of feeding elevated concentrations of monounsaturated fatty acids and vitamin E to swine on characteristics of dry-cured hams. Meat Sci. 64: 475–482.

Jerković, I., J. Mastelić and S. Tartaglia. 2007. A study of volatile flavour substances in Dalmatian traditional smoked ham: Impact of dry curing and frying. Food Chem. 104: 1030–1039.

Jiménez-Colmenero, F., J. Ventanas and F. Toldrá. 2010. Nutritional composition of dry-cured ham and its role in a healthy diet. Meat Sci. 84: 585–593.

Karolyi, D. and M. Đikić. 2013. Drniš dry-cured ham : characteristics of raw material and ripe product. Meso 15: 132–137.

Karolyi, D. and D. Guarina. 2015. Drniški pršut-Oznaka zemljopisnog podrijetla, Specifikacija proizvoda, Udruga proizvođača drniškog pršuta, Drniš.

Kemp, J.D., D.F.O. Abidoye, B.E. Langlois, J.B. Franklin and J.D. Fox. 1980. Effect of curing ingredients, skinning, and boning on yield, quality, and microflora of country hams. J. Food Sci. 45: 174–177.

Kemp, J.D., B.E. Langlois and A.E. Johnson. 1982. Effect of pre cure freezing and thawing on the microflora, fat characteristics and palatability of dry-cured ham. J. Food Prot. 45: 244–248.

Kos, I., A. Mandir and U. Toić. 2015. Dalmatinski pršut-Oznaka zemljopisnog podrijetla, Specifikacija, Udruga dalmatinski pršut, Trilj.

Lai, L., J.X. Kang, R. Li, J. Wang, W.T. Witt, H.Y. Yong, Y. Hao, D.M. Wax, C.N. Murphy, A. Rieke, M. Samuel, M.L. Linville, S.W. Korte, R.W. Evans, T.E. Starzl, R.S. Prather and Y. Dai. 2006. Generation of cloned transgenic pigs rich in omega-3 fatty acid. Nature Biotechnol. 24: 435–436.

Leak, F.W., J.D. Kemp, B.E. Langlois and J.D. Fox. 1984. Effect of tumbling and tumbling time on quality and microflora of dry-cured hams. J. Food Sci. 49: 695–698.

Maarse, H. and C.A. Visscher. 1989. Volatile Compounds in Foods. Qualitative and Quantitative Data, 6. Edition, TNO-CIVO Food Analysis Institute, The Netherlands.

Marriott, N.G., P.P. Graham, C.K. Shaer and S.K. Phelps. 1987. Accelerated production of dry-cured hams. Meat Sci. 19: 53–64.

Marriott, N.G., P.P. Graham and J.R. Claus. 1992. Accelerated dry curing of pork legs (hams): a review. J. Muscle Foods 3: 159–168.

Marriott, N.G., J.B. Tracy, R.F. Kelly and P.P. Graham. 1983. Accelerated processing of boneless hams to dry-cured state. J. Food Prot. 46: 717–721.

Martínez-Gonzalez, M.A. 2009. Consumo de jamón curado e incidencia de eventos cardiovasculares, hipertensión arterial o ganancia de peso. In: Proceeding of 5th World Congress of Dry-cured Ham. May 6–8, 2009. Aracena, Spain.

Marušić, N. 2013. Characterization of traditional Istrian and Dalmatian dry-cured ham by means of volatile compounds and quality parameters. Dissertation, University of Zagreb, Zagreb, Croatia.

Marušić, N., M. Petrović, S. Vidaček, T. Janči, T. Petrak and H. Medić. 2013. Fat content and fatty acid composition in Istrian and Dalmatian dry-cured ham. Meso 15: 307–313.

Marušić, N., M. Petrović, S. Vidaček, T. Petrak and H. Medić. 2011. Characterization of Istrian dry-cured ham by means of physical and chemical analyses and volatile compounds. Meat Sci. 88: 786–790.

Marušić, N., S. Vidaček, T. Janči, T. Petrak and H. Medić. 2014. Determination of volatile compounds and quality parameters of traditional Istrian dry-cured ham. Meat Sci. 96: 1409–1416.

Mayoral, P., C.S. Martínez, J.M. Santiago, M.V. Rodriguez, M.L. Garcia and A. Morales. 2003. Effect of ham protein substitution on oxidative stress in older Adults. J. Nutr. Health Aging. 7: 84–89.

Montgomery, R.E., J.D. Kemp and J.D. Fox. 1976. Shrinkage, palatability, and chemical characteristics of dry-cured country ham as affected by skinning procedure. J. Food Sci. 41: 1110–1115.

Motilva, M.J., F. Toldrá, M.C. Aristoy and J. Flores. 1993. Subcutaneous adipose tissue lipolysis in the processing of dry-cured ham. J. Food Biochem. 16: 323–335.

Motilva, M.J., F. Toldrá and J. Flores. 1992. Assay of lipase and esterase activities in fresh pork meat and dry-cured ham. Zeitschrift für Lebensmittel, Untersuchung und Forschung. 195: 446–450.

Motilva, M.J., F. Toldrá, M.I. Nadal and J. Flores. 1994. Pre freezing hams affects hydrolysis during dry curing. J. Food Sci. 59: 303–305.

Muriel, E., T. Antequera, M.J. Petron, A.I. Andres and J. Ruiz. 2004. Volatile compounds in Iberian dry-cured loin. Meat Sci. 68: 391–400.

Nunes, C., M.A. Coimbra, J. Saraiva and M.S. Rocha. 2008. Study of the volatile components of a candied plum and estimation of their contribution to the aroma. Food Chem. 111: 897–905.

Pastorelli, G., S. Magni, R. Rossi, E. Pagliarini, P. Baldini, P. Dirinck, F. Van Opstaele and C. Corino. 2003. Influence of dietary fat, on fatty acid composition and sensory properties of dry-cured Parma ham. Meat Sci. 65: 571–580.

Pham, A.J., M.W. Schilling, W.B. Mikel, J.B. Williams, J.M. Martin and P.C. Coggins. 2008. Relationships between sensory descriptors, consumer acceptability and volatile flavour compounds of American dry-cured ham. Meat Sci. 80: 728–737.

Pugliese, C., F. Sirtori, M. Škrlep, E. Piasentir, L. Calamai, O. Franci and M. Čandek-Potokar. 2015. The effect of ripening time on the chemical, textural, volatile and sensorial traits of *Bicep femoris* and semimembranosus muscles of the Slovenian dry-cured ham Kraškipršut. Meat Sci. 100: 58–68.

Ramirez, R. and R. Cava. 2007. Volatile profiles of dry-cured meat products from three different Iberian × Duroc genotypes. J. Agric. Food Chem. 55: 1923–1931.

Ruiz, J., J. Ventanas, R. Cava, A. Andres and C. Garcia. 1999. Volatile compounds of dry-cured Iberian ham as affected by the length of the curing process. Meat Sci. 52: 19–27.

Sabio, E., M.C. Vidal-Aragon, M.J. Bernalte and J.L. Gata. 1998. Volatile compounds present in six types of dry-cured ham from south European countries. Food Chem. 61: 493–503.

Santos, C., L. Hoz, M.I. Cambero, M.C. Cabeza and J.A. Ordoñez. 2008. Enrichment of dry-cured ham with a-linolenic acid and a-tocopherol by use of linseed oil and a-tocopheryl acetate. Meat Sci. 80: 668–674.

Sentandreu, M.A. and F. Toldrá. 2001. Dipeptidylpeptidase activities along the processing of Serrano dry-cured ham. Eur. Food Res. Technol. 213: 83–87.

Sentandreu, M.A. and F. Toldrá. 2007. Oligopeptides hydrolysed by muscle dipeptidylpeptidases can generate angiotensin I converting enzyme inhibitory dipeptides. Eur. Food Res. Technol. 224: 785–790.

Sentandreu, M.A., S. Stoeva, M.C. Aristoy, K. Laib, W. Voelter and F. Toldrá. 2003. Identification of small peptides generated in Spanish dry-cured ham. J. Food Sci. 68: 64–69.

Small, D.M., J.S. Coombes, N. Bennett, D.W. Johnson and G.C. Gobe. 2012. Oxidative stress, antioxidant therapies and chronic kidney disease. Nephrology 17: 311–321.

Stahnke, L.H. 2002. Flavour formation in fermented sausage. *In*: F. Toldrá (ed.). Research Advances in the Quality of Meat and Meat Products. Research Signpost, Trivandrum, India.

Simopoulos, A.P. 2002. The importance of the ratio of omega-6/omega-3 essential fatty acids. Biomed. Pharmacotherapy 56: 365–379.

Toldrá, F. 1998. Proteolysis and lipolysis in flavour development of dry-cured meat products. Meat Sci. 49: S101–S110.

Toldrá, F. 2006. Biochemical proteolysis basis for improved processing of dry-cured meats. *In*: L.M.L. Nollet and F. Toldrá (eds.). Advanced Technologies for Meat Processing. CRC Press, Boca Raton, Florida, USA.

Toldrá, F. 2002. Dry-cured meat products. Wiley-Blackwell, Ames, Iowa, USA.

Toldrá, F. 2007. Biochemistry of muscle and fat. *In*: F. Toldrá, Y.H. Hui, I. Astiasarán, W.K. Nip, J.G. Sebranek, E.T.F. Silveira, L.H. Stahnke and R. Talon (eds.). Handbook of Fermented Meat and Poultry. Blackwell Publishing, Ames, Iowa, USA.

Toldrá, F. and M.C. Aristoy. 2010. Dry-cured ham. *In*: F. Toldrá (ed.). Handbook of Meat Processing. Blackwell Publishing, Ames, Iowa, USA.

Ventanas, J. 2006. El jamón Ibérico, De la dehesa al paladar, Ediciones Mundi-Prensa, Madrid, Spain.

Ventanas, S., J. Ruiz, C. Garcia and J. Ventanas. 2007. Preference and juiciness of Iberian dry-cured loin as affected by intramuscular fat content, cross breeding and rearing system. Meat Sci. 77: 324–330.

Virgili, R., G. Prolari, C. Schivazzappa, C. Soresi and M. Borri. 1995. Sensory and texture quality of dry-cured ham as affected by endogenous cathepsin B activity and muscle composition. J. Agric. Food Chem. 60: 1183–1186.

WHO—World Health Organization. 2003. Diet, nutrition and the prevention of chronic diseases. WHO Technical Report Series 916, Geneva, Switzerland.

CHAPTER 3

Fish Products from South Portugal
Dried *Litão*, Tuna *Muxama*, and Canned Mackerel

Eduardo Esteves

3.1 Introduction

The Mediterranean diet, representing the dietary pattern usually consumed among the populations bordering the Mediterranean sea, has been continuously and widely reported to be a model of healthy eating for its contribution to a favourable health status and a better quality of life (Sofi et al. 2008), even if its practice in the Mediterranean region is (regrettably) diminishing (Burlingame and Dernini 2011).

Since the Seven Countries Study in the 1950s, it is now known which foods were more or less frequently consumed by rural societies in the Mediterranean area (Bach-Faig et al. 2011). Thus, the Mediterranean diet is a dietary pattern rich in plant foods (cereals, fruits, vegetables, legumes, tree nuts, seeds and olives), with olive oil as the main source of added fat, along with high to moderate intakes of fish and seafood, moderate consumption of eggs, poultry and dairy products (cheese and yoghurt), low consumption of red meat and a moderate intake of alcohol (mainly wine and usually during meals) (Bach-Faig et al. 2011, Serra-Majem et al. 2012).

It is recommended to include at least two servings of fish, particularly fatty fish, per week (Kris-Etherton 2002, Nunes et al. 2003, 2008, Lichtenstein et al. 2006, Prato and Biandolino 2015). Healthy eating trends originated an increased demand for fish and seafood which has many health benefits, namely cardiac health, since heart disease causes many deaths in the developed world. The benefits of fish oil over animal

Department of Food Engineering, Institute of Engineering and CCMAR – Centre of Marine Sciences, University of Algarve, Faro, Portugal.
Email: eesteves@ualg.pt

fats in the diet are widely recognized among consumers and this has also led to a shift away from red meat towards white meat and, more importantly, to fish and fish products. The health benefits of eating seafood (Nunes et al. 2008) have primarily been linked to marine lipids and marine ω-3 (or n-3) polyunsaturated fatty acids (PUFA) that are associated with a reduced risk of coronary heart disease (Schmidt et al. 2006) as well as a number of chronic diseases (Prato and Biandolino 2015). Furthermore, the consumption of a variety of fish shall minimize any potentially adverse effects due to environmental pollutants (Kris-Etherton 2002, Lichtenstein et al. 2006, Afonso et al. 2013). Some of these are subjected to limits for human consumption, e.g., for heavy metals Cd and Pb in the EU (Regulation (CE) no. 1881/2006). In recent years the fish consumption raised concerns, as though sometimes contaminants exceed the set legal limits for food, they do not always represent a risk for human health (Afonso et al. 2013, Copat et al. 2013, James 2013). Moreover, methods used to prepare fish should minimize the addition of saturated and trans fatty acids, as occurs with the use of cream sauces or hydrogenated fat during frying.

On the other hand, diets link environmental and human health. In fact, rising incomes and urbanization are driving a worldwide dietary transition from traditional diets to diets higher in refined sugars, refined fats, oils and meats. By 2050 these dietary trends, if ungoverned, shall become a serious contributor to an estimated 80% increase in international agricultural greenhouse gas (GHG) emissions from food production and to international land clearing. Moreover, these dietary shifts are greatly increasing the incidence of type II diabetes, coronary cardiovascular disease and other chronic non communicable diseases that lower international life expectations. Alternative diets, such as the Mediterranean diet, that supply substantial health advantages might, if widely adopted, cut back GHG emissions and land clearing and resultant species extinctions, and help prevent diet-related chronic diseases (Tilman and Clark 2014). According to Burlingame and Dernini (2011) the Mediterranean diet is an example of a sustainable diet, in which nutrition, biodiversity, local food production, culture and sustainability are strongly interconnected.

On December 4, 2013, UNESCO recognized the Mediterranean diet as an Intangible Cultural Heritage of Italy, Portugal, Spain, Morocco, Greece, Cyprus and Croatia.

In a recent study by Vanhonacker et al. (2013) the overriding healthy perception European consumers have about fish was confirmed, and contributed very strongly to the general perception consumers have about fish. Fresh fish was perceived the most healthy fish product, followed by frozen, preserved and ready fish products. Remarkably there was no single dominant barrier to fish consumption, with price, contamination risk, and smell and/or taste or preparation difficulty poorly explaining the self-reported consumption frequencies of the different fish product categories in the different countries studied. Compared to the UK, Czech Republic or Germany, for example, in the Mediterranean countries studied, Greece, Italy or Portugal, fish consumption frequency is on a very high level, independently of perceived barriers and motivational aspects, and part of the traditional Mediterranean diet (Vanhonacker et al. 2013).

Portugal and the Algarve have a centuries-old, continued relationship with the sea, fisheries and fish and fish products (Cardoso and Nunes 2013). In recent years, the levels of fish and seafood consumption in Portugal are consistently in the range 55 to 60 kg per capita per year, one of the highest in the world. Notwithstanding, fisheries and related activities (aquaculture, salt extraction, seafood industry and commerce) are not yet expressed economically, neither in terms of the number of companies (only 0.7% of the total number) nor in terms of businesses' turnover (1.5% of total) (João Cadete de Matos, Banco de Portugal, unpublished data), but this state of affairs is changing. Despite the relatively marginal production of seafood products from fisheries and aquaculture compared to other countries—according to the Eurostat (EUROSTAT 2015) in 2013 Portugal's landings represented ca. 4% of total catches in the EU, 4.8 million tons—an estimated 57 kg/capita/year of seafood products are consumed in Portugal (FAO 2014a), i.e., three times the worldwide average consumption (ca. 19 kg/capita/year). After its prime in the 1960s, during 2013 the Algarve still represents 16% (about 22,700 t) of national landings, totalling about 144,650 tons (INE 2014). In the region, the most landed fish species are mackerel (*Scomber* spp.), sardine (*Sardina pilchardus*), horse mackerel (*Trachurus* spp.); tuna and tuna-like species represent only 1% of the landings and sharks other than dogfish are included in the «diversos» category (that represented ca. 10% of the landings). Shrimps (Penaidae) and crayfish (*Nephrops* spp.) are the most representative crustaceans in the catches, while octopuses (*Octopus* spp.) and grooved carpet shells (*Ruditapes decussatus*) are the dominant molluscs in the landings and aquaculture, respectively (INE 2014). In coastal Algarve, fish consumption has historically been in the form of boiled, grilled and fried fish, namely horse mackerel, sardine, hake, sea bream, meagre, mackerel, grouper, axillary sea bream, red snapper or Lusitanian toadfish. These fish and other seafood products, e.g., clams, are also used in dishes of the traditional cuisine. In addition, specially preserved fish, e.g., Litão (blackmouth catshark *Galeus melastomus*), Muxama (tuna) or stowed sardines, are distributed across the region and are consumed throughout the year or in special occasions.

Moreover, traditional processing technologies have clearly been developed in order to preserve excess quantities of fresh fish for storage or transport. Usually these technologies are not well documented and are being lost with the trend to urbanization and convenience food. While in most countries, products of this genre are usually made from low cost raw materials, in the so-called developed world there is still a considerable demand for products such as salted and dried cod, where the raw material and the final product are of high value (James 1998, Doe 2002). This chapter will discuss the production, processing and nutritional composition of selected, traditional fish products from southern Portugal and the Algarve: dried litão, tuna muxama and canned mackerel.

3.2 Dried *litão*

Dried litão is a traditional delicatessen produced by salting and drying blackmouth catshark *Galeus melastomus* (Fig. 3.1), a small scyliorhinid catshark that is common in the northeastern Atlantic Ocean from Iceland to Senegal, including the Mediterranean Sea, and usually found over the continental slope at depths of 150 to 1,400 metres on or near muddy bottoms.

Figure 3.1 Dried litão (www.cincoquartosdelaranja.com).

In Portugal, the catches of sharks, rays and chimeras reported to FAO (FAO 2014a) declined from ca. 19,000 tons in 2010 to about 15,000 tons in 2012 of which only a minute fraction, less than 4 tons, is registered as blackmouth catshark *Galeus melastomus* (Leitão or Litão in Portugal). It is taken, in relatively high quantities, as by-catch in demersal trawl and longline fisheries that targets the Norway lobster (*Nephrops norvegicus*), red shrimp (*Aristeus antennatus*) and deepwater pink shrimp (*Parapenaeus longirostris*), and by the near-bottom longline fishery that targets European hake (*Merluccius merluccius*), conger eels (*Conger conger*) and wreckfish (*Polyprion americanus*). In both fisheries, most captured specimens are discarded (Coelho et al. 2005), as in the Cantabrian Sea for example (Olaso et al. 2004), although it is retained and utilized in some areas (see below). Most specimens are captured and returned to the sea alive, but usually with severe injuries (due to the long trawling periods or hooks) that are likely to impair their survival. An increase of 151% in catches (CPUE) was displayed in data obtained aboard commercial longline vessels operating off southern Portugal (Algarve) from 1997 and 1998 (Erzini et al. 1999) compared to 2003 (Coelho et al. 2005). Official fisheries landings statistics for this species in the Algarve show a steady increase in commercial landings—not actual capture—during the last two decades (Correia 2009, DRGM 2013). This increase may be due to a reduction in discards, since a decline in stocks of traditional target species has resulted in increased retention of this species. Despite this, off southern Portugal the population appears to be currently stable; it is assessed as Least Concern in the IUCN red list of threatened species but population trends should continue to be monitored (Serena et al. 2009). Moreover, this species is not listed as prohibited in Regulation (EU) no. 43/2014 and Regulation (EU) no. 1262/2012, amended by Regulation (EU) no. 1182/2013 wherein *G. melastomus* was excluded from the list of deep sea sharks.

G. melastomus is utilized fresh or salted-dried for human consumption, for fish meal and for leather (Gibson et al. 2006, Serena et al. 2009, Ordóñez-Del Pazo et al. 2014). In some locations in the Algarve, this species is traditionally eaten during Christmas time after being salted-dried (see below). During this period, this species can reach very high prices in the local fish markets (Serena et al. 2009). Small shark species like *G. melastomus* are usually preferred to larger species for meat as they usually have lower concentration of urea and mercury in their flesh and are also easier to process. More often, where/when there are/were no available facilities for immediate refrigeration or freezing or when there is a surplus of shark meat which cannot be sold fresh, sharks are more commonly filleted and then salted and dried, usually sun-dried, or smoked (Vannuccini 1999). As with other dried fish products, «dried litão», is rehydrated before cooking.

Curing, as a means of preserving fish, has been practiced perhaps longer than any other food preservation technique (Horner 1997b). Drying (in fact sun-drying), salting (or brining), smoking, acid curing or a combination of these methods forms the basis of preparation of traditional fish products in many countries (Sen 2005). Drying is a traditional method and "the single most common unit operation in the food industry" (Driscoll 2004) for processing and preservation of seafood products that works by decreasing water activity (a_w) and thus inhibiting microbial growth as well as undesirable chemical reactions induced by enzymes. These effects facilitate storage (at ambient temperature), transportation and consumption of products. Furthermore, unique flavour-texture combinations that are highly appreciated by consumers are also incorporated by dried products (Horner 1997b, Doe 2002, Driscoll 2004, Wang et al. 2011). Commonly, salting is a preliminary step in fish drying processes to obtain a commercial product with adequate shelf life (decreased a_w) and provide specific organoleptic and sensory characteristics to seafood products as well as better quality and higher processing yield of the final product. In addition (heavy) salting contributes to water removal and shortens subsequent dehydration time. Dried and salted fish is a very popular food item worldwide, e.g., in Europe, Asia, the Caribbean or Canada, where today the demand is driven more for the flavour of the product than for preservation purposes (Mujaffar and Sankat 2005, Guizani et al. 2008, Wang et al. 2011, Darvishi et al. 2013). A few examples of salted-dried products are stockfish (cod), bacalhau (salted-dried cod), conpoy (dried scallops), dried squid/octopus, vobla (salt-dried roach), and dried litão (dried blackmouth catshark). In what concerns catshark, in spite of the controversy about shark finning (resolved in the European Union by Regulation (EU) no. 605/2013 and not really a threat to this group of smaller, short-finned sharks) and other uses, e.g., production of leather (Ordóñez-Del Pazo et al. 2014), the flesh of sharks is highly favoured in some regions, namely in Europe wherein northern Italy and France are major consuming countries and Spain is the world's largest exporter of shark meat and Italy the largest importer (Vannuccini 1999). Notwithstanding, a number of euphemisms, e.g., petite roussette (in France) or gattucci (in Italy) have been necessary to mask the name shark to overcome consumer resistance in many countries (Vannuccini 1999). In Olhão (Algarve) there is a Confraria do Litão, a gastronomical association, founded in December 2011, to promote this local tradition.

3.2.1 Processing

The processing of blackmouth catshark to produce «dried litão» in Olhão is described next (Fig. 3.2). Fresh specimens (1) are blooded, skinned and split (2). Most fishers salt the fishes in light brine as a preparatory step (3). Specimens are then mounted (4) on inverted-T-like frames (or racks) made of giant cane that are hanged out to dry in the sun (and wind—that averages 17 to 20 km/h in the region) (5). In these conditions, the drying time can range from four days on a summer day, when air temperature and relative humidity commonly, and for continued periods, exceed 25°C and does not reach 60%, to four weeks in the winter, when ambient conditions are less beneficial to drying (12 to 16°C and >75% relative humidity). Expectedly, the method requires special care. Drying specimens must not catch rain or humidity, since these are believed to compromise the drying process. In fact sun drying is liable to contamination and insect attack (Yean et al. 1998, Sampels 2015). Moreover, dried fish should be kept in an elevated place with good air circulation to avoid further development of molds. The end-product—dried blackmouth catshark—is named «dried litão» (Fig. 3.1) and packed (6). This procedure is at the mercy

Fresh fish[1]
|
Blooding, Eviscerating,
Skinning and Splitting[2]
|
Brinning[3]
|
Mounting[4]
|
Sun-drying[5]
↓
«Packaging»[6]

Figure 3.2 Flow diagram of the traditional process of drying blackmouth catshark in the Algarve. Numbers are used as reference in the main text.

of the weather and little control is possible other than physical protection of the drying fish; notwithstanding, the traditional procedure described above implements several ways to improve the drying rate (Hall 2011b) such as increasing the surface area available for moisture loss by splitting specimens and raising the temperature of the fish and orienting the fish for maximum exposure to sun and wind.

3.2.2 Composition

In the early 1990s, Martins (unpublished data) studied the drying process of *G. melastomus* and its effect on a number of quality parameters. Recently, Rosa (unpublished data) and Lourenço (unpublished data) assessed some microbiological and physico-chemical characteristics of marketed «dried litão». Those authors and Nunes (unpublished data) carried out experiments to study the effects of process conditions upon dried product characteristics of small spotted catshark *Scyliorhinus canicula*, an akin but relatively more abundant shark species (Olaso et al. 2004). Ultimately, their aim was to evaluate the prospect of developing a salted-dried seafood delicatessen similar to «dried litão» considering that Ordóñez-Del Pazo et al. (2014) list *S. canicula* as one of several discarded species in Spanish and Portuguese métiers with valorization potential.

Expectedly drying of blackmouth catshark has significant impact in several quality characteristics of the raw material and confers the characteristics to the final food product, «dried litão». Those effects have been observed in other dried shark products, e.g., *Charcharhinus sorrah* (Guizani et al. 2008) or *C. macloti* (Al Ghabshi et al. 2012). Table 3.1 summarizes the changes in the main nutrients, other constituents and microbiological parameters compiled from several sources.

Table 3.1 Proximate composition and quality parameters of fresh and salted-dried catsharks. Values in parenthesis are the standard deviation.

Parameter	Species	Fresh	Salted-dried/ commercial	Ref.
Moisture (g/100 g)	G. melastomus	77.6 (0.38)	32.64 (3.59)	[1]
	G. melastomus	--	28.4	[2,3]
	C. sorrah	75.37 (0.37)	34.35 (1.76)	[6]
	C. macloti	50.4 (0.16)	33.0 (0.14)	[6]
	S. laticaudus	76.3 (0.20)	--	[4]
a_w	G. melastomus	--	0.675	[2,3]
	C. macloti	0.89 (0.002)	0.69 (0.004)	[6]
Protein (g/100 g)	G. melastomus	16.2 (0.20)	--	[1]
	C. macloti	22.1 (0.15)	44.72 (0.12)	[6]
	S. laticaudus	26.7 (0.75)		[4]
Lipids (g/100 g)	G. melastomus	0.30 (0.10)	--	[1]
	G. melastomus	0.55	--	[5]
	C. sorrah	0.50	--	[7]
	C. macloti	6.70 (0.03)	7.48 (0.06)	[6]
	S. laticaudus	1.35 (0.07)	--	[4]
SFA (mg/100 g)	G. melastomus	93	--	[5]
MUFA (mg/100 g)	G. melastomus	70	--	[5]
PUFA (mg/100 g)	G. melastomus	219	--	[5]
Ash (g/100 g)	G. melastomus	2.26 (0.13)	--	[1]
	C. sorrah	1.16 (0.08)	3.15 (0.33)	[7]
	C. macloti	0.92 (0.01)	27.54 (0.09)	[6]
	S. laticaudus	1.19 (0.10)	--	[4]
NaCl (g/100 g)	G. melastomus[B]	--	18.8 (0.60)	[1]
	G. melastomus	--	5.7	[2,3]
	C. macloti	0.6 (0.004)	17.34 (0.07)	[6]

Cont....

Cont.

Parameter	Species	Fresh	Salted-dried/ commercial	Ref.
L*	G. melastomus	--	46.80	[2,3]
	C. sorrah	55.5 (0.49)	61.2 (1.21)	[7]
a*	G. melastomus	--	1.38	[2,3]
	C. sorrah	12.9 (0.58)	5.1 (2.41)	[7]
b*	G. melastomus	--	8.55	[2,3]
	C. sorrah	8.8 (0.41)	14.7 (4.60)	[7]
pH	G. melastomus	--	6.77	[2,3]
TVB-N (mg/100 g)	S. laticaudus	9.98	--	[4]
	C. dussumieri	13.86	--	[8]
Microorganisms (log(cfu)/g)				
Total Viable Counts	G. melastomus	--	2.20 (0.23)	[2,3]
	C. sorrah	4.20	4.12	[7]
	C. macloti	6.19	1.48	[6]
Psycrotrophic aerobic bacteria (20°C)	G. melastomus[A]		<0.5	[1]
Psycrotrophic anaerobic bacteria (20°C)	G. melastomus[A]		<0.5	[1]
Aerobic halophilic bacteria, 15% NaCl	G. melastomus[A]		<0.5	[1]
Staphylococcus aureus	G. melastomus	--	1.51 (1.35)	[2,3]
	C. sorrah	3.02	2.27	[7]
Coliforms	G. melastomus[A]	--	ND	[1]
	C. macloti	2.65	ND	[6]
Molds	G. melastomus	--	1.39 (0.46)	[2,3]
	C. sorrah	2.12	2.61	[7]
	C. macloti[C]	2.58	1.30	[6]

[1] (Martins unpubl. data), [2,3] (Lourenço unpubl. data, Rosa unpubl. data), [4] (Mathew and Shamasundar 2002), [5] (Sirot et al. 2008), [6] (Al Ghabshi et al. 2012), [7] (Guizani et al. 2008), and [8] (Nazemroaya et al. 2011). ND—non detected. [A] After storage of dried product for 2 months at 20°C.[B]After salting. [C]Molds and yeasts.

The main objective of drying is to remove moisture (Doe 2002) and thus decrease a_w to a level at which microbial spoilage and deterioration reactions are minimized, which allows safe storage over an extended period of time (Horner 1997b, Fellows 2000, Hui et al. 2004, Vega-Gálvez et al. 2011). The moisture content of dried blackmouth catshark ranges from 28 to 33%, an important decrease from the initial ca. 77%, which corresponds to a_w <0.7. This reduction is in line with the results of Guizani et al. (2008) for spottail shark (*C. sorrah*) and Al Ghabshi et al. (2012) for hardnose shark *C. macloti*. Moreover, the measured a_w of «dried litão» is below 0.75, the level that is referred to in Derrick (2009) for dried products; only xerophilic molds are known to be able to grow at that a_w (Horner 1997b, Derrick 2009).

Appearance of foodstuffs is one of the most important factors valued by consumers. The colour of dried fish changes strikingly compared to the fresh specimens (Louka et al. 2004). The observed yellowish brown discoloration of dried fish products, observed in dried catshark (Lourenço unpublished data, Rosa unpublished data) and spottail shark (Guizani et al. 2008), is mainly due to browning (Maillard) reactions, lipid oxidation and degradation of myoglobin (Louka et al. 2004, Guizani et al. 2008, Sampels 2015). Differences in the extent of the effects might be related to drying conditions, mainly temperature. Drying in the sun might promote other effects or changes in colour due to sunlight characteristics, e.g., UV.

Drying is believed to have negative effects on the nutritional properties of food products due mainly to changes in proteins and lipids of the fish muscle. Protein denaturation/aggregation that occurs during the drying process leads to changes in amino acid composition, protein solubility and protein digestibility. Similarly, lipid oxidation can cause the degradation of fatty acids, particularly PUFAs, e.g., eicosopentaenoic acid (EPA) and docoshexaenoic acid (DHA). Furthermore, lipid oxidation products could also have a negative effect on human health since their interaction with protein constituents (i.e., amino acids, peptides) cause significant changes in proteins, affecting their solubility and nutritional value. Increased lipid oxidation in the traditional process of drying fish has been observed and attributed to light and temperature but only few studies exist (Sampels 2015). Loss of vitamins in the fish muscle also occurs during drying (Nguyen et al. 2014).

There will be changes in the concentrations of solutes within the fish, when water is removed from cured fish and moisture content is decreased (Doe 1998). Salt concentration, for example, will increase as water is removed during drying and this increase in salt molality will drive the salt molecules further into the fish (to regions of lower molality)—a phenomenon named bimodal diffusion (Doe et al. 1998). For dried catshark, the NaCl content observed (ca. 19 g/100 g), might represent a ten-fold increase as in *C. macloti* (Al Ghabshi et al. 2012). The penetration of salt by this process has beneficial implications for the microbial stability of the cured dried fish product (Doe et al. 1998) considering the pH of the product, 6.77. In fact, the abundance of microorganisms, both bacteria and molds, is quite low (Table 3.1), well below the limits suggested by the ICMSF (1986), 7 log (CFU)/g. Both the reduced a_w and increased concentration of salt are detrimental for the survival and development of the majority of microorganisms (Horner 1997b, Derrick 2009).

There are still a number of important aspects, dealing with both the conditions of the drying process and the characteristics of the final product, «dried litão», that have not been studied yet and thus preclude a more complete description of the process and product, e.g., free amino acids and fatty acids profiles.

3.3 Tuna *muxama*

Muxama (in Portugal or Mojama in Spain or Mosciame in Italy) is a Mediterranean delicacy served as thinly sliced salt cured tuna. It is made from tuna loins by curing them in salt, using a traditional procedure, other than the more usual canning of tuna or consumption as fresh steaks or sashimi.

Aníbal and Esteves (2016) give a recent account of the importance of tuna as fishing and food resources. Their work details data of catches and/or landings for the most important species for commercial and recreational fisheries worldwide and in Portugal and Spain, namely yellowfin (*Thunnus albacares*), bigeye (*T. obesus*), bluefin (*T. thynnus, T. orientalis*, and *T. macoyii*), albacore (*T. alalunga*), and skipjack (*Katsuwonus pelamis*), and briefly describes the traditional method employed to catch tunas in the south of the Iberian Peninsula (Algarve, Portugal, and Andalucia, Spain)—using the «armação» (Portugal) or «almadrava» (Spain). This is an off-shore "maze" of bottom-fixed nets to imprison, capture and hold the fish (Rebelo 2010). Gallart-Jornet et al. (2005) and Lindkvist et al. (2008) describe in detail the history of fish salting in the context of the Mediterranean diet. According to those authors, native Iberians were already drying and salting fish, mainly tuna, in pre Roman times, a practice that was further developed during Roman rule over the provinces of Hispania (i.e., modern Iberia), particularly in *Hispania Baetica* (one of three Imperial Roman provinces in Hispania) and continued during Al-Andalus (a medieval Muslim cultural domain and territory that eventually occupied most of what are today Spain and Portugal). The developments in packaging of foods in the 20th century, namely vacuum packaging (Han 2005, Perdue 2009, Arvanitoyannis 2012), contributed further to develop muxama and other tuna-related food products.

Muxama is the most important of a number of products obtainable from the 50 to 60% of a tuna that can be utilized for the manufacturing of food products or preparation of meals, viz. "Muxama", "Estupeta", "Mormos", "Rabinhos", "Faceiras" and "Orelhas", "Ventresca", "Tarantela" and "Sangacho", "Espinheta", "Tripa", "Bucho" and "Ovas", at the end of the (traditional) quartering of tunas a.k.a. "ronqueamento" (in Portugal, or "ronqueo" in Spain) (Lã and Vicente 1993, Yubero 2008).

Videos showing this particular procedure are easily found online through major web search engines and/or video sharing websites.

The word "muxama" comes from the Arab "musama" and means "dry". The production of muxama was developed by the Arabs ca. 10th century AD, together with the "almadravas" for catching tunas during their migrations between the Atlantic Ocean and Mediterranean Sea, following the Phoenicians that settled in Gdr (Gadir, Cádiz today), in the Western Mediterranean Sea, and had learned to dry tuna in sea salt to make it appropriate for trade (Godinho 2011, Gallart-Jornet et al. 2005). Muxama production has not changed much since then: the tuna loins were air dried during summer, when the weather in the south of the Iberian Peninsula was hot and dry. Nowadays, "temperature and humidity controlled rooms" allow year-round production of muxama both from locally caught tuna species, such as bigeye, yellowfin and albacore, using the almadravas and tuna that is fished elsewhere and arrives frozen to the plants (Barat and Grau 2009).

3.3.1 Processing

The preparation of muxama involves a series of steps that are succinctly described herein (Lã and Vicente 1993, Gallart-Jornet et al. 2005, Aníbal and Esteves 2016) and presented as a flow diagram in Fig. 3.3. After beheading, eviscerating/gutting and fining, tunas are quartered into four big loins (2-4), two in the tuna's dorsal (upper) side and other two in the ventral (lower) side (Fig. 3.4a). The inner, central parts in each of these loins are composed of red muscle with low lipid content, that after being properly cut into rectangle parallelograms (5-6) are used to produce muxama. The trimmed loins are generally 50 cm in length, 8 cm in width and 3 cm in height (Lã and Vicente 1993). These loins (and the muxama subsequently produced) are classified into extra or first class.

Then, the tuna loins are stacked or piled in alternated layers of sea salt for 24 to 48 hours (7). Traditionally, the piles were hard-pressed with large circular stones named «pedras de estiva» in Portugal. Later on, loins are washed several, usually six or seven times, in large tanks of tap/potable water (8). After removing all visible crystallized salts from the surface of the loins (8) and letting the loins drain for 24 to 48 hours (9)—this standing stage is carried in trays until the loins are firm enough –, they are hung and dried at 14°C and 60% relative humidity (10) for as long as 12 days, depending on their size (Lã and Vicente 1993).

Differently sized portions (11) are marketed, from 80 to 100 g to >1 kg or 3 kg, that are packed vacuum-sealed in polyethylene bags (skins) or trays (12) or alternatively immersed in vegetable (mainly olive) oil in jars or cuvettes. The former have a shelf life of six months, while the latter have only two months.

Thawed fish[1]
|
Quartering or Sectioning[2]

```
. . . . . .+. . . . . .
:    Headless,      :
:    Sectioned,     :
:     Gutted,       :
:    Quartered,     :
:    Cleaned[3–4]   :
. . . . . .|. . . . . .
```

Classifying[5]
|
Cutting[6]
|
Pile salting[7]
|
Washing or Desalting[8]
|
Standing[9]
|
Drying[10]
|
Cutting[11]
▼
Vacuum packaging[12]

Figure 3.3 Flow diagram of the salting-drying process used to produce muxama (adapted from Lã and Vicente (1993) and Gallart-Jornet et al. (2005)). Numbers are used as reference in the main text.

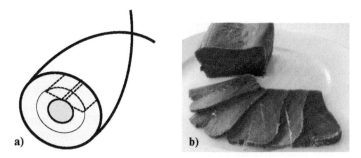

Figure 3.4 (a) Diagram representing the relative position of the tuna loins used in the production of muxama (indicated by dashed lines), (b) tuna muxama.

As expected, the curing process described above affects the proximate composition and organoleptic characteristics of the product.

Due to the salting and drying processes, both the size (in terms of volume) and weight of individual loins used for muxama (Fig. 3.4b) are significantly reduced, to ca. 46% of original size and weight according to Lã and Vicente (1993). This is a significant shrinkage that is mainly due to water removed due to brining and during subsequent drying of tuna loins (Nguyen et al. 2014). Furthermore, Lã and Vicente (1993) found that a_w is reduced from 0.96 in the thawed muscle of tuna used for processing to 0.79 in the muxama. This is relatively lower than the a_w measured by Gómez et al. (1991) in samples of muxama from Spain, 0.851. In contrast, Lã and Vicente (1993) found that pH increased from 5.72 to 7.10. While the initial pH is in line with the range of values published for fresh fish (i.e., 5.2 to 7.1 or 7.3), the final value obtained by Lã and Vicente (1993) was relatively high compared to the pH of 6.14 measured by Gómez et al. (1991) and the pH of 5.8 reported by Gallart-Jornet et al. (2005) for commercial samples from Spain and warranted attention since it is within the pH values that are optimal for microorganism growth (Gram and Huss 1996, Derrick 2009, Fernandes et al. 2009). In addition, Rahman et al. (2000) concluded that drying temperature of 50°C or below is not lethal to the microbiota in minced tuna. Notwithstanding, the product is considered stable because of the relatively low a_w, that is expected to inhibit the development of a number of microorganisms particularly pathogens (Hall 2011b), and the salt content, ca. 10.4% (Lã and Vicente 1993), that further creates an unsuitable medium for microorganisms development. Recently, Barat and Grau (2009) studied the simultaneous thawing and salting of frozen tuna loins by using dry salt or brine and observed a clear shortening of the processing time required to achieve the same NaCl concentration as in the commercial samples. In a recent set of experiments, Esteves and Aníbal (unpublished data) describe the dynamics of important physico-chemical parameters such as moisture content, a_w, NaCl concentration, pH and colour, during the salting and drying stages of muxama production and tested the effects of changes in the traditional processing conditions followed in southern Portugal (Algarve), aiming at optimizing the production procedure.

The storage stability of packed muxama has been confirmed by the results of physico-chemical and microbiological analyses reported in the technical specifications sheets of a few producers that comply with national and European legislation, namely Reg. (CE) no. 2073/2005 amended by Reg. (CE) no. 1019/2013 and altered by Reg. (CE) no. 1441/2007, Reg. (CE) no. 2074/2005 amended by Reg. (CE) no. 1022/2008 and Reg. (CE) no. 1881/2006. The reported TVB-N content, 28 ± 2 mg N/100 g, is within the range of limit values stipulated for unprocessed fish and fish products (even though of different species) of 25 to 35 mg N/100 g. On the other hand, the hazards related to heavy metals contamination and histamine content are under control since reported values are <0.10 mg/kg for Cd, <0.30 mg/kg for Pb and 0.99 to 1.00 mg/kg for Hg and <100 to 200 ppm for histamine (below the m = 200 mg/kg and M = 400 mg/kg limits set by Reg. 1019/2013), respectively. In terms of microbiological parameters, counts of TVC below 10^4–10^5 cfu/g, abundance of Enterobacteriaceae and *Staphylococcus aureus* less than 100 cfu/g, and non-detected *Salmonella* or *Listeria monocytogenes* per 25 g reflect muxama's safety. The worrying values observed by Lã and Vicente (1993) for molds and yeasts, 1.9×10^4 and 4.0×10^6 cfu/g respectively, have not been reproduced elsewhere.

3.3.2 Composition

Drying is supposed to have negative effects on the nutritional properties of food products due mainly to changes in proteins and lipids of the fish muscle. Tables 3.2 and 3.3 aggregate information on raw, fresh tuna and muxama compiled from a number of sources. Protein denaturation/ aggregation that takes place during the drying process leads to changes in amino acid composition, protein solubility and protein digestibility (Nguyen et al. 2014). Similarly, lipid oxidation can cause the degradation of fatty acids, particularly PUFA, e.g., EPA and DHA. Apparently the results compiled herein (Table 3.3) provide evidence of this effect in bluefin tuna but less so in yellowfin tuna. Furthermore, lipid oxidation products could also have a negative effect on human health. The interactions between lipid oxidation products and protein constituents (i.e., amino acids, peptides) cause significant changes in proteins, affecting their solubility and nutritional value (Nguyen et al. 2014). While studying the effects of drying on protein and fatty acids of sand smelt *Atherina boyeri* Smida et al. (2014) found that, from a nutritional point of view, drying conditions of 50°C, 30% relative humidity and 2 m/s air-flow are not completely detrimental for the preservation of sand smelt—the authors observed a decrease in total fatty acids from 4.9 to 2.8 mg/100 g. The ω3 PUFA were much less sensitive to heat than the ω6 PUFA. Loss of vitamins in the fish muscle also occurs during drying (Nguyen et al. 2014). Unfortunately there is no published data about these changes specifically for muxama.

Muxama has the general organoleptic characteristics of any product subjected to the combined processes of salting and drying. It presents a rigid, typical consistency of a dried product, with a dark red-brown coloring with pleasant odour and flavour, exempt from rancid off-taste/ odour, and smooth texture and with a salt content appropriate for the product in question. Muxama is usually served in extremely thin slices with olive oil and chopped tomatoes and/or roasted almonds (cf. Gallart-Jornet et al. 2005); sometimes, this appetizer is aromatized with rosemary.

Two municipalities in Andalucia (Spain), Barbate and Isla Cristina, that represent ca. 75% of the muxama production in that region (Ministerio de Agricultura, Alimentación y Medio Ambiente de España 2014), have very recently filed applications for Protected Geographical Indication(s) (PGI)—Mojama de Barbate and Mojama de Isla

Table 3.2 Proximate composition (per 100 g) of raw, fresh tuna (bluefin, *T. thynnus*, and yellowfin, *T. albacares*) and muxama (different producers) gathered from available sources. Values in parenthesis are the standard error (±SE).

Bluefin

Proximate composition	Raw				Muxama	
	[1]	[2]	[3]	[6]	[4]	[6]
Energy (kcal/kJ)	137.5 / 575.2	144 / 602	143 / 597	177.0–200.7 / 741–840[A]	199.5–204.4 / 833.9–854.4	-- / --
Moisture (g)	69.7	68.1 (0.78)	--	65–71.3	--	49.6
Protein (g)	24.8	23.3 (0.38)	18.8–25.0	21.1–29.9	49.2–48.4	38–42
Lipids (g)	3.5	4.90 (0.97)	5.3–14.6	6.28–13	0.3–1.2	0.95
Ash (g)	1.7	1.2 (0.02)				
NaCl[B] (g)	0.113	0.098	0.113	0.1–0.3	7–8	7–8

Yellowfin

Proximate composition	Raw				Muxama		
	[4]	[5]	[3]	[6]	[4]	[6]	[5]
Energy (kcal/kJ)	109 / 454	115.0 / 481.0	143 / 597	103.9–139.3 / 434.8–582.8	187.4–194.7 / 784[A]–814	-- / --	47.6 / 199.4
Moisture (g)	74.0 (0.92)	70.5 (1.49)		71.1		49.6	46.6 (1.19)
Protein (g)	24.4 (0.74)	26.5	24.3–26.8	23.4	41.7–48	38–42	
Lipids (g)	0.49 (0.11)	1.0	0.73–3.05	1–5.0	2.3–0.3	0.95	
Ash (g)	1.6 (0.07)	0.01					0.001
NaCl (g)				0.1–0.3	7–8	7–8	10.4

Legend: [1] (Ruano et al. 2012), [2] USDA (2016), [3] (Torry Research Station 1989), [4] «Salazones Herpac, S.L.», [5] (Lã and Vicente 1993), [6] (Gallart-Jornet et al. 2005). [A]Estimated from protein and lipid contents. [B]Except for values reported by [5] and [6], NaCl content was estimated from sodium concentration ([Na] × 2.5).

Table 3.3 Fatty acids (per 100 g) of raw tuna and muxama compiled from available sources. Values in parenthesis are the standard deviation.

Fatty acids (g)	Raw (bluefin) [1]	(bluefin) [2]	Muxama (bluefin) [4]	Raw (yellowfin) [2]	Muxama (yellowfin) [4]
Saturated	1.73	1.257	<0.1–0.4	0.172	<0.1–0.7
Monounsaturated	1.77	1.600	<0.1–0.3	0.116	<0.1–0.7
Polyunsaturated	0.79	1.433	0.2–0.5	0.147	0.2–0.9
ω3	0.69	1.298[A]		0.104[A]	
EPA	0.05	0.283		0.012 (0.004)	
DHA	0.42	0.890		0.088 (0.022)	
ω6	0.11			0.008[B]	

Legend: [1] (Bandarra et al. 2004, INSA 2010), [2] USDA (2016), [4] «SalazonesHerpac, S.L.»; [A]Sum of EPA, DPA and DHA. [B]Sum of 18:2 n-6 c,c, 18:3 n-6 c,c,c, 20:2 n-6 c,c and 20:3 n-6.

Cristina—to the European Union's quality schemes for agricultural products and foodstuffs (Regulation (EU) no. 1151/2012).

3.4 Canned mackerel

Heating is the oldest and most reliable method of preservation of food products including seafood (Venugopal 2010). Heat processed fishery products may be grouped into three categories, depending on the intensity of thermal energy applied: conventional or traditional processes that depend on heat treatment below 100°C (e.g., curing); products that receive thermal treatment at temperatures above 100°C (such as those produced by canning, retort pouch packaging, etc.); and, more recently, those products involving a combination of mild heating with other processes such as chilling, uses of food additives including antimicrobial and antioxidant compounds, packaging, etc. (Venugopal 2010).

Thermal sterilization of fish on a commercial scale was initiated with the invention of canning. The technology for preserving foods in cans was developed at the beginning of the 19th century by a Frenchman, Nicolas Appert (Warne 1988). Canning, i.e., the thermal treatment (at temperatures above 100°C) of fish in sealed (metallic) cans, eliminates bacterial as well as autolytic spoilage, giving products with shelf lives of one to two years or more at ambient temperatures (Venugopal 2010). The main species of fish to be considered for canning are the oily (fat) fish (Aubourg 2001, Garthwaite 2010, Hall 2011a), namely mackerel, *Scomber* spp. (Fig. 3.5a and b).

In European waters, the catches of Atlantic and chub mackerel have been increasing since 2006 (ca. 540,000 tons) to about 890,000 to 930,000

Figure 3.5 (a) Fresh mackerel, (b) Portuguese canned mackerel fillets.

tons in 2011–2012 (FAO 2014b). In Portugal, reported landings of *Scomber* spp. have steadily increased to about 37,000 t in 2012. Only a fraction of these fish are used in the Portuguese canning industry considering its production level in 2012 of ca. 2,600 t (INE 2014).

It is well-known and for a long time (Seno 1974) that the processing and utilization of mackerel (*Scomber* spp.) present a number of technical problems, such as the large variations in the fat content of the fish (a typical range throughout the year is 6 to 23% (Keay 2001)), the high enzyme activity and the large catches. However, as the fish are usually caught in nearby waters and catches/stocks are (seemingly) increasing, they have formed a very attractive raw material for processing and utilization. Besides chilling, several methods of processing mackerel are reported, such as salting and/or drying, dried mackerel sticks ('fushi'), mackerel paste, smoking (ordinary, electric or liquid), and canning. The latter is described and discussed herein/below.

Raw fish for canning must be as fresh as possible (Seno 1974) as the quality of the raw material is the key factor which governs the final (canned fish) product quality (Seno 1974, Garthwaite 2010, Sampels 2015) and the microbiological content of the raw material is crucial for the effective preservation or stability of the end product (Gaze 2010). Mackerel meat has stronger enzymatic activity compared with other fish and decomposition of ATP-related substances is rapid (Seno 1974). In fact, although it is recognized as a healthy food, it remains underutilized in addition because of its short chilled shelf life (up to 10 days), that is also related to lipid hydrolysis and oxidation (Sanjuás-Rey et al. 2012). Fatty fish like mackerel contain high levels of unsaturated fatty acids, which are susceptible to attack by atmospheric oxygen leading to rancidity (Garthwaite 2010). On the other hand, when frozen fish is used for canning, thawing conditions are important. Thawing of frozen (fish) blocks using water, e.g., by immersion in moving water, is commonly used, inexpensive and simple method (where an ample supply of clean water is available) and is more rapid than air thawing due to higher rates of heat transfer between the product and water. Nevertheless, it may cause the fish to lose quality in terms of flavour and appearance, and if water is absorbed during the thawing process this may cause problems in subsequent operations during the canning process (Garthwaite 2010).

The nutritive value of fresh mackerel and its products is generally as high (or even higher) as that of tuna, bonito, and sardine (Bandarra et al. 2004, INSA 2010), while canned mackerel in brine has a higher value than that of other canned fish because its subcutaneous fat does not come off the body during preheat preparation (Seno 1974). Similarly, canned mackerel in (olive) oil has comparable nutritional value to that of canned tuna and sardine (see next). In general, the nutritional value of the fish and shellfish is not seriously damaged during thermal processing (Méndez and Abuín 2012).

3.4.1 Processing

The unit operations for finfish canning include: skinning; filleting; separation of fish parts after evisceration and trimming of fins, scales and other inedible parts; brining; cooking; exhausting; hot filling; and sealing (Horner 1997a). Warne (1988) uses different, more-general designations for these operations and elaborates on the later, key stages in the process, namely: retorting (for cans or for glass); post process handling (incl. chlorination and cooling water quality, and post-process hygiene and sanitation); and the final operations (incl. assessing of container damage during handling and storage, and control of cooling rate, and temperature of storage). In fact, relatively few unit operations are unique to one particular manufacturing process (Warne 1988). The customary process of canning mackerel carried out in the south of Portugal is described below (and presented diagrammatically in Fig. 3.6) and closely follows international, standardized practices outlined, e.g., in Seno (1974), Warne (1988), Horner (1997a) and CAC (2012). Readers are also referred to Fellows (2000), Ababouch (2002), Bratt (2010), Sun (2012), and Clark et al. (2014) for further readings on fish canning processing and safety and heat treatment applied to food products, respectively.

In Portugal and in the Algarve, most plants are located near fishing ports thus fresh fish is used but frozen fish is occasionally utilized. At reception (1) of mackerel, freshness is assessed considering appearance/colour (brightness), onset/resolution of rigor mortis, odour and temperature. Upon lot acceptance, fishes are transported directly to the production or stored at 0 to 4°C awaiting processing. When frozen fish is used, thawing of blocks is carried out in refrigerated chambers (overnight) or in agitated water. Following washing (4) in light brine (the specification of its strength is species-specific), specimens are beheaded, eviscerated and trimmed (5) and bled (6) in running water before being transferred to the cooking tank wherein a mixture 1:1 of water and brine (4 to 6° Baumé, i.e.,

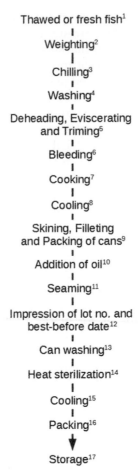

Thawed or fresh fish[1]
|
Weighting[2]
|
Chilling[3]
|
Washing[4]
|
Deheading, Eviscerating
and Triming[5]
|
Bleeding[6]
|
Cooking[7]
|
Cooling[8]
|
Skining, Filleting
and Packing of cans[9]
|
Addition of oil[10]
|
Seaming[11]
|
Impression of lot no. and
best-before date[12]
|
Can washing[13]
|
Heat sterilization[14]
|
Cooling[15]
|
Packing[16]
↓
Storage[17]

Figure 3.6 Flow diagram of the canning process used to produce canned mackerel fillets. Numbers are used as reference in the main text.

ca. 40 to 60 g NaCl per litre of water) has been previously heated to 100°C. The fish are cooked (7) for 15 to 25 minutes depending on specimens' sizes (more time for larger fishes). Apart from the modification of the sensory characteristics, this process causes the denaturation of proteins and loss of water, which otherwise would be released during the retorting process. Hence, the proteinaceous exudate, which forms an unsightly curd in the surrounding liquor during processing, is minimized (Horner 1997a). Subsequent cooling (8) (eventually in a refrigerated room/chamber) allows the flesh to set and become manageable, and the excess water is released. In fact, one use of pre-cooking is in the production of boneless mackerel fillets: in the cooked form the skin and visible spines must be painstakingly removed, usually by hand, but it is easier to separate the two fillets from either side of the fish skeleton, and then place them directly into cans (9), ca. 4 to 5 fillets per can, or place onto modified conveyors leading to automatic can filling machinery (Garthwaite 2010).

Oil (olive oil or sunflower) is added (10) to the cans while they are transported on a conveyor to the seamers. Both filling and sealing are operations that require close control to ensure the integrity of the final seam (Horner 1997a). Failure to form hermetically sealed containers means that product safety and shelf stability is at risk (Warne 1988). The re-contamination of the contents after processing, usually by ingress through the double seam, is at the origin of the majority of incidences of canned food spoilage by microorganisms (Horner 1997a); accounting for as much as 60 to 80% of the cases (Ababouch 2002). This is called 'leaker spoilage'. Nondestructive visual and tactile examination of can seams may detect a fault of the several possible in the seaming operation (Horner 1997a). Besides the abundant literature available from packaging material and sealing machine suppliers recommending methods of seal formation and criteria for their evaluation, in several countries regulatory authorities have published procedures for the evaluation of seal adequacy (Florêncio and Albuquerque (1996) and Florêncio (1996) in Portugal); the purpose of these regulations is to ensure that not only local manufacturers have guidelines to follow, but also that foreign manufacturers can comply with the requirements of the country to which they are exporting to (Warne 1988). Thence it is important that sealed containers be indelibly coded (12) with details of the production date and time, product codes, the manufacturing plant and any other information that is necessary to identify the origin and nature of the product.

After washing the (sealed) cans (13), they are placed in crates and then subjected to the heat-sterilization step (14), during which a combination of increased temperature (and pressure) inactivates most microorganisms that are hazardous for human consumption. Their initial numbers are usually small and the health risk low unless they increase in number, which is usually the result of poor handling and inappropriate storage conditions. *Clostridium botulinum* has been chosen as the organism which must be destroyed to a high level of probability because it is spore-forming, anaerobic and its toxin is particularly toxic to humans (Piggott and Tucker 1990, Hall 2011a, Ababouch and Karunasagar 2014, Skipnes 2014). The severity of the heat-sterilization step applied to low acid foods (>pH 4.5), that include most canned fish products which are canned at about neutral pH, requires a full heat treatment and rapid cooling to prevent the germination of surviving spores (Hall 2011a, Skipnes 2014). In the fish canning industry, a 12 log reduction in spore counts (i.e., a 12D process) is used as a minimum heat process to be applied. For proteolytic strains of *C. botulinum* with higher resistance, this would be $12 \times 0.25 = 3.0$ min at 121.1°C and a $Z = 10$°C (also called F_0); in commercial practice, higher

F-values (e.g., 5.0) are often used to produce botulism-safe canned food (Ryder et al. 2014). The most common medium in use for heat processing of canned foods (and fish) to commercial sterility, is saturated steam under pressure (Horner 1997a, Gaze 2010). Frequently used conditions in conventional retorts are 1.5 to 2.0 bar pressure and 111.4 to 120.2°C (condensation) temperature (Myrseth 1985, Horner 1997a, Gaze 2010). In the processing line that is being described herein, a temperature of 118°C is maintained for 45 minutes to sterilize canned mackerel. Other factors, such as the heating and cooling curve, the heat transfer characteristics of the contents and the size and shape of the can can be considered to define the conditions. Moreover, as the central operation in canning, the nature of the equipment used for heat sterilization and the mode of operation will have profound effects on the energy/water usage of the overall canning process (Hall 2011a).

After the cold spot of products has received a heat treatment adequate for commercial sterilization, it is usually desirable to cool it as rapidly as possible to avoid too great a degree of over-cooking. Additionally, a long, slow cooling process (by, for example, allowing the can contents to cool naturally after removing the heating medium) can lead to thermophilic spores (surviving a process designed to eliminate C. *botulinum*) germinating and multiplying to cause spoilage (Warne 1988, Horner 1997a, Gaze 2010).

Although canning is designed to preserve as much as possible all of the nutritional constituents present in the initial fish for human nutrition, the extensive heat treatment involved in the cooking and sterilization steps substantially alters the nature of the raw material so that, in effect, a product with different, but not necessarily bad, characteristics is formed (Naczk and Arthyukova 1990, Scheller et al. 1999, Sikorski et al. 2000, Aubourg 2001, Awuah et al. 2007, Sampels 2015).

During the canning process, (pre)cooking is employed to reduce excess moisture, so that the total exudate released in the canned product is minimum (Naczk and Arthyukova 1990, Méndez and Abuín 2012), the sensory, physical and chemical qualities of the product are improved and the shelf life is prolonged (Méndez and Abuín 2012). The length of cooking is dependent on fish species, size/weight and initial temperature (Garthwaite 2010, Méndez and Abuín 2012). After cooking and before being placed in cans, the fish is cooled by allowing it to stand at room temperature (12 to 18°C), eventually overnight (Aubourg 2001).

3.4.2 Composition

The heat degradation of nutrients, the oxidation of vitamins and lipids, the leaching of water-soluble vitamins, minerals and proteins and the toughening and drying of fragile fish protein have been pointed out

(Aubourg 2001, Sampels 2015) as damage pathways for the cooking treatment in terms of the nutritional and sensory qualities, especially if over processing is carried out.

Leaching of water soluble vitamins and soluble proteins into cooking liquors is increasingly recognized as the major source of loss of nutrients (Horner 1997a, Venugopal 2010). Notwithstanding, since endogenous enzymes are inactivated and microbial development is stopped by heat, most attention during cooking has been given to lipid oxidation and further interaction of oxidized lipids with other constituents, especially proteins. At the same time, proteins are heat denatured as a result of the thermal treatment and turn into more reactive molecules (Horner 1997a). The most labile proteins are the sarcoplasmic proteins and myosin, and their denaturation is sufficient to ensure the observed textural changes (Méndez and Abuín 2012). The protein denaturation during fish precooking also implies texture and binding changes that allow, for example, easier separation of muscle from bone. During cooking of tuna, the moisture falls about 10%, the protein increases about 9%, and the fat about 61% (Gallardo et al. 1984 in Méndez and Abuín 2012).

According to the literature review by Aubourg (2001), a wide range of experiments showed no differences in the content of PUFA as a result of cooking even at different cooking conditions despite the increased risk of oxidation due to the application of heat (Sampels 2015). However, lipolysis, i.e., free fatty acids (FFA) formation, and carbonyl compounds (TBA value) have been observed to develop when comparing raw and cooked samples. Moreover, compounds resulting from the interaction of oxidized lipids and proteins have shown to be produced. PUFA in sardine were stable to cooking but unstable to oxidation during the subsequent refrigerated storage with a sharp increase in TBA value with increasing storage time. In contrast, mild heating treatment during cooking might have a positive role on lipid hydrolysis and oxidation, thru enzyme denaturation/inactivation, as suggested for herring (Aubourg 2001). Notwithstanding, most cooking has a negligible effect on fatty acid composition of fish (Venugopal 2005).

On the other hand, the heat sterilization process destroys not only viable microorganisms but also affects the desirable nutritive and organoleptic properties of food components such as proteins, sugars, fats, vitamins as well as texture, flavour, colour, etc. (Naczk and Arthyukova 1990, Sampels 2015). The reduction of nutrient content in thermally processed foods is dependent on the time/temperature treatment used, the rate of heat transfer into the product, and how fast the canned product is cooled after processing (Piggott and Tucker 1990). The effect of heat on the constituents of foods is generally deleterious to the overall quality. Effects include the degradation of vitamins, the softening of texture, loss of colour, development of off-flavours, and destruction of enzymes. Some of these are desirable, e.g., softening of texture. All these reactions, chemical or physical, have different kinetics to microbial inactivation. Bacterial spores have Z-values[1] between 7 and 12°C whereas other constituents have values up to 50°C. This means that if high processing temperatures are used for short times (HTST-processes) there will be less damage of thermolabile components in contrast to longer processes (Holdsworth 2009). The goal of the canning industry is to produce canned foods which are microbiologically safe and have high nutritional value (Naczk and Arthyukova 1990). In general, there is no important change in the proximate composition of canned mackerel (Table 3.4). The values reported herein for mackerel energy content, ca. 110 to 300 kcal (460 to 1210 kJ) per 100 g, which are mainly attributable to the protein and fat contents (Venugopal 2005), remain stable after canning. Notwithstanding, a more indepth, nutrient level look at published data shows a different image.

Fish lipids are highly concentrated in PUFA, especially EPA and DHA (Table 3.5). This composition leads to a high nutritional value of fish products associated with the well-known beneficial effects of ω3 PUFA on human health (Nunes et al. 2003, 2008, Prato and Biandolino 2015). Hydrolysis and oxidation are the major alterations of fish lipids during thermal processing (Naczk and Arthyukova 1990, Méndez and Abuín 2012). Lipid degradation is often focused on the reactivity of PUFA since they can produce a significant number of compounds, including volatile derivatives, by action of heat particularly sterilization treatments involving high temperatures and long times, e.g., 110°C and 130°C during 120 minutes, that lead to a significant decrease of ω3 fatty acids (Méndez and Abuín 2012), via the process of lipid oxidation. In this way, the highly unsaturated fatty acid composition renders fish flesh extremely susceptible to oxidation and rapid degradation during processes that involve thermal treatments. In addition, oxidation of ω3 fatty acids and the effect of oxidized lipids on proteins and amino acids in fish muscle cause loss of nutritional value (Naczk and Arthyukova 1990, Méndez and Abuín 2012). Even though the fatty acids are relatively well preserved (Méndez and Abuín 2012), the retention of PUFA during canning varies with species (Piggott and Tucker 1990). Hale and Brown (1983) found no significant changes in the PUFA contents or lipid class distributions due to heat processing in sealed cans (F_0 range of 13.3–18.8) of Spanish sardine, *Sardinella aurita*, thread herring, *Opisthonema oglinum*, and chub mackerel, *Scomber japonicus*. Canned fish, frequently packed in vegetable oil, not only

---

[1] Z value is the temperature increment required to reduce the D-value by 90%; and D is the heat treatment time (min) required to reduce the number or concentration of spores, microorganisms or quality factors by 90%.

Table 3.4 Proximate composition of raw, fresh and canned mackerel, *Scomber* spp., compiled from available sources.

	Raw [1,2]	Raw (*S. scombrus*) [3]	Raw (*S. japonicus*) [4]	Raw [5]	Canned fillets (*S. japonicus* in olive oil) [1,2]	Canned (*S. scombrus* in natural juices) [3]	Canned (*S. japonicus* in light brine) [4]	Canned (*Scomber* spp. in brine) [5]
Energy (kcal/kJ)	202/844	289/1210	112ᵃ/472	111/464	182/763	225/943	128ᴬ/544	190/795
Moisture (g)	64.3	56.6 (50.0–68.0)	73.73	76.0	63.7	62.3 (58.9–65.0)	69.77	65.7
Protein (g)	20.3	18.1 (14.6–21.0)	21.08	18.0	24.0	19.8 (18.0–21.4)	24.79	17.4
Total fat (g)	13.4	24.4 (7.0–34.7)	3.07	4.0	9.6	16.4 (13.2–21.4)	3.30	13.4
Ash (g)	1.42	1.2 (1.0–1.6)	2.66	1.3	2.40	1.6 (1.1–2.3)	3.03	3.5

Legend: [1,2] (Bandarra et al. 2004, INSA 2010), [3] (Saxholt et al. 2008) [4] (Hale and Brown 1983) [5] (Seno 1974). ᴬusing the following relationship: 17 kJ/g protein, 37 kJ/g fat, 17 kJ/g carbohydrate, 8 kJ/g dietary fiber and 29 kJ/g alcohol (cf. Regulation (EU) no. 1169/2011, *Official Journal of the European Union* L304/18).

Table 3.5 Fatty acids (g/100 g) of raw and canned mackerel (whole/fillets), compiled from available sources.

	Raw [1,2]	Raw (S. scombrus) [3]	Raw (S. scombrus) [4]	Raw (S. japonicus) [5]	Canned fillets (S. japonicus) in olive oil [1,2]	Canned (S. scombrus) [3]	Canned (S. scombrus) in natural juices [4]	Canned (S. japonicus) in light brine [5]
Lipids	13.4	7.07	24.4	3.07	9.6	13.2	16.4	3.30
Fatty acids								
Saturated	3.61	1.87	5.7	0.75	1.48	3.73	3.61	0.73–0.74
Monounsaturated	3.67	1.44	9.8	0.48	5.99	1.79	6.51	0.41–0.48
Polyunsaturated	4.66	2.85	6.6	0.92	1.23	5.98	4.57	1.01–0.95
LA18:2ω6	0.248	0.15	0.80	0.03	0.504	0.35	0.26	0.02–0.03
EPA	1.22	0.66	1.22	0.11	0.136	0.79	0.94	0.10–0.13
DHA	2.13	1.40	3.28	0.45	0.424	1.99	2.25	0.45–0.51
ω3	4.14	2.59	5.78	0.65	0.683	4.35	4.31	0.70–0.74
ω6	0.53	0.26	0.79	0.15	0.554	1.59	0.26	0.15
Cholesterol	14/45		80		35		80	

Legend:[1, 2] (Bandarra et al. 2004, INSA 2010), [3] (Sirot et al. 2008), [4] (Saxholt et al. 2008), [5] (Hale and Brown 1983).

increases in calories content but also may vitiate the beneficial effects of ω3 PUFA (Venugopal 2005). The values compiled herein indicate that canning mackerel is somewhat detrimental to PUFA content, more so in ω3 PUFA. However, there is also an exchange of ω3 PUFA from fish flesh to filling oil, i.e., the dilution of the natural lipids of fish muscle by triacylglycerols from packing oils (Hale and Brown 1983) resulting in the oil enriched with ω3 PUFA, that is not quantified when analyzing fish muscle. Reportedly, there is decrease in the ω3/ω6 ratio in canned fish with added plant oils (Sampels 2015). Moreover, oxidation can be minimized by employing packing media containing natural antioxidants such as extra virgin olive oil rich in polyphenols or certain vegetable oils rich in tocopherols (Hale and Brown 1983, Méndez and Abuín 2012). This advises the consumption of both fish and sauce.

Fish is a source of vitamins, in which concentration depends on species, anatomical part of the fish, and age. Most of the vitamins are degraded to some extension during heating. Thus, vitamins are among the most sensitive food component to be affected by heat sterilization (Awuah et al. 2007, Méndez and Abuín 2012, Skipnes 2014). Vitamin degradation during heat treatment is not simple and is dependent on other agents such as oxygen, light and water/fat solubility. In addition, vitamin degradation depends on pH and may be catalyzed by chemicals present, metals, other vitamins and enzymes (Awuah et al. 2007). Relevant heat sensitive vitamins are the fat soluble vitamins A (in the presence of oxygen), D, E and beta carotene, and water-soluble vitamins B1 (thiamine), B2 (riboflavin) in acid environment, and B3 (niacin) (Naczk and Arthyukova 1990, Piggott and Tucker 1990, Awuah et al. 2007). Losses of 25 to 55% (Naczk and Arthyukova 1990) or even about 70% (Piggott and Tucker 1990) of thiamine can occur during canning. Thiamine is the most thermolabile of the complex-B vitamins, and its losses during precooking and after conventional sterilization are reported at about 42% and 90%, respectively (Méndez and Abuín 2012). Niacin and riboflavin showed higher heat stability than thiamine, and the retention in canned albacore was about 60 to 70%, respectively (Méndez and Abuín 2012). Data compiled herein (Table 3.6) confirms these estimates. Processing at HTST periods and immediate cooling can significantly reduce damage to these components (Piggott and Tucker 1990). Moreover, the extent of degradation of vitamins depends on the species and fat content of seafoods and also on the sterilization parameters used (Naczk and Arthyukova 1990, Skipnes 2014).

The major group of nutrients in fish muscle are proteins (Skipnes 2014). The muscle of fish mainly consists of myofibrillar proteins that represent 65 to 75% of the total muscle proteins. They are contractile proteins as myosin and actin and regulatory proteins as tropomyosin and troponin,

Table 3.6 Vitamin profiles (per 100 g) of raw and canned mackerel (whole/fillets), compiled from available sources.

	Raw [1,2]	Raw [3]	Raw [4]	Canned fillets (*S. japonicus*) in olive oil [1,2]	Canned *S. scombrus* in natural juices [3]	Canned *S. scombrus* in brine [4]
A (mg)	28	50.5 (21.0–88.0)	15	n.a./23	22.3 (5.0–33.0)	0
D (ug)	2.4	5.45 (2.10–18.9)		0.4/<0.70	2.71 (1.60–4.70)	
E (mg)	1.3	1.3 (0.730–1.64)		1.9	0.9 (0.70–1.03)	
B1 (mg)	0.13	0.14 (0.138–0.147)	0.15	0.04	0.06	0.02
B2 (mg)	0.23	0.35 (0.220–0.484)	0.20	0.20	0.16 (0.158–0.168)	0.10
B3 (mg)	22	8.0 (4.10–11.4)	8.0	15.8	6.6	–
B6 (mg)	1.0	0.610 (0.544–0.662)		0.26	0.25 (0.230–0.311)	
B12 (ug)	14	8.40 (2.60–14.0)		17	7.70 (5.34–9.60)	
Folate (ug)	14	8		17	10	

Legend: [1,2] (Bandarra et al. 2004, INSA 2010), [3] (Saxholt et al. 2008), [4] (Seno 1974).

and other minor proteins. Other proteins of fish muscle are sarcoplasmic proteins or water-soluble proteins, which account for 20 to 30% of the total proteins. They are mostly enzymes and are involved in postmortem biochemical changes. Fish flesh also contains connective tissue proteins, about 3%, which are related with the texture of fish fillets (Nunes et al. 2008, Méndez and Abuín 2012). Muscle cellularity is one major determinant of texture together with lipid and water content (Cheng et al. 2014) since the lipid and water content may reduce the structural cohesion of the muscle, lowering its mechanical strength (Méndez and Abuín 2012).

The effects of thermal processing on proteins can be divided (Awuah et al. 2007) into those responsible for altering the secondary, tertiary and quaternary structure of proteins; and those that alter the primary structure. The former, unfolds the proteins and improves their bioavailability since peptide bonds become readily accessible to digestive enzymes, while the later, on the other hand, may lower digestibility and produce proteins that are not biologically available. Heat processing brings about denaturation of proteins which results in loss of solubility and enzymatic activity. The latter is a beneficial effect brought about by sterilization/canning. The increased protein content in canned mackerel observed in the data compiled (Table 3.4) might be an artifact of water loss due to precooking/canning.

Free amino acids and creatine are the major components of the total non protein nitrogen (NPN) compounds in teleost muscle. They are associated with autolysis and bacterial actions in the earlier stages of alteration and recognized as contributors to the flavour (Méndez and Abuín 2012). Free amino acids can be significantly affected by heating. Losses in amino acids can be observed after short, <60 to 90 minute periods of heating (Naczk and Arthyukova 1990). Thermal processing below 100°C will not alter the composition of amino acids, though their functional properties may be changed and may give enhanced digestibility (Skipnes 2014). At higher, sterilization temperatures, 110 to 130°C, detrimental changes in free amino acids in canned tuna (and supposedly tuna-like species) are not significant only if standardized conditions (F in the range of six to 32 minutes) are used (Méndez and Abuín 2012). Kinetics studies (e.g., Banga et al. 1992) indicate that the "classical" methods of sterilization of tuna, i.e., temperatures of retort under 125°C and F values under 12 minutes with $Z = 10$°C, have only a slight effect on availability ($\leq$10 percent loss) of lysine—the most sensitive of the amino acids and its availability serves as index of damage. Conversely, over-processing, at retort temperatures but longer periods—in the range of hours—causes considerable negative change in amino acids content namely histidine (Aitken and Connell 1979). The heating at 115°C causes a loss in cystine plus cysteine which has been associated with corrosion/discolouration of cans and formation of an offensive off-flavour when the can is opened (Méndez and Abuín 2012). Data compiled (Table 3.7) herein gives no evidence of significant losses in amino acids concentrations and reinforces the adequacy of traditional, conventional processing conditions.

The prevention of oxidation in canning is also important for conservation of the many minerals, both macro minerals and trace elements (Table 3.8), for which fish is an important source (Skipnes 2014). Fish contains an important amount of potassium (around 300 mg/100 g), phosphorous (ca. 200 mg/100 g), magnesium (about 25 mg/100 g), and calcium (ca. 15 mg/100 g). Moreover, fish contributes with trace elements as iron, zinc, copper, and other elements such as iodine, selenium, manganese, copper, cadmium, lead, vanadium, cobalt, and boron, which are important for human health. A number of studies have reported the effect of the heat treatments on these elements but found little or no effect on the minerals after thermal processing. In canned mackerel major increases have been observed in sodium, most probably due to brining, and in calcium, as a result of bones becoming edible (Piggott and Tucker 1990) and/or the detrimental effect of heating upon the ability of proteins

Table 3.7 Amino acids profiles (mg per 100 g) of raw and canned mackerel.

	Raw [1,2]	Raw [3]	Canned fillets (*S. japonicus*) in olive oil [1,2]	*S. scombrus* in natural juices [3]
Aspartic acid	2.1	1.7	2.7	1.8
Threonine	1.1	0.78	1.3	0.89
Serine	0.9	0.73	1.1	0.85
Glutamic acid	3.1	2.1	3.8	2.2
Proline	0.5	0.61	1.5	0.73
Glycine	1.1	0.84	1.4	1.1
Alanine	1.5	1.1	1.5	1.1
Cystine+Cysteine	n.a.	0.14	n.a.	0.12
Valine	1.3	1.1	1.6	1.2
Methionine	0.5	0.52	0.8	0.54
Isoleucine	1.1	0.9	1.3	0.95
Leucine	1.8	1.4	2.3	1.5
Tyrosine	0.8	0.64	1.1	0.76
Phenylalanine	1.0	0.73	1.2	0.79
Histidine	1.4	0.84	1.8	0.98
Lysine	2.0	1.6	2.6	1.7
Arginine	1.5	1	1.7	0.98
Tryptophan (mg)	3.8	0.19	4.5	0.21

Legend: [1,2] (Bandarra et al. 2004, INSA 2010), [3] (Saxholt et al. 2008).

to bind calcium (Naczk and Arthyukova 1990). The compiled data (Table 3.8) gives examples of these effects.

Although some changes may be desirable, the rather severe temperature for an extended period of time would trigger chemical reactions, resulting in loss of nutrients and sensory characteristics such as appearance, colour, flavour and texture (Awuah et al. 2007, Sampels 2015). The effect of thermal processing on the sensory quality in marine products is difficult to predict because of intra and interspecific variability of fish species and factors such as appearance, odour, colour, flavour, and texture (Méndez and Abuín 2012). However, the changes in the sensory quality originated by the thermal processing are closely related to chemical reactions, which usually have a temperature dependence that can be described by a Z value of 33°C (Lund 1977). Kinetics of thermal degradation of surface colour in canned tuna was reported by Banga et

Table 3.8 Concentration (per 100 g) of minerals in raw and canned mackerel (whole/fillets).

	Raw [1,2]	Raw [3]	Raw [4]	Canned fillets (*S. japonicus*) in olive oil [1,2]	Canned [3] *S. scombrus* in natural juices	Canned [4]
Ash (g)	1.42	1.2 (1.0–1.6)	1.3	2.40	1.6 (1.1–2.3)	3.5
Na (mg)	78	70 (44.5–101)		623	590	
K (mg)	360	256 (219–277)		383	256 (219–277)	
Ca (mg)	39.0	15 (5.00–20.0)	5	9.5	15 (14.0–21.0)	290
P (mg)	282	267 (148–778)	190	249	220	260
Mg (mg)	37	24 (23.1–25.3)		35	24 (23.1–25.3)	
Fe (mg)	1.1	0.75 (0.37–1.17)	1.8	0.6	1.1 (0.920–1.15)	1.6
Mn (mg)	0.02			<0.02		
Cu (mg)	0.17	0.09 (0.03–0.20)		0.10	0.11 (0.079–0.120)	
Zn (mg)	2.2	0.10		0.7	1.6 (1.10–2.30)	
Se (ug)		31.9 (20.1–52.8)			31.9 (20.1–52.8)	
I (ug)		0.423			21.9	

Legend: [1,2] (Bandarra et al. 2004, INSA 2010), [3] (Saxholt et al. 2008), [4] (Seno 1974).

al. (1993). The Z-value found by these authors was Z = 44°C that is much higher than that found by Ohlsson (1980) for the temperature dependence of sensory quality changes during the thermal processing in fish (Z = 25°C).

The conventional, traditional canning of mackerel carried out in Portugal and the Algarve produces a nutritionally rich fish product that utilizes an abundant but less-valued fish species. Notwithstanding, despite the abundance of scientific literature on fish canning, the changes triggered by the process are still not completely elucidated.

3.5 Conclusion

In summary, I agree with Venugopal (2005): Seafood products contribute significantly to provide various nutrients. Some of the nutritional losses due to processing need not pose concerns and consumption of processed seafood does not necessarily compromise their nutritional value. Nevertheless, care in applications of the processing methods can prudently minimize the (almost) inevitable nutrient losses. It needs to be emphasized that apart from the advantages of value addition to otherwise unused, undervalued species, processing, in general, has a number of nutritional benefits such as enhancing palatability of the product, digestibility of proteins, and bioavailability of nutrients including fat, vitamins and minerals, resulting in the overall increase in the health benefits. Moreover, there is no need to use added preservatives in the processed fish described in this chapter.

Since «dried litão» is still traditional and localized, a number of aspects related to its production and characteristics, not yet studied, constitute an interesting subject not only in terms of food science and technology but also in terms of ethnology. On the other hand, muxama is a high-valued, relatively well-studied food product obtained from tuna. Notwithstanding, a few details related with its production, mostly changes in its characteristics are still to be clarified. Finally, the canning of mackerel is established and has been fairly studied, but the suspected changes arising from the process are not yet fully detailed.

In general, there are no specified, detailed methods or recipes for the production of traditional products. This lack makes product quality control and uniformity a near impossibility (Yean et al. 1998). Besides attempting to compile information about the fish products considered herein and divulging their benefits in terms of nutrition and health, this chapter exposes the absence of information dealing with the changes in physico-chemical and microbiological parameters as well as in nutritional composition due to processing that might be specific of those foodstuffs.

References

Ababouch, L. 2002. HACCP in the fish canning industry. *In*: H.A. Bremner (eds.). Safety and Quality Issues in Fish Processing. Woodhead Publishing Ltd., Cambridge, UK.

Ababouch, L. and I. Karunasagar. 2014. Developments in food safety and quality systems. *In*: J. Ryder, I. Karusanagar and L. Ababouch (eds.). Assessment and Management of Seafood Safety and Quality: Current Practices and Emerging Issues. FAO Fisheries and Aquaculture Technical Paper No. 574, Rome, Italy.

Afonso, C., C. Cardoso, H.M. Lourenço, P. Anacleto, N.M. Bandarra, M.L. Carvalho, M. Castro and M.L. Nunes. 2013. Evaluation of hazards and benefits associated with the consumption of six fish species from the Portuguese coast. J. Food Compos. Anal. 32: 59–67.

Aitken, A. and J.J. Connell. 1979. Fish. *In*: R.J. Priestley (ed.). Effects of Heating on Foodstuffs. Applied Science Publishers Ltd., London, UK.

Aníbal, J. and E. Esteves. 2016. Muxama and Estupeta: Traditional food products obtained from tuna loins in south Portugal and Spain. *In*: K. Kristbergsson and J. Oliveira (eds.). Traditional Food Products: General and Consumer Aspects. Springer Science+Business Media, New York, USA.

Arvanitoyannis, I. 2012. Modified atmosphere and active packaging technologies. CRC Press, Boca Raton, Florida, USA.

Aubourg, S.P. 2001. Review: loss of quality during the manufacture of canned fish products. Food Sci. Technol. Int. 7: 199–215.

Awuah, G.B., H.S. Ramaswamy and A. Economides. 2007. Thermal processing and quality: principles and overview. Chem. Eng. Process. Process Intensif. 46: 584–602.

Bach-Faig, A., E.M. Berry, D. Lairon, J. Reguant, A. Trichopoulou, S. Dernini, F.X. Medina, M. Battino, R. Belahsen, G. Miranda and L. Serra-Majem. 2011. Mediterranean diet pyramid Today. Science and cultural updates. Public Health Nutr. 14: 2274–2284.

Bandarra, N.M., M.A. Calhau, L. Oliveira, M. Ramos, M.G. Dias, H. Bártolo, M.R. Faria, M.C. Fonseca, J. Gonçalves, I. Batista and M.L. Nunes. 2004. Composição e valor nutricional dos produtos da pesca mais consumidos em Portugal. Instituto Nacional das Pescas e do Mar, Lisboa, Portugal.

Banga, J.R., A.A. Alonso, J.M. Gallardo and R.I. Pérez-Martín. 1992. Degradation kinetics of protein digestibility and available lysine during thermal processing of tuna. J. Food Sci. 57: 913–915.

Banga, J.R., A.A. Alonso, J.M. Gallardo and R.I. Pérez-Martín. 1993. Kinetics of thermal degradation of thiamine and surface colour in canned tuna. Z. Lebensm. Unters For. 197: 127–131.

Barat, J.M. and R. Grau. 2009. Thawing and salting studies of dry cured tuna loins. J. Food Eng. 91: 455–459.

Bratt, L. 2010. Fish canning handbook. Wiley-Blackwell. Chichester, UK.

Burlingame, B. and S. Dernini. 2011. Sustainable diets: The Mediterranean diet as an example. Public Health Nutr. 14: 2285–2287.

CAC. 2012. Code of practice for fish and fishery products. 2nd ed. World Health Organization (WHO) and Food and Agriculture Organization (FAO) of the United Nations. Rome, Italy.

Cardoso, C. and M.L. Nunes. 2013. A Importância do consumo de produtos da pesca em Portugal. *In*: J.L. Santos, I. Carmo, P. Graça and I. Ribeiro (eds.). O futuro da alimentação: ambiente, saúde e economia. Fundação Calouste Gulbenkian, Lisboa, Portugal.

Cheng, J.H., D.-W. Sun, Z. Han and X.A. Zeng. 2014. Texture and structure measurements and analyses for evaluation of fish and fillet freshness quality: a review. Compr. Ver. Food Sci. Food Safe. 13: 52–61.

Clark, S., S. Jung and B. Lamsal. 2014. Food processing: principles and applications. John Wiley & Sons, Ltd., Chichester, UK.

Coelho, R., K. Erzini, L. Bentes, C. Correia, P.G. Lino, P. Monteiro, J. Ribeiro and J.M.S. Gonçalves. 2005. Semi-pelagic longline and trammel net elasmobranch catches in southern Portugal: catch composition & catch rates and discards. J. Northw. Atl. Fish. Sci. 35: 531–537.

Copat, C., G. Arena, M. Fiore, C. Ledda, R. Fallico, S. Sciacca and M. Ferrante. 2013. Heavy metals concentrations in fish and shellfish from the eastern Mediterranean sea: consumption advisories. Food Chem. Toxicol. 53: 33–37.

Correia, J.P.S. 2009. Pesca comercial de tubarões e raias em Portugal. Tese de Doutoramento, Universidade de Aveiro, Portugal.

Darvishi, H., M. Azadbakht, A. Rezaeiasl and A. Farhang. 2013. Drying characteristics of sardine fish dried with microwave heating. J. Saudi Soc. Agric. Sci. 12: 121–27.

Derrick, S. 2009. Cured, smoked and dried fish. *In*: R. Fernandes (ed.). Microbiology Handbook: Fish and Seafood. Leatherhead Food International Ltd. and Royal Society of Chemistry, Surrey and Cambridge, UK.

Doe, P.E. 1998. Fish Drying & Smoking: Production and Quality. Technomic Publishing Company, Inc., Lancaster, USA.

Doe, P.E. 2002. Fish drying. *In*: H.A. Bremner (ed.). Safety and Quality Issues in Food Processing. Woodhead Publishing Ltd., Cambridge, UK.

Doe, P.E., Z. Sikorski, N. Haard, J. Olley and B.S. Pan. 1998. Basic principles. *In*: P.E. Doe (ed.). Fish Drying and Smoking: Production and Quality. Lancaster, USA: Technomic Publishing Company, Inc.

DRGM. 2013. Publicações—Estatística. Direção-Geral de Recursos Naturais, Segurança e Serviços Marítimos, Lisboa.

Driscoll, R. 2004. Food dehydration. *In*: J.S. Smith and Y.H. Hui (eds.). Food Processing: Principles and Applications. Blackwell Publishing Ltd., USA.

Erzini, K., K.I. Stergiou, L. Bentes, J.M.S. Economidis, P.S. Gonçalves, P.G. Lino, D. Moutopoulos, G. Petrakis, J. Ribeiro and P. Vulgaridou. 1999. Comparative fixed gear selectivity studies in Portugal and Greece. final report. Commission of the European Communities, Project DG XIV Ref. 96/065.

EUROSTAT. 2015. Agriculture, Forestry and Fishery Statistics. 2014 Edition. Eurostat, Luxembourg.

FAO. 2014a. FAOSTAT. Food Balance Sheet. Retrieved January 18, 2015 (URL: http://data. fao.org/ref/48dc9161-53e2-4883-93c0-8f099e5e67ab.html?version=1.0).

FAO. 2014b. FISHSTAT. Global Capture Production (Dataset). Retrieved January 19, 2015 (URL: http://data.fao.org/ref/af556541-1c8e-4e98-8510-1b2cafba5935. html?version=1.0).

Fellows, P. 2000. Food Processing. Principles and Practice. 2nd ed. Woodhead Publishing Limited, Abington, UK.

Fernandes, R. 2009. Microbiology Handbook: Fish and Seafood. Leatherhead Food International Ltd. and Royal Society of Chemistry, Surrey and Cambridge, UK.

Florêncio, T.J. 1996. Manual de Identificação e Classificação de Defeitos Visuais Exteriores nas Embalagens Metálicas de Conservas de Pescado. Relat Técn.-Cient. Inst. Port. Invest. Mar: 16. 47 pp.

Florêncio, T.J. and M.M. Albuquerque. 1996. Parâmetros de Cravação. Relat Técn.-Cient. Inst. Port. Invest. Mar. 20: 18.

Gallart-Jornet, L., I.E. Roberto and P.F. Maupoei. 2005. La Salazón de Pescado, Una Tradición en la Dieta Mediterránea. Universidad Politécnica de Valencia, Valencia, Spain.

Garthwaite, T. 2010. Fish quality. *In*: L. Bratt (ed.). Fish Canning Handbook. Wiley-Blackwell, Chichester, UK.

Gaze, J. 2010. Principal causes of spoilage in canned fish products. *In*: L. Bratt (ed.). Fish Canning Handbook. Wiley-Blackwell, Chichester, UK.

Al Ghabshi, A., H. Al-Khadhuri, N. Al-Aboudi, S. Al-Gharabi, A. Al-Khatri, N. Al-Mazrooei and P.S. Sudheesh. 2012. Effect of the freshness of starting material on the final product quality of dried salted shark. Adv. J. Food Sci. Technol. 4(2): 60–63.

Gibson, C., S.V. Valenti, S.L. Fowler and S. Fordham. 2006. The conservation status of northeast atlantic chondrichthyans. Report of the Shark Specialist Group Northeast Atlantic Regional Red List Workshop, Peterborough, UK.

Godinho, M. 2011. A Muxama. Centro de Investigação e Informação do Património de Cacela |Projeto TASA. Retrieved January 15, 2015 (http://www.projectotasa.com/2011/01/a-muxama/).

Gómez, R., M.A. Carmona and J. Fernández-Salguero. 1991. Estudio de los Alimentos de Humedad Intermedia Españoles. I. Actividad del Agua y pH. II Jornadas Científicas sobre "Alimentación Española", Córdoba: 124–130.

Gram, L. and H.H. Huss. 1996. Microbiological spoilage of fish and fish products. Int. J. Food Microbiol. 33(1): 121–137.

Guizani, N., A.O. Al-Shoukri, A. Mothershaw and M.S. Rahman. 2008. Effects of salting and drying on shark (*Carcharhinus sorrah*) meat quality characteristics. Drying Technol. 26(6): 705–713.

Hale, M.B. and T. Brown. 1983. Fatty acids and lipid classes of three underutilized species and changes due to canning. Mar. Fish. Ver. 45(5-6): 45–48.

Hall, G.M. 2011a. Canning of fish and fish products. *In*: G.M. Hall (ed.). Fish Processing: Sustainability and New Opportunities. Wiley-Blackwell, Chichester, UK.

Hall, G.M. 2011b. Preservation by curing (drying, salting and smoking). *In*: G.M. Hall (ed.). Fish Processing: Sustainability and New Opportunities. Wiley-Blackwell, Chichester, UK.

Han, J.H. 2005. Innovations in food packaging. Elsevier Science/Academic Press, Amsterdam.

Holdsworth, S.D. 2009. Principles of thermal processing: sterilization. *In*: R. Simpson (ed.). Engineering Aspects of Thermal Food Processing. CRC Press Inc., Boca Raton, Florida, USA.

Horner, W.F.A. 1997a. Canning of fish and fish produtcts. *In*: G.M. Hall (ed.). Fish Processing Technology. Blackie Academic and Professional/Chapman & Hall, London, UK.

Horner, W.F.A. 1997b. Preservation of fish by curing (drying, salting, and smoking). *In*: G.M. Hall (ed.). Fish Processing Technology. Blackie Academic and Professional/Chapman & Hall, London, UK.

Hui, Y.H., M.H. Lim, W.K. Nip, J.S. Smith and P.H.F. Yu. 2004. Principles of food processing. *In*: J.S. Smith and Y.H. Hui (eds.). Food Processing Principles and Applications. Blackwell Publishing Ltd., Ames, Iowa, USA.

ICMSF. 1986. Microorganisms in Foods. 2. Sampling for Microbiological Analysis: Principles and Specific Applications. 2nd ed. University of Toronto Press, Toronto, Canada.

INE. 2014. Estatísticas da Pesca 2013. Instituto Nacional de Estatística e Direção-Geral de Recursos Naturais, Segurança e Serviços Marítimos, Lisboa, Portugal.

INSA. 2010. Tabela da Composição de Alimentos. Instituto Nacional de Saúde Dr. Ricardo Jorge, Lisboa, Portugal.

James, D. 1998. Production, consumption, and demand. *In*: P. Doe (ed.). Fish Drying and Smoking: Production and Quality. CRC Press, Boca Raton, Florida, USA.

James, D. 2013. Risks and benefits of seafood consumption. Globefish Research Programme, FAO, Rome, Italy.

Keay, J.N. 2001. Handling and processing mackerel. Torry Advisory Note 66 (Revised). FAO and Support unit for International Fisheries and Aquatic Research (SIFAR), Rome, Italy.

Kris-Etherton, P.M. 2002. Fish consumption, fish oil, Omega-3 fatty acids, and cardiovascular disease. Circulation 106(21): 2747–2757.

Lã, A. and L. Vicente. 1993. O Atum Esquecido. Universidade do Algarve, Faro, Portugal.

Lichtenstein, A.H., L.J. Appel, M. Brands, M. Carnethon, S. Daniels, H.A. Franch, B. Franklin, P. Kris-Etherton, W.S. Harris, B. Howard, N. Karanja, M. Lefevre, L. Rudel, F. Sacks, L. Van Horn, M. Winston and J. Wylie-Roset. 2006. Diet and lifestyle recommendations revision 2006: A scientific statement from the American Heart Association Nutrition Committee. Circulation 114(1): 82–96.

Lindkvist, K.B., L. Gallart-Jornet and M.C. Stabell. 2008. The restructuring of the Spanish salted fish market. Can Geogr-Geogr Can 52(1): 105–120.

Louka, N., F. Juhel, V. Fazilleau and P. Loonis. 2004. A novel colorimetry analysis used to compare different drying fish processes. Food Control 15(5): 327–334.

Lund, D.B. 1977. Maximizing nutrient retention. Food Technol. 31: 71–78.

Mathew, S. and B.A. Shamasundar. 2002. Effect of ice storage on the functional properties of proteins from shark (*Scoliodon Laticaudus*) meat. Die Nahrung 46(4): 220–226.

Méndez, M.I.M. and J.M.G. Abuín. 2012. Thermal processing of fishery products. *In*: D.-W. Sun (ed.). Thermal Food Processing: New Technologies and Quality Issues. CRC Press, Boca Raton, Florida, USA.

Ministerio de Agricultura, Alimentación y Medio Ambiente de España. 2014. Indicación Geográfica Protegida. Retrieved December 11, 2015 (URL: http://www.magrama. gob.es/es/alimentacion/temas/calidad-agroalimentaria/calidad-diferenciada/dop/ pesca_molus_crus/MojamaIslaCristina.aspx#).

Mujaffar, S. and C.K. Sankat. 2005. The air drying behaviour of shark fillets. Can. Biosyst. Eng. 47: 3.11–3.21.

Myrseth, A. 1985. Planning and Engineering Data 2. Fish Canning. FAO Fish. Circ. 784: 77 p.

Naczk, M. and A.S. Arthyukova. 1990. Canning. *In*: Z.E. Sikorski (ed.). Seafood: Resources, Nutritional Composition, and Preservation. CRC Press, Boca Raton, Florida, USA.

Nazemroaya, S., M.A. Sahari and M. Rezaei. 2011. Identification of fatty acid in mackerel (*Scomberomorus Commersoni*) and shark (*Carcharhinus Dussumieri*) fillets and their changes during six month of frozen storage at –18°C. J. Agric. Sci. Technol. 13: 553–566.

Nguyen, M.V., S. Arason and T.M. Eikevik. 2014. Drying of fish. *In*: I.S. Boziaris (ed.). Seafood Processing: Technology, Quality and Safety. John Wiley & Sons Ltd., Chichester, UK.

Nunes, M.L., I. Batista, N.M. Bandarra, M.G. Morais and P.O. Rodrigues. 2008. Produtos da pesca: valor nutricional e importância para a saúde e bem-estar dos consumidores. Public. Avulsas IPIMAR 18: 77.

Nunes, M.L., N.M. Bandarra and I. Batista. 2003. Fish products: contribution for a healthy food. Electron. J. Environ. Agric. Food Chem. 2(4): 453–457.

Ohlsson, T. 1980. Temperature dependence of sensory quality changes during thermal processing. J. Food Sci. 45(4): 836–839.

Olaso, I., F. Velasco, F. Sánchez, A. Serrano, C. Rodríguez-Cabello and O. Cendrero. 2004. Trophic Relations of lesser spotted catshark (*Scyliorhinus Canicula*) and blackmouth catshark (*Galeus Melastomus*) in the Cantabrian sea. J. Northw. Atl. Fish. Sci. 35: 481–494.

Ordóñez-Del Pazo, T., L.T. Antelo, A. Franco-Uría, R.I. Pérez-Martín, C.G. Sotelo and A.A. Alonso. 2014. Fish discards management in selected Spanish and Portuguese métiers: identification and potential valorisation. Trends Food Sci. Technol. 36(1): 29–43.

Perdue, R. 2009. Vacuum packaging. *In*: K.L. Yam (ed.). The Wiley Encyclopedia of Packaging Technology. 3rd ed. John Wiley & Sons, New Jersey, USA.

Piggott, G.M. and B. Tucker. 1990. Seafood: Effects of Technology on Nutrition. Marcel Dekker Inc., New York, USA.

Prato, E. and F. Biandolino. 2015. The contribution of fish to the Mediterranean diet. *In*: V.R. Preedy and R.R. Watson (eds.). The Mediterranean Diet: An Evidence-based Approach. Academic Press, UK.

Rahman, M.S., N. Guizani, M.H. Al-Ruzeiki and A.S. Al Khalasi. 2000. Microflora changes in tuna mince during convection air drying. Drying Technol. 18(10): 2369–2379.

Rebelo, M.J.F. 2010. As Indústrias da Pesca e Conservas de Atum no Algarve do Século XX. Dissertação de Mestrado, Universidade do Algarve, Faro, Portugal.

Ruano, F., P. Ramos, M. Quaresma, N.M. Bandarra and I.P. da Fonseca. 2012. Evolution of fatty acid profile and condition index in mollusc bivalves submitted to different depuration periods. Revista Port. Ciênc. Vet. 111: 75–84.

Ryder, J., I. Karunasagar and L. Ababouch. 2014. Assessment and management of seafood safety and quality: current practices and emerging issues. FAO Fish. Aquac. Tech. Pap. 574: 432 p.

Sampels, S. 2015. The effects of processing technologies and preparation on the final quality of fish products. Trends Food Sci. Technol. 44(2): 131–146.

Sanjuás-Rey, M., J.M. Gallardo, J. Barros-Velázquez and S.P. Aubourg. 2012. Microbial activity inhibition in chilled mackerel (*Scomber Scombrus*) by employment of an organic acid-icing system. J. Food Sci. 77(5): 264–269.

Saxholt, E., A.T. Christensen, A. Møller, H.B. Hartkopp, K. Hess Ygil and O.H. Hels. 2008. Danish Food Composition Databank, Revision 7. Department of Nutrition, National Food Institute, Technical University of Denmark. Retrieved February 12, 2015 (http://www.foodcomp.dk/).

Scheller, R.M., V.M. Snarski, J.G. Eaton and G.W. Oehlert. 1999. An analysis of the influence of annual thermal variables on the occurrence of fifteen warmwater fishes. Trans. Am. Fish. Soc. 128: 257–264.

Schmidt, E.B., L.H. Rasmussen, J.G. Rasmussen, A.M. Joensen, M.B. Madsen and J.H. Christensen. 2006. Fish, marine n-3 polyunsaturated fatty acids and coronary heart disease: A minireview with focus on clinical trial data. Prostag. Leukotr. Ess. Fatty Acids 75(3): 191–195.

Sen, D.P. 2005. Advances in fish processing Technology. Allied Publishers, New Delhi, India.

Seno, Y. 1974. Processing and utilization of mackerel. *In*: R. Kreuzer (ed.). Fishery Products. Food and Agriculture Organization (FAO) of the United Nations and Fishing News (Books) Ltd., Surrey, UK.

Serena, F., C. Mancusi, N. Ungaro, N.R. Hareide, J. Guallart, R. Coelho and P. Crozier. 2009. *Galeus melastomus* (blackmouth catshark). The IUCN Red List of Threatened Species. Version 2014.3. Retrieved February 3, 2015 (http://www.iucnredlist.org/details/161398/0).

Serra-Majem, L., A. Bach-Faig and B. Raidó-Quintana. 2012. Nutritional and cultural aspects of the Mediterranean diet. Int. J. Vitam. Nutr. Res. 82(3): 157–162.

Sikorski, Z.E., A. Gildberg and A. Ruiter. 2000. Productos pesqueros. *In*: A. Ruiter (ed.). El pescado y los productos de la pesca, composicion, propriedades nutritivas y estabilidade. Editorial Acríbia SA, Zaragoza, Spain.

Sirot, V., M. Oseredczuk, N. Bemrah-Aouachria, J.L. Volatier and J.C. Leblanc. 2008. Lipid and fatty acid composition of fish and seafood consumed in France: CALIPSO Study. J. Food Compos. Anal. 21(1): 8–16.

Skipnes, D. 2014. Heat processing of fish. *In*: I.S. Boziaris (ed.). Seafood Processing: Technology, Quality and Safety. John Wiley & Sons Ltd., Chichester, UK.

Smida, M.A.B., A. Bolje, A. Ouerhani, M. Barhoumi, H. Mejri, M. El Cafsi and R. Fehri-Bedoui. 2014. Effects of drying on the biochemical composition of *Atherina boyeri* from the Tunisian Coast. Food Nutr. Sci. 5(14): 1399–1407.

Sun, D.-W. 2012. Thermal Food Processing: New Technologies and Quality Issues. 2nd ed. CRC Press Inc., Boca Raton, Florida, USA.

Tilman, D. and M. Clark. 2014. Global Diets Link Environmental Sustainability and Human Health. Nature 515(7528): 518–522.

Torry Research Station. 1989. Yield and nutritional value of the commercially more important fish species. FAO Fish. Tech. Pap. 309: 187 p.

USDA. 2016. USDA National Nutrient Database for Standard Reference. Release 28. Retrieved January 18, 2016 (URL: http://www.ars.usda.gov/nea/bhnrc/ndl).

Vanhonacker, F., Z. Pieniak and W. Verbeke. 2013. European consumer perceptions and barriers for fresh, frozen, preserved and ready-meal fish products. Brit. Food J. 115(4): 508–525.

Vannuccini, S. 1999. Shark Utilization, Marketing and Trade. FAO Fish. Tech. Pap. 389: 470 p.

Vega-Gálveza, A., M. Miranda, R. Clavería, I. Quispe, J. Vergara, E. Uribe, H. Paez and K. Di Scala. 2011. Effect of air temperature on drying kinetics and quality characteristics of osmo-treated jumbo squid (*Dosidicus Gigas*). LWT-Food Sci. Technol. 44(1): 16–23.

Venugopal, V. 2005. Seafood Processing: Adding Value Through Quick Freezing, Retortable Packaging and Cook-Chilling. CRC Press, Boca Raton, Florida, USA.

Venugopal, V. 2010. Heat Treated Fishery Products. *In*: J. Ryder, L. Ababouch and M. Balaban (eds.). Second International Congress on Seafood Technology on Sustainable, Innovative and Healthy Seafood. Food and Agriculture Organization (FAO) and The University of Alaska, Anchorage, USA.

Wang, Y., M. Zhang and A.S. Mujumdar. 2011. Trends in Processing technologies for dried aquatic products. Drying Technol. 29(4): 382–394.

Warne, D. 1988. Manual on Fish Canning. FAO Fish. Tech. Pap. 285: 71 p.

Yean, Y.S., R. Pruthiarenun, P. Doe, T. Motohiro and K. Gopakumar. 1998. Dried and smoked fish products. *In*: P.E. Doe (ed.). Fish Drying and Smoking: Production and Quality. Technomic Publishing Company Inc., Lancaster, USA.

Yubero, I.D. 2008. Sabores de Andalucía. Distrib. Consum. 101: 116–125.

CHAPTER 4

Egyptian Legumes and Cereal Foods
Traditional and New Methods for Processing

Mohammed Hefni[1,*] and *Cornelia M. Witthöft*[2]

4.1 Introduction

Legumes and cereals play an important role in the traditional diet in several regions of the world (Messina 1999). In Egypt, cereals occupy the first place in the human diet as a source of calories, with proteins and legumes as the second (FAO 2011). Public health authorities around the world recommend the consumption of cereals and legumes because of health benefits deriving from their chemical composition, e.g., a low content of saturated fat and a high content of essential nutrients and phytochemicals (Anderson 2004, Messina 2014).

Legumes belong to the family *Leguminosae* and are divided into two major groups: oil seeds and pulses. The group of oil seeds, which includes, e.g., soybeans (*Glycine max*), peanuts (*Arachis hypogaea*) and alfalfa (*Medicago sativa*), is mainly used for oil production and feed.

Pulses (dried seeds) are mainly used for human consumption in the forms of canned legumes, germinated sprouts, and/or noodles. The production and use of legumes date back to ancient cultures in Asia, the Middle East, South America, and North Africa.

There are about 1300 species of legumes, of which only about 20 are consumed by humans. The most abundant legume species in the world are chickpeas (*Cicer arietinum*), faba beans (*Vicia faba*), pigeon peas (*Cajanus cajan*), lentils (*Lens culinaris*), mung beans (*Vigna radiata*), soybeans (*Glycine max*), winged beans (*Psophocarpus tetragonoloba*), cowpeas (*Vigna unguiculata*), peas (*Pisum sativum*), peanuts (*Arachis hypogaea*), and black

[1] Food Industries Department, Faculty of Agriculture, Mansoura University, Egypt.
[2] Department of Chemistry and Biomedical Sciences, Faculty of Health and Life Sciences, Linnaeus University, Kalmar, Sweden.
* Corresponding author: mohammed.hefni@mans.edu.eg

gram (*Vigna mungo*). The recommendation to consume legumes is based on them being good sources of protein, minerals, dietary fibre, B-vitamins and polyphenols (Messina 2014). Results from several intervention and prospective studies suggest that regular consumption of legumes is associated with several health benefits, e.g., reduction of low density lipoprotein cholesterol, which is a risk factor for the metabolic syndrome, ischemic heart disease and type 2 diabetes (Anderson et al. 1999, Dahl et al. 2012, Jukanti et al. 2012, Messina 2014). In addition have isoflavones, found in dried beans, received considerable attention in recent years for their potential role in the prevention of certain forms of cancer and a number of chronic diseases (Dahl et al. 2012, Jukanti et al. 2012, Messina 1999).

Cereals belong to the family *Poaceae*. The most abundant cereal species in the world are wheat (*Triticum sativum*), corn (*Zea mays*), rice (*Oryza sativa*), rye (*Secale cereale*), barley (*Hordeum vulgare*), oats (*Avena sativa*), sorghum (*Sorghum bicolor*), fenugreek (*Trigonella foenumgraecum*), and pearl millet (*Pennisetum glaucum*).

Cereals are staple foods in several countries of the world. They are good sources of carbohydrates, protein-, dietary fibre, B-vitamins and minerals. Cereals contain other phytochemicals which may have health promoting effects (Okarter and Lu 2010). In the past ten years, more focus has been given to whole grain cereals since results from epidemiological studies have shown that regular consumption of whole grains is associated with a reduced risk for chronic diseases such as cardiovascular diseases (Anderson 2003, Okarter and Lu 2010), type 2 diabetes (Liu et al. 2000, Tapola et al. 2005) and some cancers (Jacobs et al. 1998, Haas et al. 2009). Cereals are, however, deficient in the amino acid lysine (Šramková 2009). In contrast, legumes contain low amount of the sulphur-amino acids methionine and cysteine (Iqbal et al. 2006) but have a considerable amount of lysine. When legumes and cereals are eaten together in the same meal, which is the case in many Egyptian dishes, they could therefore provide a balanced supply of essential amino acids.

Legumes and cereals have since long been an important part of the traditional Egyptian diet, providing about 60% of the energy intake (FAO 2011). Among the legumes most frequently consumed in Egypt are faba beans or broad beans (*Vicia faba* L.). Several traditional and industrial processing methods are used to produce a number of faba bean dishes and low cost protein meals (see 4.4).

Cereal grains are processed into a wide variety of ingredients or foods for humans and animal feed. Among cereals, wheat is in Egypt frequently consumed in the form of baladi bread with each meal (FAO 2011).

This chapter focuses on the different traditional and industrial processing methods which are applied in Egypt to faba beans and wheat grains. It also describes the most frequently consumed foods and their chemical composition.

4.2 Egyptian food consumption data

The traditional Egyptian diet is based on carbohydrate rich foods. The daily protein intake is comprised 75% from plant origin and only 25% from animal origin (FAO 2011). The majority of the daily per capita energy intake (about 60% energy) originates from cereals and legumes, followed by sugar (sucrose) and sweeteners (16%) and vegetables (12%). Fruits, milk, meat and oil crops supplied only 12% of the energy (around 3% for each commodity) (Fig. 4.1).

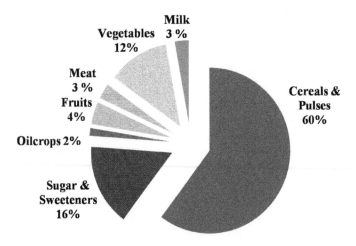

Figure 4.1 Daily per capita energy intake (%) in Egypt from different food commodities (source: FAO 2011).

The food groups that are considered the pillars of the traditional Egyptian diet include cereals, legumes and vegetables (Fig. 4.1). Dried faba beans are commonly consumed (approx. 6.3 kg/capita/year, FAO 2011) as stews, canned or germinated, usually with bread, for breakfast. Fresh green faba beans are commonly cooked in tomato sauce and eaten with rice. Wheat is consumed daily (approx. 363 g/capita/day) as baladi bread.

4.3 Processing methods for legumes and cereals

In Egypt, the faba beans are harvested either in the green stage or after field drying on the plant. The term "pulses" refers to the dried seeds of legumes to differentiate them from the fresh beans.

Wheat is commonly milled into flour and is almost completely used for bread making. Conventional food processing methods used for cereal and legumes in Egypt include soaking, germination, and fermentation and are frequently used during household processing. Freezing and canning are used during industrial processing as briefly described below.

4.3.1 Soaking

Soaking is the step of almost all industrial or household processing methods for pulses (Fig. 4.2). Soaking softens the thin outer shell of the seeds which reduces the cooking time. Soaking has furthermore been reported to reduce several antinutritional factors (e.g., phytate) and to increase the protein digestibility (Khalil and Mansour 1995). Soaking is carried out by adding fluid to the dried seeds (Table 4.1). Thereafter the water is discarded and the seeds are rinsed. The soaking medium is typically water during household preparation, and brine solution (0.25–1%) with or without sodium bicarbonate or acetic acid during industrial processing. The soaking time varies for different pulses. Soaking is usually carried out at ambient temperature (25 to 30°C) overnight, but also longer time periods of 10, 12, and 14 hours have been used (Khalil and Mansour 1995, Hefni and Witthöft 2014). The use of heated water for soaking to assist moisture penetration has also been reported (Sattar et al. 1989).

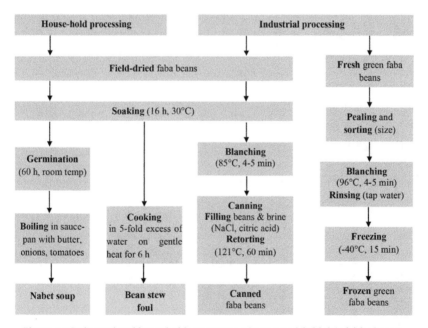

Figure 4.2 Industrial and household processing of green and field dried faba beans.

4.3.2 Germination

Germination is one of the important household processing methods in Egypt and is used to prepare several traditional and contemporary dishes. Germination is a biological process to obtain a typical flavour

Table 4.1 Industrial and household processing of legume and cereal foods and possible changes in their chemical composition.

Process	Description	Food	Application	Product	Changes in chemical composition	References
Soaking	Dried seeds are soaked in tap water over night (1:5 beans:water)	Dried pulses	Industrial processing and household preparation	Preliminary step to all cooking and processing methods	- Reducing antinutritional compounds - Increasing protein digestibility	Admassu and Kumar 2005 Martín-Cabrejas et al. 2009
Germination	Soaked seeds are germinated in the dark at room temperature for 3 days	Dried faba beans	Household preparation	Nabet soup	- Increasing protein digestibility - Increasing antioxidant activity - Reducing antinutritional compounds	López-Amorós et al. 2006 Khalil and Mansour 1995
Fermentation	The wheat flour is mixed baker's yeast, salt, little sugar, and fermented. Fermented dough is divided into portions, shaped into round flat form and baked	Wheat flour	Industrial processing and household preparation	Bread	- Increasing protein digestibility - Increasing antioxidant activity - Reducing antinutritional compounds	Reyes-Moreno et al. 2004 Meignen et al. 2001 Mehta et al. 2012
Freezing	Green beans are blanched and frozen	Green faba beans	Industrial processing	Frozen beans	- Insignificant losses of soluble and unstable nutrients such as minerals and vitamins during blanching and storage	Rickman et al. 2007 Hefni et al. 2015
Canning	Soaked seeds are blanched and autoclaved	Dried pulses	Industrial processing	Ready to eat canned pluses	- Reducing antinutritional compounds - Losses of some nutrients such as minerals, vitamins and protein - Significant decline in total phenolic content	Khalil and Mansour 1995 Drumm et al. 1990 Rickman et al. 2007

and texture in foods (Urbano et al. 2005, López-Amorós et al. 2006, Vernaza et al. 2012). The process usually includes a soaking and an incubation step (Fig. 4.2). Soaking is performed as described in the previous section. Subsequent incubation is done at different conditions with respect to time and temperature. At the household scale, soaked seeds are placed between thick layers of wet cotton cloths to allow germination in the dark at room temperature for three days. At the industrial scale, the process is carried out using macro or micro malting equipment at defined temperature and humidity (Table 4.1). Different germination times (up to 10 days) have been reported for legumes (Shohag et al. 2012, Hefni et al. 2015).

Cereals are also germinated after a shorter time of soaking (five hours). This process is still used today in Egypt for producing fenugreek sprouts which are freshly consumed as a snack. Total germination time varies depending on cereal species and the desired product; germination of cereals between two and seven days has been reported (Kariluoto et al. 2006, Jägerstad et al. 2005, Koehler et al. 2007, Hefni and Witthöft 2011).

During germination, the protein and the carbohydrates in the seeds are degraded by enzymes (Mbithi et al. 2001, Koehler et al. 2007). These changes positively affect the physicochemical characteristics of the legumes and cereals (Chang and Harrold 1988, Vidal-Valverde et al. 2002, Martín-Cabrejas et al. 2008). Germinated legume seeds are of nutritional interest, as this process is reported to on one hand increase the content of soluble dietary fibre (Koehler et al. 2007, Martín-Cabrejas et al. 2008, Aguilera et al. 2013), bioactive phenolic compounds and the antioxidant activity (Aguilera et al. 2013, López-Amorós et al. 2006) and on the other hand, decrease the level of several antinutritional factors (Vidal-Valverde et al. 2002), as shown in Fig. 4.3.

4.3.3 Fermentation

Fermentation of cereals and legumes, spontaneously or with selected microorganisms, is a common food preparation method in many African and Asian countries.

In Egypt, bread is one of the most widespread cereal products fermented by yeasts (*Saccharomyces cerevisiae*) (regarding the method of preparation, see section 4.4.5). During fermentation, raw materials undergo overall changes in composition, flavour and textural properties (Thiele et al. 2002, Loponen et al. 2004). Fermentation does not only enhance the nutrient content of foods due to biosynthesis of vitamins, essential amino acids, and proteins, it also improves the protein quality and fiber digestibility (Mehta et al. 2012). Furthermore, fermentation improves micronutrient bioavailability and supports the degradation of antinutritional factors (Reyes-Moreno et al. 2004).

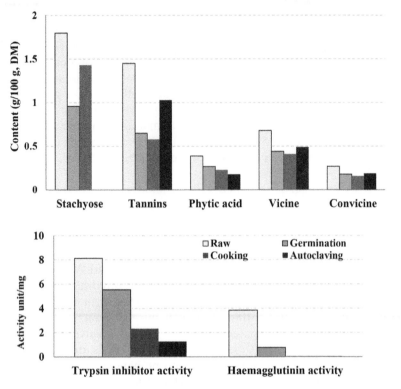

Figure 4.3 Effect of food processing on the content of antinutritional compounds in faba beans (compiled from data from Khalil and Mansour 1995). Processing conditions: **Soaking:** done prior to all cooking, by incubating the dried seeds in water for 12 hours. **Germination:** The soaked beans are incubated between cotton layers in the dark at ambient temperature for three days. **Cooking:** soaked beans are cooked in tap water on a hot plate until tenderness. **Autoclaving:** soaked faba beans are autoclaved at 121°C for 30 minutes (Source: Khalil and Mansour 1995).

4.3.4 Canning

The canning of foods has been practiced for almost 200 years to make the foods sterile. Canning is a heat sterilization process applied to foods to increase the palatability (by cooking) and the shelf life by inactivation or destruction of oxidizing enzymes or pathogenic bacteria. The canned food products therefore will remain shelf-stable as long as the container is intact.

The conventional Egyptian industrial canning of legumes comprises of three major steps: soaking, blanching and thermal processing (Table 4.1). Soaking is carried out as described above (see 4.3.1) to reduce the cooking time and to ensure product tenderness. Blanching (at 85°C for 4 to 5 minutes) is still today used by the industry to inactivate any oxidative enzymes which might produce off-flavours, and also to soften the product. Blanched seeds are filled in the tins with different brine solutions depending on the type of product, but in general the brine solutions contain sodium chloride, citric acid and ethylene diamine tetra acetic acid. Thereafter, the tins are sealed and autoclaved at 121.6°C for an hour. Conditions for heat sterilization of canned beans, which is a low-acid food, are defined to ensure that all spores of *Clostridium botulinum* are destroyed and to prevent the spoilage of the product by heat resistant, non pathogenic organisms.

Canning improves the nutritional value of canned faba beans by reducing the content of several antinutritional factors (Fig. 4.3). However, thermal processing during canning can result in losses of some nutrients such as minerals, vitamins and protein, which are partly destroyed (oxidized) and partly leach into the canning medium (Drumm et al. 1990, Hefni and Witthöft 2014). As in Egypt the canning medium is habitually consumed with the beans, the leached nutrients are not lost for consumption.

There are several types of canned bean products available in the Egyptian market, which are produced with different brine solutions or spices; for example, plain beans or beans with hot pepper, vegetable oil, tahina (sesame paste), chickpeas and olive oil or tomato sauce.

4.3.5 Freezing

The immature green pods of many legumes (faba beans, *Vicia faba*, green beans, *Phaseolus vulgaris*) are in Egypt freshly cooked or industrially frozen. Freezing facilitates the distribution and further processing of vegetables around the year, independent of the season and the place of growing. Common processing steps for frozen green faba beans are outlined (Fig. 4.2) and include blanching and freezing. Blanching prior to freezing may cause losses of water-soluble B-vitamins and minor destruction of phenolic compounds depending on the cultivar (Rickman et al. 2007).

4.4 Traditional Egyptian foods

Egypt has a rich tradition in food preparation. Many of the traditional foods of animal or plant origin are consumed over the whole country (Hassan-Wassef 2004). They play an important role in the nutritional status of the Egyptian population because they are frequently consumed.

Fermentation, soaking and germination are used till today in both household and industrial food production. Dried faba beans are commonly consumed after canning as "foul", germinated and blanched as soup (nabet) or deep-fried as paste balls or falafel (Bakr and Bayomy 1997).

A short description of the preparation and chemical composition of these foods is given in the section below.

4.4.1 Bean stew foul

Bean stew is one of the predominant dishes in Egypt. The stew is prepared at both household and industrial scale, as outlined in Fig. 4.2. In both methods the seeds are soaked overnight. At industrial scale, the soaked seeds are blanched and autoclaved, while at household scale, the seeds are cooked in a pan on gentle heat for approximately 6 hours or until they become soft.

Foul (Fig. 4.5) is usually served for breakfast with fried falafel and baladi bread. The dish bean stew consists of cooked faba beans which are mashed with vegetable oil, garlic and lemon juice prior to serving. Some other ingredient might be added into the dish as, e.g., butter, tomato sauce, tahini and spices (salt, cumin).

The nutritional value of the dish is highly dependent on how it is cooked and served. The chemical analysis of the dish foul shows that the presence of the other ingredients (such as oil, garlic, lemon juice, tomato sauce, tahini, parsley and coriander) can significantly increase the content of vitamins, minerals and dietary fibre (Table 4.2).

4.4.2 Falafel

Falafel is traditionally consumed in Egypt either together with bean stew or with baladi bread (as a sandwich) topped with salad (tomato, onion, cucumber and parsley), pickled vegetables and tahini sauce.

Falafel is made from grounded decorticated faba beans (Fig. 4.4) after overnight soaking. The soaked beans are minced with other ingredients such as onions, parsley, garlic, cumin, baking powder and salt. The dough is shaped by hand into small balls and deep-fried.

4.4.3 Nabet soup

Germinated bean soup or nabet soup is prepared from germinated faba beans, which are boiled in water after the addition of other ingredients (Fig. 4.2). Nabet is a traditional Egyptian dish which is only prepared in the household. The soup (Fig. 4.5) is eaten spiced with lemon juice and cumin, and pieces of roasted baladi bread.

4.4.4 Koshari

This traditional dish is considered the Egyptian national dish. It is rich in carbohydrates and consists of rice, lentils, pasta and tomato sauce. Other ingredients are caramelized onions, garlic and chickpeas. The dish is prepared by cooking the main ingredients, rice, lentils, pasta and chickpeas, separately. These are then mixed directly before being served

Table 4.2 Chemical composition (per 100 g fresh weight) of traditional Egyptian foods, cooked, as eaten.

Nutrient	Bean stew (Foul)	Foul sandwich	Bean falafel, fried	Falafel sandwich	Germinated bean soup (nabet)	Cooked fresh green beans	Rice koshari	Baladi bread
Moisture, g	74	56	33	53	70	77	63	35
Energy, kcal	98	173	355	205	114	79	157	254
Protein, g	6	5	11	6	9	6	5	9
Fat, g	0.7	2	20	6	0.6	0.4	4	1
Ash, g	0.8	1.8	2.2	1.6	1.2	1.0	1.6	1.1
Fibre, g	2.0	1.5	1.5	1.3	1.0	2.4	0.9	1.3
Carbohydrates, g	17	33	33	32	18	13	26	53
Na, mg	24	669	524	587	330	6	401	338
K, mg	218	300	258	273	218	250	185	236
Ca, mg	37	43	24	33	37	31	8	42
P, mg	183	-	118	-	166	110	27	134
Mg, mg	-	29	41	30	-	30	-	14
Fe, mg	2.4	2.1	2.7	2.6	2.4	1.1	0.8	2.9
Zn, mg	1	1.2	1.2	1.4	1	0.8	1.1	1.8
Cu, mg	-	0.1	0.4	0.1	-	0.3	0.1	0.1
Vitamin A, µg	t	-	-	-	t	43	-	nd
Vitamin C, mg	nd	-	6	-	nd	27	-	nd
Thiamin, mg	-	-	-	-	-	0.2	-	0.3
Riboflavin, mg	-	-	-	-	-	0.2	-	0.1

t: traces, -: not analysed, nd: not detected. Source: Egyptian food composition data base 2005.

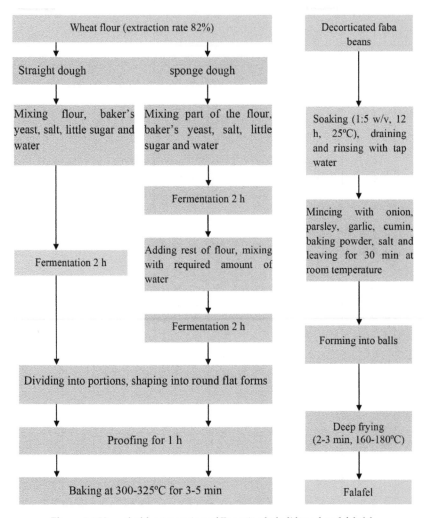

Figure 4.4 Household preparation of Egyptian baladi bread and falafel.

and topped with caramelized onions, garlic sauce and tomato sauce (Fig. 4.5). This dish is the most popular lunch meal in Egypt over the last century.

4.4.5 Baladi bread

Wheat flour is in Egypt almost completely consumed in form of baladi bread. This bread is a circular, 15 to 20 cm in diameter and 1 to 2 cm thick, flat loaf consisting of two layers with almost no crumbs (a type of pita bread).

Figure 4.5 Traditional Egyptian foods commonly consumed (a–e) and a novel food (f).

It is produced from wheat flour from a high extraction rate at industrial and household (hand-made) scale. The bread is baked using two different methods as straight or sponge dough (Fig. 4.4). For the straight dough method, the wheat flour and all the other ingredients (yeast (*Saccharomyces cerevisiae*), salt, little sugar and the required amount of water) are mixed at the same time. For the sponge method, all the ingredients are mixed with a small portion of the flour and allowed to ferment for two hours, thereafter the rest of the flour is added as outline in Fig. 4.4. This bread (Fig. 4.5) is a main component in all meals.

4.5 Chemical composition

Faba beans are one of the most important ingredients of Egyptian dishes. The protein content of raw dried faba beans is between 25 and 40%. Most pulses are very low in fat, generally containing <5% of energy as fat, except chickpeas and soy beans, which contain <15% and 47% fat, respectively (reviewed by Messina 2014). Pulses are good source of dietary fibre, have a low glycemic index (48, 49) and are shown to lower serum cholesterol in hypercholesteremic individuals (Foster-Powell and Miller 1995). Beans also are good sources of micronutrients such as folate, iron, zinc, and calcium (reviewed by Messina 2014).

Data in Table 4.2 show the chemical composition of several typical Egyptian foods as eaten. The highest protein content (11 g/100 g fresh weight, FW) is found in falafel followed by germinated nabet beans and baladi bread (9 g/100 g fresh weight). In other foods the protein content is around 5 g/100 g food as eaten.

With the exception of falafel which contains 20 g/100 g (fresh weight) fat because it is deep-fried, the fat content in the other foods ranges from 0.7 to 4 g/100 g fresh weight. The mineral and fibre content is similar in all foods ranging from 0.8 to 2.2 g/100 g FW for ash and from 1 to 2.4 g/100 g FW for fibre. One serving of these foods, bean stew (200 g), falafel (100 g), germinated nabet soup (150 g), koshari (200 g) or baladi bread (120 g), provides 10 to 14 g protein which is around 20% of the current Recommended Dietary Allowance (RDA) (Institute of Medicine 1998). These foods are also good sources of dietary fibre. One portion of each provides 1.5 to 4 g dietary fibre. One portion of those foods provides furthermore 20 to 70 µg dietary folate; a good body status of this vitamin is reported to reduce the risk of neural tube defects and prevents megaloblastic anemia (Boushey et al. 1995, Berry et al. 1999).

4.6 Development of novel functional cereal and legume ingredients

In Egypt, micronutrient deficiencies are widespread. Anemia is a major public health problem with a prevalence of 30 to 50% among children in the age below 5, of 30% among women of reproductive age and of 45% among pregnant women (EDHS 2005, FFI 2009). Egyptian authorities decided to introduce wheat flour fortified with iron and folic acid for baking of subsidized baladi bread to reduce the widespread micronutrient deficiencies (GAIN 2009). However, complementary approaches to folic acid fortification are required; e.g., by promoting the consumption of naturally folate-rich foods and by increasing the folate content in foods, e.g., by bio-processing.

Germination is a bio-processing technique which has been reported to increase in cereals and legumes the folate content (Jägerstad et al. 2005, Kariluoto et al. 2006, Kariluoto et al. 2006, Koehler et al. 2007, Hefni and Witthöft 2011, Shohag et al. 2012) and other macro and micro nutrients such as soluble and insoluble dietary fibres (Martín-Cabrejas et al. 2008, Aguilera et al. 2013, Rumiyati et al. 2013), riboflavin, thiamin, biotin, pantothenic acid and tocopherols (Plaza et al. 2003). Germination has been shown also to increase the antioxidant activity (Aguilera et al. 2013, López-Amorós et al. 2006) (Table 4.3).

Table 4.3 Effect of germination on the chemical composition of cereal and legume foods.

Compound	Food	Raw	Germinated	Germination condition	References
Phenolic compounds					
p-Hydroxybenzoic aldehyde (µg/100 g DM)	White bean, Phaseolus vulgaris	nd	48–144	6 days, 20°C	López-Amorós et al. 2006
	Peas, Pisum sativum	nd	19–22		
	Lentil, Lens culinaris	nd	47–54		
Dietary fibre					
Insoluble dietary fibre (%, DM)	Oats, Avena sativa	22	31	144 h, 20°C	Hübner et al. 2010
	Barley, Hordeum vulgare	14	16		
Soluble dietary fibre (%, DM)	Oats, Avena sativa	3.4	1.4		
	Barley, Hordeum vulgare	3	1.4		
Dietary fibre (mg/g, DM)	Wheat, Triticum sativum	150	190	102 h, 25°C	Kohler et al. 2007
Vitamins					
Ascorbic acid (mg/100 g, DM)	White beans, Phaseolus vulgaris	3	14	5 days, 25°C	Sangronis and Machado 2007
	Black beans, Phaseolus vulgaris	9	12		
	Pigeon beans, Cajanus cajan	4	14		
Thiamine (mg/100 g, DM)	White beans, Phaseolus vulgaris	0.8	0.9		
	Black beans, Phaseolus vulgaris	0.7	0.9		
	Pigeon beans, Cajanus cajan	0.8	0.9		
Folate (µg/100 g, DM)	Wheat, Triticum sativum	23–28	105–145		Hefni and Witthöft 2011
	Rye, Secale cereale	31–39	90–165	72 h, 25°C	
	Faba beans, Vicia faba	110	304		
	Chickpeas, Cicer arietinum	226	775		
	Soybeans, Glycine max	200–230	759–815	96 h, 25°C	Shohag et al. 2012
	Mungbeans, Vigna radiata	141–170	640–696		

DM: dry matter

Flour from germinated cereals and legumes is used to produce novel Egyptian foods with enhanced folate content (Hefni and Witthöft 2011, 2014, Hefni et al. 2015). A short description of the preparation and food composition data for folate are given in the following section.

4.6.1 Cookies with increased folate content

The aim of the product development was to produce folate-rich cookies using flour from germinated chickpeas. The cookies were produced at industrial scale. Briefly, chickpeas are germinated for 48 hours, dried, milled and sieved. Germination (48 hours) of chickpeas almost doubled the folate content (Fig. 4.6). Wheat flour used for baking cookies is substituted with 50% sieved chickpeas flour. The folate content in the novel cookies with germinated chickpea flour is as high as 85 µg/100 g fresh weight compared to 15 µg/100 g in cookies made entirely from wheat flour (Fig. 4.6). A serving of 100 g could provide 85 µg folate which is approximately 20% of the RDA of 400 µg/day.

4.6.2 Baladi bread with increased folate content

The aim was to produce Egyptian baladi bread with high folate content using germinated wheat flour. Wheat grains are germinated, dried and milled. Germination increased folate content in wheat grains up to four times (Fig. 4.6). Bread is baked from wheat flour which is substituted

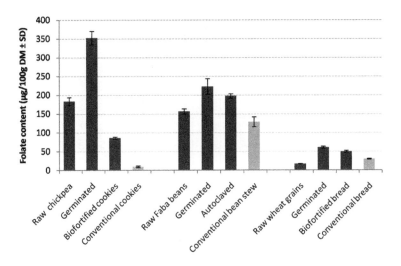

Figure 4.6 Folate content (µg/100 g ± SD, DM) in vitamin enhanced foods with ingredients from germinated chickpeas, faba beans and wheat as compared with conventional products. DM: dry matter, SD: standard deviation.

with germinated wheat flour (50%) (Hefni and Witthöft 2011). Using 50% germinated wheat flour, baladi bread acceptable with respect to colour and the separation into two layers is prepared. The folate content in this bread is as high as 52 µg/100 g DM compared to 31 µg/100 g DM in conventional bread. Based on an average consumption of four loafs per day, eating this novel bread would increase the average daily folate intake by approximately 75 µg.

This approach of increasing the folate content in a bakery product by partial replacement of wheat flour with germinated cereal flour is suitable for both household and commercial production of bread. The application of this approach on other bakery products will allow the development of further novel foods with increased content of folate and other nutrients.

4.6.3 Canned germinated faba beans—foul with increased folate content

This product was also produced at industrial scale with the aim to provide foul with increased folate content from canned faba beans. In this novel industrial canning method, soaked faba beans are first germinated for 48 hours. Thereafter, the germinated faba beans are filled in tins and autoclaved. Germination of faba beans increased the folate content in the canned beans up to 55% (Fig. 4.6). The overall process of industrial canning of germinated faba beans resulted in a net folate increase of 30% compared to the folate content in the raw material. The new product germinated canned faba beans contained 200 µg folate/100 g ± SD, dry matter, which is 52% higher than in a conventional product. One serving (200 g) of this foul (including the canning medium, as commonly consumed in Egypt), provides 85 µg folate.

In summary, several Egyptian traditional and industrial processing methods for faba beans and wheat grains were presented. Furthermore, the most frequently consumed foods and their chemical composition, and a novel approach using traditional bio-processing methods to develop foods with increased folate content was described. These novel ingredients could also be useful in other countries, where mandatory folic acid fortification is not practiced, to improve dietary folate intake; and foods with increased content of other nutrients could also be produced.

References

Admassu, S.E. and R.S. Kumar. 2005. Antinutritional factors and *in vitro* protein digestibility of improved haricot bean (*Phaseolus vulgaris* L.) varieties grown in Ethiopia. Int. J. Food Sci. Nutr. 55(6): 377–387.

Aguilera, Y., M.F. Díaz, T. Jiménez, V. Benítez, T. Herrera, C. Cuadrado, M. Martín-Pedrosa and M.A. Martín-Cabrejas. 2013. Changes in nonnutritional factors and antioxidant activity during germination of nonconventional legumes. J. Agric. Food Chem. 61(34): 8120–8125.

Anderson, J.W. 2003. Whole grains protect against atherosclerotic cardiovascular disease. Proc. Nutr. Soc. 62(1): 135–142.

Anderson, J.W. 2004. Whole grains and coronary heart disease: the whole kernel of truth. Am. J. Clin. Nutr. 80(6): 1459–1460.

Anderson, J.W., B.M. Smith and C.S. Washnock. 1999. Cardiovascular and renal benefits of dry bean and soybean intake. Am. J. Clin. Nutr. 70(3 Suppl): 464S–474S.

Bakr, A.A. and M.F.F. Bayomy. 1997. Effect of selected Egyptian cooking methods on faba bean nutritive value and dietary protein utilization 2: Ability of faba bean products to support hemoglobin response in rats. Plant Foods Hum. Nutr. 50(1): 81–91.

Berry, R.J., Z. Li, J.D. Erickson, S. Li, C.A. Moore, H. Wang, J. Mulinare, P. Zhao, L.Y. Wong, J. Gindler, S.X. Hong and A. Correa. 1999. Prevention of neural-tube defects with folic acid in China. China-U.S. Collaborative Project for Neural Tube Defect Prevention. N. Engl. J. Med. 341(20): 1485–1490.

Boushey, C.J., S.A. Beresford, G.S. Omenn and A.G. Motulsky. 1995. A quantitative assessment of plasma homocysteine as a risk factor for vascular disease. Probable benefits of increasing folic acid intakes. J.A.M.A. 274(13): 1049–1057.

Chang, K.C. and R.L. Harrold. 1988. Changes in selected biochemical components, *in vitro* protein digestibility and amino acids in two bean cultivars during germination. J. Food Sci. 53(3): 783–787.

Dahl, W.J., L.M. Foster and R.T. Tyler. 2012. Review of the health benefits of peas (*Pisum sativum* L.). Br. J. Nutr. 108(Suppl 1): S3–10.

Drumm, T.D., J.I. Gray, G.L. Hosfield and M.A. Uebersax. 1990. Lipid, saccharide, protein, phenolic acid and saponin contents of four market classes of edible dry beans as influenced by soaking and canning. J. Sci. Food Agric. 51(4): 425–435.

Egypt Demographic and Health Survey (EDHS). 2005. National Population Council. Cairo, Egypt.

Flour Fortification Initiative (FFI). 2009. FFI country information: Egypt. Overview of progress towards fortification.

Food and Agricultural Organization (FAO). 2011. Food balance sheet.<http://www.fao.org>. Accessed December 28, 2014.

Foster-Powell, K. and J.B. Miller. 1995. International tables of glycemic index. Am. J. Clin. Nutr. 62(4): 871S–890S.

Global Alliance for Improved Nutrition—GAIN. 2009. WFP, MOSS and GAIN celebrate start of flour fortification in Egypt to reduce widespread anemia by 28%. http://www.gainhealth.org/press-releases/wfp-moss-and-gain-celebrate-start-flour-fortification-egypt-reduce-widespread-anemia. Accessed 27 November 2009.

Haas, P., M.J. Machado, A.A. Anton, A.S. Silva and A. de Francisco. 2009. Effectiveness of whole grain consumption in the prevention of colorectal cancer: meta-analysis of cohort studies. Int. J. Food Sci. Nutr. (6): 1–13.

Hassan-Wassef, H. 2004. Food habits of the Egyptians: newly emerging trends. East Mediterr. Health J. 10(6): 898–915.

Hefni, M.E. and C.M. Witthöft. 2011. Increasing the folate content in Egyptian baladi bread using germinated wheat flour. LWT-Food Sci. Technol. 44(3): 706–712.

Hefni, M.E. and C.M. Witthöft. 2014. Folate content in processed legume foods commonly consumed in Egypt. LWT-Food Sci. Technol. 57(1): 337–343.

Hefni, M.E., M.T. Shalaby and C.M. Witthöft. 2015. Folate content in faba beans (*Viciafaba* L.)-effects of cultivar, maturity stage, industrial processing, and bioprocessing. Food Sci. Nutr.: DOI: 10.1002/fsn1003.1192.

Hübner, F., T. O'Neill, K. Cashman and E. Arendt. 2010. The influence of germination conditions on beta-glucan, dietary fibre and phytate during the germination of oats and barley. Eur. Food Res. Technol. 231(1): 27–35.

Institute of Medicine. 1998. Dietary Reference Intakes for Thiamin, Riboflavin, Niacin, Vitamin B6, Folate, Vitamin B12, Pantothenic Acid, Biotin, and Choline. National Academies Press, Washington DC, USA.

Iqbal, A., I.A. Khalil, N. Ateeq and M. Sayyar Khan. 2006. Nutritional quality of important food legumes. Food Chem. 97(2): 331–335.

Jacobs, D.R., Jr., K.A. Meyer, L.H. Kushi and A.R. Folsom. 1998. Whole-grain intake may reduce the risk of ischemic heart disease death in postmenopausal women: the Iowa Women's Health Study. Am. J. Clin. Nutr. 68(2): 248–257.

Jägerstad, M., V. Piironen, C. Walker, G. Ros, E. Carnovale, M. Holasova and H. Nau. 2005. Increasing natural food folates through bio-processing and biotechnology. Trends Food Sci. Technol. 16(6-7): 298–306.

Jukanti, A.K., P.M. Gaur, C.L. Gowda and R.N. Chibbar. 2012. Nutritional quality and health benefits of chickpea (*Cicer arietinum* L.)—a review. Br. J. Nutr. 108(Suppl 1): S11–26.

Kariluoto, S., K. Liukkonen, O. Myllymäki, L. Vahteristo, A. Kaukovirta-Norja and V. Piironen. 2006. Effect of germination and thermal treatments on folates in rye. J. Agric. Food Chem. 54(25): 9522–9528.

Khalil, A.H. and E.H. Mansour. 1995. The effect of cooking, autoclaving and germination on the nutritional quality of faba beans. Food Chem. 54(2): 177–182.

Koehler, P., G. Hartmann, H. Wieser and M. Rychlik. 2007. Changes of folates, dietary fibre, and proteins in wheat as affected by germination. J. Agric. Food Chem. 55(12): 4678–4683.

Liu, S., J.E. Manson, M.J. Stampfer, F.B. Hu, E. Giovannucci, G.A. Colditz, C.H. Hennekens and W.C. Willett. 2000. A prospective study of whole grain intake and risk of type 2 diabetes mellitus in US women. Am. J. Public. Health. 90(9): 1409–1415.

López-Amorós, M.L., T. Hernández and I. Estrella. 2006. Effect of germination on legume phenolic compounds and their antioxidant activity. J. Food Compos. Anal. 19(4): 277–283.

Loponen, J., M. Mikola, K. Katina, T. Sontag-Strohm and H. Salovaara. 2004. Degradation of HMW glutenins during wheat sourdough fermentations. Cereal Chem. 81(1): 87–93.

Martín-Cabrejas, M.A., M.F. Díaz, Y. Aguilera, V. Benítez, E. Mollaand and R.M. Esteban. 2008. Influence of germination on the soluble carbohydrates and dietary fibre fractions in non-conventional legumes. Food Chem. 107(3): 1045–1052.

Mbithi, S., J. Van Camp, R. Rodriguez and A. Huyghebaert. 2001. Effects of sprouting on nutrient and antinutrient composition of kidney beans (*Phaseolus vulgaris* var. *Rose coco*). Eur. Food Res. Technol. 212(2): 188–191.

Mehta, B.M., A. Kamal-Eldin and R.Z. Iwanski. 2012. Fermentation, effects on food properties, CRC Press, Taylor and Francis Group, Boca Raton, Florida, USA. ISBN: 978-1-4398-5334-4.

Messina, V. 1999. Legumes and soybeans: overview of their nutritional profiles and health effects. Am. J. Clin. Nutr. 70(3): 439s–450s.

Messina, V. 2014. Nutritional and health benefits of dried beans. Am. J. Clin. Nutr. 100(1): 437S–442S.

Okarter, N. and R.H. Liu. 2010. Health benefits of whole grain phytochemicals. Crit. Rev. Food Sci. Nutr. 50(3): 193–208.

Plaza, L., B. de Ancos and P. Cano. 2003. Nutritional and health-related compounds in sprouts and seeds of soybean (*Glycine max*), wheat (*Triticum aestivum* L.) and alfalfa (*Medicago sativa*) treated by a new drying method. Eur. Food Res. Technol. 216(2): 138–144.

Reyes-Moreno, C., E.O. Cuevas-Rodríguez, J. Milán-Carrillo, O.G. Cárdenas-Valenzuela and J. Barrón-Hoyos. 2004. Solid state fermentation process for producing chickpea (*Cicer arietinum* L.) tempeh flour. Physicochemical and nutritional characteristics of the product. J. Sci. Food Agric. 84(3): 271–278.

Rickman, J.C., D.M. Barrett and C.M. Bruhn. 2007. Nutritional comparison of fresh, frozen and canned fruits and vegetables. Part 1. Vitamins C and B and phenolic compounds. J. Sci. Food Agric. 87(6): 930–944.

Rumiyati, V.J. and A.P. James. 2013. Total phenolic and phytosterol compounds and the radical scavenging activity of germinated Australian sweet lupin flour. Plant Foods Hum. Nutr. 68(4): 352–357.

Sangronis, E. and C.J. Machado. 2007. Influence of germination on the nutritional quality of *Phaseolus vulgaris* and *Cajanus cajan*. LWT-Food Sci. Technol. 40(1): 116–120.

Sattar, A., S.K. Durrani, F. Mahmood, A. Ahmad and I. Khan. 1989. Effect of soaking and germination temperatures on selected nutrients and antinutrients of mungbean. Food Chem. 34(2): 111–120.

Shohag, M.J.I., Y. Wei and X. Yang. 2012. Changes of folate and other potential health-promoting phytochemicals in legume seeds as affected by germination. J. Agric. Food Chem. 60(36): 9137–9143.

Šramková, Z., E. Gregová and E. Šturdík. 2009. Chemical composition and nutritional quality of wheat grain. Acta. Chimica. Slovaca. 2(1): 115–138.

Tapola, N., H. Karvonen, L. Niskanen, M. Mikola and E. Sarkkinen. 2005. Glycemic responses of oat bran products in type 2 diabetic patients. Nutr. Metab. Cardiovasc. Dis. 15(4): 255–261.

Thiele, C., M.G. Gänzle and R.F. Vogel. 2002. Contribution of sourdough Lactobacilli, yeast, and cereal enzymes to the generation of amino acids in dough relevant for bread flavour. Cereal Chem. 79(1): 45–51.

Urbano, G., P. Aranda, A. Vílchez, C. Aranda, L. Cabrera, J. Porres and M. López-Jurado. 2005. Effects of germination on the composition and nutritive value of proteins in *Pisumsativum*, L. Food Chem. 93(4): 671–679.

Vernaza, M.G., V.P. Dia, E.G. de Mejia and Y.K. Chang. 2012. Antioxidant and antiinflammatory properties of germinated and hydrolysed Brazilian soybean flours. Food Chem. 134(4): 2217–2225.

Vidal-Valverde, C., J. Frias, I. Sierra, I. Blazquez, F. Lambein and Y. Kuo. 2002. New functional legume foods by germination: effect on the nutritive value of beans, lentils and peas. Eur. Food Res. Technol. 215(6): 472–477.

CHAPTER 5

Slovenian Dairy Products

Andreja Čanžek Majhenič[1],* and Petra Mohar Lorbeg[2]

5.1 Introduction

Although Slovenia is a small country on the sunny side of the Alps, its cuisine is very rich and diverse, influenced by the Slovenian landscape, climate, history and neighbouring cultures. The heritage and originality of each nation are not designated merely by its towns, nature and citizens; an important contribution is also derived from traditional foods and beverages. Although a small country, Slovenian ethnologists have recently divided Slovenia into 23 gastronomic regions.

Slovenia has a very long tradition of farmhouse cheesemaking that in certain regions of the country, has been influenced by foreign cheesemaking technologies. The Swiss master cheesemaker Hitz, that came to Slovenia in the late 19th century, and the well known Montasio cheese—produced in regions of neighboring Italy—had previously influenced the technology of Slovenian cheeses such as Tolminc, Bohinj cheese, Bovec cheese, and Mohant. Other dairy products such as skuta and sour milk, were and still are typical fermented dairy products of the highlands, which besides cheesemaking, also offered another possibility for milk preservation.

Among consumers, there is a growing interest for artisanal dairy products. This is mainly due to the uniqueness of such products which is a result of the climate, vegetation and consequently, the activities of raw milk microbiota. Traditional cheeses and other dairy products are synonymous with natural and artisanal dairy products that have been made for centuries from raw milk from different dairy animals and following traditional protocols. The unique organoleptic characteristics

[1] Chair of Dairy Science, Biotechnical Faculty, University of Ljubljana, Ljubljana, Slovenia.
[2] Institute of Dairy Science and Probiotics, Biotechnical Faculty, University of Ljubljana, Ljubljana, Slovenia.
* Corresponding author: andreja.canzek@bf.uni-lj.si

of traditional dairy products reflect properties of milk combined with the fermentation and ripening processes caused by strains of lactic acid bacteria (LAB), which are determined by the cheesemaking environment and induced by the cheese production technology. Awareness of the importance to preserve unique traditional dairy products together with an increasing distrust of consumers towards additives, preservatives, and new food-transmitted diseases forces producers and scientists to find new, safe and, above all, natural ways to fulfil consumers' requests for safe and authentic traditional dairy products.

This chapter describes the best known Slovenian PDO (Protected Designation of Origin) cheeses, as well as some other historically and locally important dairy products.

5.2 Mohant cheese (PDO)

The strong smelly, semi soft cheese called Mohant originates from the Bohinj region, nestled in the Julian Alps. This geographically separated area enabled the formation of a unique cheese, typical in its production, taste, smell and texture. The first written records date to the 19th century, when Mohant was produced on mountain farms and in villages in Bohinj, initially to preserve the surplus milk. For many decades, Mohant was produced on a farmhouse level as a main source of animal proteins and was consumed with either bread or potato. Since 1997 many activities had been taking place with the intention to preserve the originality of the Mohant cheese; in 2013, it was finally registered by the EU as a foodstuff with PDO (EC No.: SI-PDO-0005-0424-29.10.2004). Nowadays, Mohant is produced by local farmers and in small dairies in villages such as Podjelje, Nemški Rovt, Brod, Srednja Vas and Češnjica. In the summer, it is also produced on Alpine dairy farms Zajamniki, Krstenica and Laz.

5.2.1 Processing

Mohant cheese is traditionally manufactured from the raw milk of the autochthonous cattle breed called Cika (Fig. 5.1), but milk from other cattle breeds may also be used. Evening milk is skimmed using the ladle (or a skimmer) and whole morning milk is added. The milk is heated to 32 to 35°C and after the rennet addition, curding of the milk takes place in about 30 to 45 minutes. The coagulum is cut to the size of a hazelnut. Traditionally it was cut using a wooden piece of equipment called trnač. After scalding at about 38°C, the curd rests under the whey for about 15 minutes. The whey is removed using cheesecloth and the curd is drained. At this stage the curd can be either transferred into wooden or plastic containers, called deže, where it is first salted, then pressed and finally weighed, or it can be placed into the molds and pressed. Cheese in containers is then fermented for 12 hours at 20 to 22°C, causing the remaining whey to excrete, which leads to the formation of anaerobic conditions. Cheese in the molds is pressed for 24 hours at 20 to 22°C, then dry salted and finally moved to the containers and pressed to achieve anaerobic conditions. Fermentation is followed by the three-step ripening process: the cold ripening at 12 to 16°C for 15 to 20 days, the warm ripening at 19 to 23°C for 15 to 20 days (where cheeses undergo the most drastic changes in texture and taste) and the concluding cold ripening at 12 to 16°C for upto two months. During fermentation time the secreted whey is frequently removed and the surface of the cheese is regularly cleaned. Matured Mohant is kneaded thoroughly, then packed into glass or plastic jars and kept at refrigeration temperatures of 3 to 5°C.

Figure 5.1 Cika cattle: autochthonous Slovenian cattle breed (Courtesy: Dr. Mojca Simčič).

5.2.2 *Characteristics*

Mohant is semi soft cheese, packed into plastic or glass jars. Because of special ripening process in anaerobic conditions and at relatively high temperatures, the Mohant has a strong smell which makes it unique and recognizable. The typical Mohant is light yellow, plastic, smooth or slightly cloddy and hardly spreadable (Fig. 5.2). The flavour is sharp, pungent and can be slightly bitter as the result of proteolysis and lipolysis. A typical chemical composition might be: 17 to 26% protein, 23 to 29% fat, about 50% moisture and 0.5 to 2.5% salt (Juhant 2009).

5.3 Bohinj cheese

Bohinj cheese production started in 1873 on the Alpine pasture Bitenjska planina where Swiss master cheesemaker Hitz started to teach people from the Bohinj region the basics of cheese technology. Some years later, several small dairies were established under Hitz supervision including the dairy in village Srednja vas (which is still active), and the dairy in Stara Fužina, where the Alpine Dairy Farming Museum today presents the history

Figure 5.2 Matured Mohant cheese.

of cheesemaking in the Bohinj region. The Bohinj cheese technology is therefore based on the technology of Swiss Emmental cheese, that with certain modifications through its history, has resulted in a unique cheese with golden yellow rind, numerous large eyes and a nutty flavour. In the past, Bohinj cheese was produced in all Bohinj Alpine pastures and valleys but only the dairy Bohinjska sirarna in Srednja vas has preserved the tradition of typical Bohinj cheese production.

5.3.1 Processing

Bohinj cheese is traditionally made from raw or low pasteurized cow milk originating from the Bohinj region. The milk can be standardized to 3 to 3.3% of fat content. Warm milk (31 to 33°C) is inoculated with thermophilic (e.g., *Streptococcus thermophilus* and *Lactobacillus helveticus*), and propionibacterium starter culture. Rennet is added to give a firm coagulum in 20 to 50 minutes; calcium chloride (20 g/100 litres) may be added to ensure the sufficient coagulum strength. The coagulum is gradually cut into small pieces. During stirring, the curd is heated to temperature optimal for starter culture (44 to 46°C) for 20 to 50 minutes. Further scalding at 50 to 52°C for 20 to 60 minutes gives the curd grain the size of a wheat grain. When the curd grains are firm enough they are forced, by the use of the cheesecloth or mechanical pump, into round molds 30 to 60 cm in diameter and 10 to 16 cm in height. The cheeses are then pressed for 24 hours at 20 to 25°C. During pressing, the cheeses are turned frequently to ensure sufficient whey removal and a symmetrical round form. Over the next three to four days, they are salted in brine at 15 to 17°C. After draining and surface drying, they are removed to an initial ripening room with 15 to 18°C for about two weeks. A subsequent longer ripening period for four to six weeks at higher temperature (22 to 24°C) allows the formation of the characteristic large eyes. The final stage is maturation at 12 to 14°C for two to four months that encourages flavour development; with longer maturation period the cheese acquires a more distinctive and piquant aroma (Perko 2003).

5.3.2 Characteristics

Bohinj cheese is a hard, cylindrical cheese which can be slightly convex. A whole cheese weighs 45 to 55 kg, with measures of around 60 cm in diameter and 8 to 15 cm in height. The rind is firm, dry and yellow while the cheese body is ivory to yellow with a high number of eyes. The eyes are regular, large (5 to 15 mm), smooth and rather equally scattered (Fig. 5.3). The flavour is mild and aromatic with a nutty taste. Cheeses with longer maturation time can be more aromatic and slightly piquant. Typical chemical composition may be: 21 to 29% protein, 32 to 36% fat, about 32 to 38% moisture, 48 to 55% fat-in-dry-matter and 1 to 2% salt (Perko 2003).

Figure 5.3 Bohinj cheese (Courtesy: Bohinjska sirarna, d.o.o.).

5.4 Tolminc cheese (PDO)

The first written records on cheesemaking in the Tolmin region date back to the 13th and 14th centuries when cheese was used to pay taxes to the nobility. The name Tolminc cheese (Fromaggio di Tolmino) was first written in 1756 on a price list in the Italian town Udine. Tolminc had a low price compared to other cheeses on that price list; probably because it was made from skimmed milk as the upper Soča Valley was known for its production of high quality butter. Even though the 19th century saw Swiss cheesemaking expert Hitz teach the people of Tolmin the technology of Swiss cheese, the technology of Tolminc remained largely unchanged. The cheese Montasio, produced in the neighbouring province of Friuli in Italy might have had some impact on Tolminc technology. Nowadays, Tolminc cheese is registered as PDO (EU 187/2012) and it can be produced only in the strictly defined area around Tolmin, the town located in the foothills of the Alps next to Italian border. It is made by local farmers, small dairies and in the summer time, also on Alpine pastures. Tolmin has specific

Alpine flora that represents an important part of the feed to milking cows on the pastures and has an important impact on the cheese aroma and flavour, together with natural milk microbiota. Natural microbiota plays a fundamental role in fermentation and is one of the most important factors influencing cheese quality and characteristic flavour.

5.4.1 Processing

Tolminc is made from whole or partly skimmed cow milk. Generally, raw milk is used but it is also allowed to use thermized milk (30 minutes at 64°C or 15 seconds at 68°C). In both cases, the maturation of the milk is an important phase of the technology. Traditionally, the evening milk rests overnight in a vat at room temperature to obtain partly skimmed milk by the gravity separation of fat. At this time the fermentation by indigenous milk microbiota starts the natural acidification of milk. In the morning, the upper layer of cream is separated using a ladle and fresh morning milk is added. Milk is warmed to 32 to 34°C and at this stage natural or commercial thermophilic starter cultures can be added. Natural starter culture can be made from milk or whey from the previous day which is thermized and then fermented at 40 to 45°C for at least 12 hours. After addition of commercial rennet, milk coagulates to a solid curd in 30 minutes. The curd is gradually cut using a curd knife and a cheese harp to the size of a hazelnut and stirred gently during the subsequent scalding at 44 to 48°C. When the curd grains are dried, the stirring is stopped and the curd is allowed to settle down under the whey. The curd is transferred into the molds and pressed for 6 to 12 hours at room temperature. Meanwhile, cheeses are turned a few times to ensure sufficient whey removal and for proper formation of the surface. Cheeses are salted in saturated brine for 24 to 48 hours. Brining is carried out at 10 to 14°C so the curd is cooled down rapidly (to stop further syneresis and to slow down the bacterial growth). The cheeses are removed from the brine and allowed to dry; then they are placed on wooden shelves for ripening at 15 to 18°C. During ripening, cheeses are turned and the surface is brushed or washed using brine or salted water to keep surfaces clean. After two months of ripening, the cheeses are mature but more pronounced flavour may be achieved with prolonged maturation up to one year (Specification for Tolminc cheese 2012). The detailed study of Tolminc microbial population revealed predominance of lactobacilli and lactococci (Mohar Lorbeg 2008), where enterococci play an important part (Čanžek Majhenič et al. 2005).

5.4.2 Characteristics

Tolminc is a hard type, flat, round cheese weighing 3 to 5 kg. Each cheese is 25 to 30 cm in diameter and 8 to 9 cm in height, with a smooth and pale yellow rind. The body is yellow, semi firm and smooth, with a few eyes each the size of a lentil or pea (Fig. 5.4). Young cheeses have a sweet and milky aroma that with longer maturation become more intensive, sweet to tangy and slightly nutty (Specification for Tolminc cheese 2012). Typical chemical composition might be: 25 to 27% of protein, 31 to 35% of fat, about 35% of moisture and 1.5 to 2.0% of salt (Podoreh 2009).

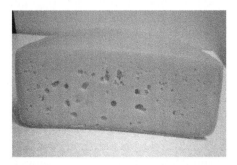

Figure 5.4 Tolminc (Courtesy: Teja Tomažinčič).

5.5 Nanos cheese (PDO)

Nanos is the name of the plateau in the Vipava Valley where cheesemaking was mentioned by land registers of the Vipava Manor as early as the 16th century. Historical notes from the 19th century reported on the vast grazing of sheep on the pastures of the Nanos plateau, and described the first cheesemaking protocol of Nanos cheese. Before World War II, sheep breeding as well as cheesemaking started to decline and completely diminished a few years after the end of the war. The production of Nanos cheese was revived by the Vipava Dairy in 1986 but due to the shortage of sheep milk nowadays it is made from cow milk. The milk from this area is distinguished for its high content of beta carotene that gives the cheese an intensive yellow colour. During the two months of its ripening, a characteristic spicy taste is given to Nanos cheese by the specific microbiota of the wine-growing region where the cheese ripening depot is located. PDO status was implemented by EU 987/2011 regulation.

5.5.1 Processing

At least 80% of the milk for Nanos cheese production comes from a brown cow breed that must graze on pastures of Vipava Valley, Trnovo plateau, surroundings of Karst region, and Postojna Gate. Besides high content of beta carotene, milk from this Karst region also contains more minerals: especially calcium, an important trait from the cheesemaking point of view. The starting material for Nanos cheese production in

Vipava Dairy is raw milk that is thermally treated at 68°C for 15 seconds, partially skimmed, and left at 2 to 10°C overnight in storage tanks. After adjustment of the temperature to 32 to 34°C, the milk is inoculated with thermophilic starter culture and lysozyme, and stirred for about 15 to 40 minutes. The addition of starter culture is an important step since the natural microbiota is partially destroyed during the milk heat treatment. The addition of standard rennet ensures that coagulation occurs within 30 to 40 minutes, a process that may well be aided by enzyme activity introduced from the remaining natural microbiota as well as added starter culture. The curd is cut into 2.5 cm cubes, and stirring with subsequent scalding results in pieces of pea or corn size. Scalding over 40 to 45 minutes brings the temperature to 45 to 50°C, while the final firming of the curd—with constant stirring—takes place for the next 30 to 45 minutes. The content of the vat is transferred to the press vat for total pressing of the curd and after 20 to 60 minutes, the resultant curd mass is cut into cubes of appropriate dimensions that fit to molds of 35 cm in diameter and 15 cm in height. During 2.5 to four hours of pressing, cheeses are turned three times and each time the pressure is gradually increased. On removal from the molds, cheeses are left overnight at 12 to 18°C or until the pH of 5 to 5.4 is reached, and then salted in brine baths for three days. The concentration, temperature and pH of brine are 18 to 20%, 12 to 18°C and 5 to 5.4, respectively. When the cheese surface is dry enough, the cheeses are transferred into the ripening room with 83 to 88% relative humidity and 14 to 16°C. Within the first 14 days cheeses are turned and wiped every three days, and then once a week, to the minimal maturation period of 60 days. However, prolonged maturation to upto 12 months gives rise to stronger flavour and harder texture. Concerning the maturation of Nanos cheese there is a strict demand that cheeses mature in the ripening room that is positioned in the upper part of the wine-growing district of Vipava Valley at about 500 metres above sea level. Vipavska burja, the famous north wind of Vipava, is equally important for the Nanos cheese maturation due to its positive selection in favour of *Penicillium* molds, as they represent between 60 to 80% of microbial population on the surface of Nanos cheese (Specification for Nanos cheese 2013).

5.5.2 Characteristics

Nanos cheese is a hard type cheese, and the cheese wheel is about 8 to 10 cm in height, 34 cm in diameter and between 8 and 10 kg in weight. The rind is smooth, firm and dry, yellow and brick reddish in colour. The cheese body is yellowish to intensive yellow in colour, flexible and plastic, with no visible fractures, although rarely scattered pea-sized eyes are allowed (Fig. 5.5). The typical piquant flavour of Nanos cheese originates from the presence of noble molds that are specific to the wine-growing

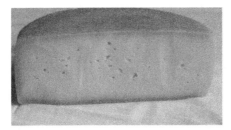

Figure 5.5 Nanos cheese (Courtesy: Dr. Iva Boltar).

region of the Vipava Valley and are manually cleaned from the cheese surface during maturation. Yeasts are equally important microbes on the surface of Nanos cheese and although they diminish by the end of the ripening period, their proteolytic and lipolytic enzymes significantly contribute to the typical aroma of Nanos cheese, which is piquant and slightly nutty (Specification for Nanos cheese 2013). Recently, the thorough research concerning the volatile profile of Nanos cheese was published (Boltar et al. 2015). The typical chemical composition of Nanos cheese at 60 days ripening is: at least 60% of dry matter, at least 45% fat in dry matter, about 35 to 40% moisture and 1.5 to 2% salt.

5.6 Bovec cheese

This cheese was named after the town of Bovec and is mentioned in the 14th century. In the past, Bovec cheese was used as a means of payment and was valued much higher than other cheeses. Even today, it continues to maintain its special status and symbolizes a testimony to the rich farming and cheesemaking heritage around Bovec. Its taste and smell are well rounded and pleasantly harmonized with a slightly spicy note. It is also registered as a PDO cheese (EU 753/2012). Today, Bovec cheese is produced on farms and on two highland Alpine dairies in the municipality of Bovec and its surroundings, which represents the area of protection.

5.6.1 Processing

Bovec cheese is a hard type, full fat sheep cheese, but varieties made with upto 20% of cow and/or goat milk are allowed as well. In the traditional system, raw milk drawn on the evening before production is poured into a large vat and left overnight at 9 to 15°C. The next morning fresh milk is added and the mix is poured into the vat. After adjustment of the temperature to 35 to 36°C, rennet consisting of chymosin-pepsin is added to successfully coagulate the milk in 30 to 45 minutes. The resultant coagulum is cut into beans or peas-sized pieces, while stirring during the subsequent scalding stage results in the size of wheat grain (Fig. 5.6). Scalding over the

Figure 5.6 Stirring and scalding of the coagulum
for Bovec cheese (Courtesy: Davorin Koren).

next 15 to 30 minutes brings the final temperature to 44 to 49°C, which is maintained until the cheese grain is firm. After, the curd is allowed to settle to the bottom of the vat and is gently pressed by hand. The curd mass is cut with the copper wire into coarse lumps; each lump is removed from the whey and placed into a mold. Alternatively, the entire curd mass may be scooped into the cloth drawn across the bottom of the vat with the help of a stainless steel rod and, by pulling together the corners of the cloth, lifted out. After transferring to a special draining table the curd is cut manually into portions of appropriate dimensions and transferred to molds that are often cloth-lined. The cheeses are pressed over the period of four to six hours, with frequent turning, before being brined for 24 to 43 hours at 14 to 16°C, or being dry salted for upto six days with sea salt. Maturation at 14 to 19°C and humidity of 75 to 80% follows. The minimum ripening time is usually 60 days, but it can be extended upto two years. During this time the cheeses are frequently turned and, if necessary, wiped with brine-soaked cloth (Specification for Bovec cheese 2014).

5.6.2 Characteristics

Bovec cheese wheel is about 8 to 12 cm in height, 20 to 26 cm in diameter and between 2.5 and 4.5 kg in weight. The rind is smooth and grey-brown in colour, with a flat upper side and slightly convex peripheral side. At the minimum ripening time of two months, the cheese will develop compact and homogenous texture with typical "shell" break that is liable to fracture but not crumble. Rarely scattered eyes of the size of a lentil or pea are evident, with a few natural fissures allowed (Fig. 5.7). The flavour becomes aromatic, intensive to mild-piquant, and the interior will be grey to pale yellow. A typical chemical composition of Bovec cheese is about: 23 to 26% of protein, 35 to 45% fat, 50 to 55% of fat-in-dry-matter, 30 to 40% moisture and 1.5 to 2.5% salt (Specification for Bovec cheese 2014).

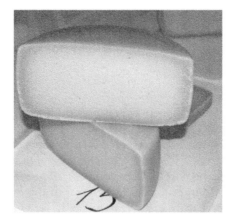

Figure 5.7 Bovec cheese (Courtesy: Davorin Koren).

5.7 Karst ewe cheese

Karst ewe cheese is a hard type cheese that is a result of typical Karst soil and climate, vegetation and the autochthonous ewe breed called Istrian Pramenka. Its PDO designation is registered on the national level (Pravilnik o Kraškem ovčjem siru z zaščiteno označbo porekla, UL RS 29/2008, 21.3.2008).

5.7.1 Processing

Traditionally, raw sheep milk was employed as a starting material but the use of thermized milk (64°C, 30 minutes) is also allowed. During the season, cheese is made from the mix of the evening and morning milk, or rarely, immediately after each milking. After adjustment of temperature to 31 to 33°C, the bulk milk is inoculated with either 0.4 to 0.6% of "kisava" (fermented secondary whey), or thermophilic starter bacteria in DVS form. A chymosin or mix of chymosin-pepsin is added in amounts that encourage gel formation within 30 to 40 minutes. The curd is cut into wheat-sized granules. The temperature is slowly raised of the stirred curd/whey mixture to 38 to 42°C within 10 to 15 minutes for firm pieces of curd, that rapidly settle from the whey on cessation of stirring. After 10 minutes, the curd mass is collected into cloths and placed in molds. Alternatively, immediately after scalding, the entire content of the vat is transferred into perforated molds on a cheese table to allow the whey to drain off. Collected either way, the molds are subjected to increasing pressure for 12 to 18 hours. The cheeses are turned two to three times. After removal from the molds, Karst ewe cheese is dry salted for three to six days or brined in 20% brine at 15 to 18°C for 24 hours. Further, the cheeses are placed into a maturation store with temperatures of 15 to 18°C and humidity of 75 to 85% where they mature for a few months, and exceptionally, for up to one year (Specification for Karst ewe cheese 2008).

5.7.2 Characteristics

Between 5.8 and 6.6 kg of sheep milk is needed for the production of one kg of Karst ewe cheese, where the wheel is about 8 to 12 cm in height, 20 to 26 cm in diameter and between 2.5 and 4.5 kg in weight. The rind is smooth and flat, with a slightly convex peripheral side and yellow brownish in colour. The cheese has a compact and uniform texture that is liable to fracture but not crumble. The colour of the body is pale yellow to brownish. None or only a few rarely scattered eyes of the size of a small lentil are visible on cutting. A few natural fissures are also allowed (Fig. 5.8). At the minimum ripening time of two months, the flavour is aromatic, intensive to mild piquant, and the interior a grey to straw colour. With the passage of time the colour of the interior darkens and the flavour becomes more piquant. From the microbiological point of view, the *Lactobacillus* community in Karst ewe cheese was investigated by Čanžek Majhenič et al. (2007). The typical chemical composition of Karst ewe cheese is as follows: 22 to 26% of protein, 28 to 36% of fat, 47 to 57% of fat-in-dry-matter, 35 to 40% of moisture and 1.5 to 3% salt.

Figure 5.8 Karst ewe cheese.

5.8 Dolenjski ewe cheese

Dolenjski is the least known sheep cheese in Slovenia, originating from the eastern part of the country. It is a semi hard and full fat type of cheese.

5.8.1 Processing

Milk drawn on the evening before production is kept overnight at 4°C, mixed with fresh morning milk, and thermized at 63°C for 30 minutes, or used raw. Either way, after the temperature is adjusted to 32°C, the milk is inoculated with selected thermophilic starter cultures that act as the main starter culture or as a supportive starter culture to the indigenous milk microbiota, and is coagulated by the use of chymosin or chymosin-pepsin mix to ensure coagulation within 30 to 45 minutes. The resultant coagulum is cut into pieces of hazelnut size and with subsequent scalding over the next 30 minutes at 42 to 45°C, the size of the cheese grain is reduced to the size of a pea. When the curd is sufficiently firm, small cheese producers leave the curd to rapidly settle from the whey on the bottom of the vat, and then collect the curd mass into cloths and place it in molds, whereas bigger cheese producers transfer the entire content of the vat into perforated molds on the cheese table to allow the whey to drain off. Pressing takes 10 to 14 hours with the weight of 3 kg/kg of cheese. Dolenjski cheese can either be brined (22%) for 24 hours or dry salted at 15°C for two days. Maturation takes place at 13 to 15°C and humidity of 75 to 80% for at least 60 days (Havranek et al. 2012).

5.8.2 Characteristics

The rind is smooth and grey-brownish in colour, with upper and peripheral sides slightly convex. The interior of the cheese may be grey or pale straw in colour, with rarely scattered eyes of the size of lentils, where the presence of small holes is not too pronounced (Fig. 5.9). The texture is hard, firm and compact but not crumbly. The taste and smell are aromatic, clean and fully rounded with a slightly piquant note. The typical chemical composition of Dolenjski cheese at about two months is: 25 to 26% protein, 25 to 35% fat, 45 to 55% fat-in-dry-matter, 35 to 40% moisture and 1.5 to 2.5% salt.

Figure 5.9 Dolenjski ewe cheese.

5.9 Trnič cheese

Trnič is a special type of hard cheese which was traditionally made by the herdsmen on Velika planina, plateau in the Kamnik-Savinja Alps in the

19th and early 20th centuries. Only a few individuals have preserved the knowledge of making Trnič today. Trnič was mainly made by the herdsmen in summer during the grazing season. At first it was made to preserve the milk surplus. Milk was left at room temperature to sour, cream was removed and soured milk was poured into a special clay pot. The pot was left on the fireplace to heat and the mass was drained using the wooden strainer. After salting, some cream was added and the curd was shaped into clumps. Clumps were partially dried on the shelves above the fireplace. Afterwards, they were shaped and decorated using carved wooden seals, called pisava. Every herdsman carved his own pisava to mark the cheeses he made (Fig. 5.10). Finally, the cheeses were left on shelves above the fireplace to dry and to smoke. Cheeses were always made in pairs (both with the same ornaments) as its shape symbolizes the breasts of a woman. At the end of the grazing season, the herdsman would keep one and present the other to his beloved as proof of his love and fidelity—and sometimes as an engagement symbol. If his beloved accepted the gift, it showed that she agreed to his courtship. For several years, this cheese was recognized as an authentic souvenir from Velika planina since it looked attractive and could be kept unchanged for several years without preservation. Nowadays, Trnič is also used as a food ingredient. Young Trnič can be grated on risotto, pasta and salad; it can also be added to meat or fish dishes or simply cut into slices and served with the addition of honey, pepper, olive or pumpkin oil. Aged Trnič becomes harder and the taste becomes sharper and nobler; because it is very dry it can be soaked in hot milk to make cutting easier. Today, Trnič is produced by some farmers located near the town of Kamnik and can be found in several tourist spots in the summer time, as well as on Velika planina, Mala planina and Gojška planina—mountain plateaus situated above Kamnik.

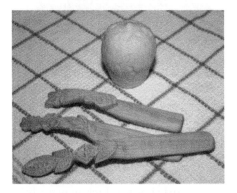

Figure 5.10 Trnič and pisava.

5.9.1 Processing

Trnič is made from raw cow milk which is fermented at room temperature with the addition of mesophilic starter culture. After milk coagulation, the separated fat from the surface is removed and the coagulum is transferred to the vat. Scalding is done by indirect heating while the mass is stirred. In the summer, the scalding temperature is higher (around 60°C) than in the winter (52 to 55°C). The curd is separated from the whey using cheesecloth. The curd in cheesecloth is hung for 12 hours at room temperature to drain. Afterwards, the drained curd is spread over dry cheesecloth placed on trays and left at room temperature until it is dry enough for shaping (about 24 hours). After salting (two g of salt per 100 g of curd is added), the curd is kneaded and then manually formed first into a small ball, which is further shaped to distinctive form and decorated using a carved wooden seal. Trnič is then left at room temperature to dry for about four weeks while its size is reduced by half (Fig. 5.11). To make one Trnič of 120 g, about two litres of milk is needed (H. Kropivšek personal communication).

Figure 5.11 Freshly made and matured Trnič.

5.9.2 Characteristics

Trnič is a dried cheese, with a cup-shaped body and a nipple on top, decorated with relief ornaments. Its size depends on the size of the cheesemakers' hands and usually is 5 to 6 cm in height and 4 to 5 cm in diameter. The colour of the cheese depends on the season: winter cheeses are pale yellow and summer cheeses are dark yellow to brownish. It is hard, dry and crumbly, with a salty and slightly acidic taste. With prolonged maturation, the cheese becomes harder and more piquant. Its typical chemical characteristics might be: 40 to 42% protein, 31 to 42% of fat, 83 to 87% solids and 4 to 5% salt.

5.10 Other dairy products

5.10.1 Pregreta smetana, škrlupec (overheated cream)

Overheated cream is a little-known traditional dairy product, only produced on farm level. It is typical for the Posavje, Bela Krajina and Dolenjska regions in Slovenia. The manufacturing protocols were and are still passed from generation to generation by oral tradition and personal notes that are based mostly on empirical experiences. Consequently, many variations for overheated cream production exist, but two main processes predominate.

5.10.1.1 Processing

Immediately after milking, the milk is poured in shallow terracotta pots (latvice) and left in cold storage overnight in the fridge or pantry to allow fat separation. Then, the latvice are placed in the oven until the separated layer on the milk is crusty and golden brownish in colour. The formed crust is occasionally immersed in milk to obtain a thicker layer of cream as well as to prevent burning. After overheating, which usually lasts for three to four hours, the contents are left to cool and are then strained. According to the oral tradition, the effective straining of overheated cream is of great importance for its further use. The remaining milk does change in colour and flavour and is usually unsuitable for further processing and therefore often fed to animals or discarded. For one kg of overheated cream about 12 litres of raw cow milk is needed.

Another variation differs in the first stage of manufacture where overheated cream is made from a skin-like layer (or crust) formed on the surface of boiled milk. After being heated to a boiling point, the milk is left for a few minutes and then poured into wide and shallow containers to allow the formation of a fat/serum protein surface layer. Normally, upto three days are required for fat/protein aggregation. Prior to overheating, the fat/protein mass (crust) is cut into small pieces, transferred into a baking tin and overheated with occasional stirring until a golden brown colour and crust patches are formed. Overheated cream is then left overnight to cool down and packed into glass jars.

The growing interest among consumers in the revival of overheated cream production, as well as variations in its manufacture (and consecutively in sensory characteristics), led us to introduce a compromised technological procedure for overheated cream production that would also unify its sensorial characteristics. Therefore, overheated cream was made in four different batches from 30 litres of milk that was boiled, left for a few minutes and then poured into wide containers. Separated fat/protein mass was overheated. Milk heat treatment and fat/protein separation were identical in all batches; overheating was accomplished with the changing of some technological parameters as follows: the size of crust patches (whole pieces of crust, cut crust), the temperature (150°C to 170°C) the time (up to four hours) of overheating and the mode of stirring (no stirring, stirring every half hour, stirring every hour). The cream was overheated to a golden colour with dark brown patches of crust, cooled and filled into glass containers. The results of chemical and sensory analysis of the samples revealed that the optimal technological process was as follows: prior overheating fat/protein mass (crust) was cut into smaller pieces and overheated for three hours at 170°C with intermediate stirring every half hour (Rudolf 2013).

5.10.1.2 Characteristics

The taste and smell of overheated cream is soft, creamy and caramelized. The texture is buttery, smooth to slightly crumbly with crust patches that are golden brown to brownish in colour and evenly marbled (Fig. 5.12). Overheated cream is to be kept in a dark and cold place and it should be used within a few days of production (two weeks is the maximum shelf life). Pregreta smetana can be used as a topping, as a stuffing ingredient in traditional Bela Krajina rolled cake, as well as an addition to everyday and special occasion/festive dishes (potica roll, štruklji dough roll, soups, gravies). A typical chemical composition of overheated cream is about: 5.8% of protein, 74% of fat, 87% of fat-in-dry-matter and 15% of moisture.

Figure 5.12 Pregreta smetana in earthenware (with the remaining milk in the cup) and potica roll made with pregreta smetana (Courtesy: Špela Rudolf).

5.10.2 Skuta

Skuta refers to fresh cheeses made by several technological procedures and with differing characteristics. Traditionally, skuta was made without the addition of rennet; the combination of acid and heat was used for milk coagulation. Skuta is one of the most widely produced cheeses in Slovenia;

it is produced in small dairies, by local farmers and during grazing season on highland farms. Different types of skuta also have specific names: sour skuta is quark type cheese, sweet skuta is made with the addition of acidic compound to the milk and sirarska skuta is ricotta type whey cheese.

5.10.2.1 Sour skuta processing

Sour skuta is manufactured with natural fermentation in combination with heat. It is generally made of raw cow milk which is left at room temperature to sour naturally. Some sour milk or whey from a previous batch may be added to promote fermentation. When coagulum is firm, fat is removed using a spoon. Coagulum is cut into strips and slowly heated until the curd is separated from the whey (50 to 60°C). The curd is left under the whey to cool to room temperature and then it is transferred to a colander or a cheesecloth and left to strain for about 12 hours.

5.10.2.2 Sweet skuta processing

Sweet skuta is made with the addition of sour component to whole milk; usually apple or wine vinegar, as well as fermented whey, lemon juice or citric acid is used. Milk is heated to a high temperature (80 to 90°C) and after the addition of sour component, precipitation of denatured proteins begins. After a short time, the precipitate can be separated from the whey using a colander or it may be left to cool before straining to improve consistency.

5.10.2.3 Sirarska skuta processing

Sirarska skuta can be made of cow, goat or ewe whey retained after cheese production. Some whole milk may be added to improve the yield and texture of the cheese. Whey left in the vat is heated to 90°C and acid component is added afterwards; the addition of fermented whey or citric acid is most common. Precipitated proteins segregated on the surface are removed after a short time and drained using cheese molds or cheesecloth. To extend stability, some salt can be added to the product.

5.10.2.4 Characteristics

Sour skuta is a white, soft, creamy and somewhat crumbly fresh cheese with an acidic taste (Fig. 5.13). Sweet skuta is white, more compact and somewhat elastic and chewy, with a slightly acidic and milky taste. Sirarska skuta is soft and creamy, white and slightly sweet in taste.

Figure 5.13 Sour skuta.

All types of skuta can be used in various ways: as a spread (some herbs, spices or aromatic oils can be added), or as an ingredient to typical dishes like štruklji, prekmurska gibanica (layer cake) and other dishes, like pancakes. The typical chemical composition of sour skuta might be: around 5% of fat, 13% of protein and 77% of moisture. The typical chemical composition of sweet skuta might be: around 26% of fat, 17% of protein and 55% of moisture. The typical chemical composition of sirarska skuta might be: around 10% of fat, 12% of protein and 77% of moisture.

5.10.3 Kislo mleko (sour milk)

Kislo mleko is a traditional product made from cow milk.

5.10.3.1 Processing

The starting material for kislo mleko is raw milk that is put either in wooden or earthen pots, inoculated with the small amount of the previous day's kislo mleko, and left overnight at the ambient temperature for fermentation. The fermentation can also be aided by the naturally present microbiota of raw milk, without any addition of previously fermented product.

5.10.3.2 Characteristics

Kislo mleko is meant for direct consumption and due to its acidic taste, is a refreshing dairy product. The coagulum is white in colour and very tender; it is covered with a layer of separated fat. It is fermented with mesophilic lactic acid bacteria that contribute to its nicely rounded taste and aroma.

References

Boltar, I., A. Čanžek Majhenič, K. Jarni, T. Jug and M. Bavcon Kralj. 2015. Volatile compounds in Nanos cheese: their formation during ripening and seasonal variation. J. Food Science Technol. 52(1): 608–623.

Čanžek Majhenič, A., I. Rogelj and B. Perko. 2005. Enterococci from Tolminc cheese: Population structure, antibiotic susceptibility and incidence of virulence determinants. Int. J. Food Microbiol. 102: 239–244.

Čanžek Majhenič, A., P. Mohar Lorbeg and I. Rogelj. 2007. Characterisation of the *Lactobacillus* community in traditional Karst ewe cheese. Int. J. Dairy Technol. 60, 3: 182–190.

Havranek, J., B. Perko, A. Trmčić, A. Čanžek Majhenič and I. Rogelj. 2012. Cheeses of Slovenia. *In*: N. Antunac and N. Mikulec (eds.). An Atlas of Sheep Cheeses of the Countries of the Western Balkans. Faculty of Agriculture, University of Zagreb, Zagreb, Croatia.

Juhant, G. 2009. Production process of Mohant cheese—cheese of geographical origin. BSc Thesis, University of Ljubljana, Ljubljana, Slovenia.

Mohar Lorbeg, P. 2008. Phenotypic and genotypic diversity of enterococci from traditional Slovenian cheeses. PhD Thesis, University of Ljubljana, Ljubljana, Slovenia.

Perko, B. 2003. Slovenski avtohtoni-siri z geografskim poreklom. University of Ljubljana, Rodica, Slovenia.

Podoreh, T. 2009. Execution of technological process in making Tolminc cheese with geographical designation of origin. BSc Thesis, University of Ljubljana, Ljubljana, Slovenia.

Rudolf, Š. 2013. Technological procedure of overheated cream (pregreta smetana) production. MSc Thesis, University of Ljubljana, Ljubljana, Slovenia.

Specification for Bovec cheese. 2014. (http://www.mkgp.gov.si/fileadmin/mkgp.gov.si/pageuploads/podrocja/Varna_in_kakovostna_hrana_in_krma/zasciteni_kmetijski_pridelki/Specifikacije/Bovski_sir_specifikacija.pdf).

Specification for Karst ewe cheese. 2008. (http://www.mkgp.gov.si/fileadmin/mkgp.gov.si/pageuploads/podrocja/Varna_in_kakovostna_hrana_in_krma/zasciteni_kmetijski_pridelki/Specifikacije/KRASKI_OVCJI_SIR.pdf).

Specification for Mohant cheese. 2012. (http://www.mko.gov.si/fileadmin/mko.gov.si/pageuploads/podrocja/Varna_in_kakovostna_hrana_in_krma/zasciteni_kmetijski_pridelki/Specifikacije/MOHANT_SP_EU.pdf).

Specification for Nanos cheese. 2013. (http://www.mkgp.gov.si/fileadmin/mkgp.gov.si/pageuploads/podrocja/Varna_in_kakovostna_hrana_in_krma/zasciteni_kmetijski_pridelki/Specifikacije/NANOSKI_-_spec._junij_2013.pdf).

Specification for Tolminc cheese. 2012. (http://www.mkgp.gov.si/fileadmin/mkgp.gov.si/pageuploads/podrocja/Varna_in_kakovostna_hrana_in_krma/zasciteni_kmetijski_pridelki/Specifikacije/TOLMINC_januar_2012.pdf).

CHAPTER 6

Spanish Dry-cured Hams and Fermented Sausages

Jorge Ruiz,[1] María J. Pagán,[2] Purificación García-Segovía[2],* and Javier Martínez-Monzó[2]

6.1 Introduction

Production of dry-cured meat products in the Mediterranean can be traced back to several centuries before Christ. In this area, the environmental conditions throughout the year, with cold winters and hot and dry summers, allowed the production of self-stable meat products when industrial chilling technologies were not available. There is an enormous variety of different regional dry-cured meat products.

Meat products manufactured in Spain reached 1,300,000 Mt in 2012 (Mercasa 2013). The amount of dry-cured hams and dry fermented sausages is around 19% and 15%, respectively of total manufactured meat products.

6.2 Dry-cured ham

From an historical perspective, meat curing may be defined as the addition of salt to meats for the purpose of preservation. Together with sodium chloride, nitrate was added as an impurity in salt. This allowed for carryover of meat from times of plenty to times of scarcity (Martin 2001). The combined effect of curing salts and dehydration of the pieces during the drying process play a key role in, first, stabilizing the product by reducing water activity, and secondly, allowing the occurrence of chemical and biochemical reactions leading to development of the sensory features of dry-cured ham.

[1] Department of Food Science, University of Copenhagen, Frederiksberg, Denmark.
[2] Department of Food Technology, Polytechnic University of Valencia, Valencia, Spain.
* Corresponding author: pugarse@tal.upv.es

Even though the main goal of dry-curing technology was the preservation of food products, allowing storage and consumption of meat throughout the year, nowadays the main purpose of this technology is to obtain products showing attractive and exclusive sensory properties, especially flavour and texture (Martin 2001). The characteristics and chemical composition of the raw material and the procedures and conditions followed during the processing determine the features and sensory attributes of dry-cured ham. In this sense, numerous studies have shown the influence of pig breed (Muriel et al. 2004a), pig feeding (Cava et al. 2000), fresh ham freezing (Pérez-Palacios et al. 2011), amount of salt (Andrés et al. 2005), processing length (Ruiz et al. 1999b), or ripening temperature (Andrés et al. 2004b) on chemical parameters and sensory features of dry-cured products.

Among them, perhaps the most valuable and representative one is dry-cured ham, especially Parma and San Danielle hams in Italy, Bayonne ham in France, Presunto Barranco in Portugal and Serrano and Iberian hams in Spain.

6.2.1 Types of Spanish dry-cured hams

In Spain there are two main types of dry-cured hams: Serrano and Iberian. Serrano hams are made with thighs from selected pig breeds, without any special attention to their feeding and with different ripening times. Nevertheless, within this type of hams, there are different categories: high quality ones, made out of high intramuscular fat Duroc cross-breeds, slaughtered at 120 kg, in which the feeding specifically excludes polyunsaturated fat and ripened for almost a year; and also lower quality ones, from very lean pigs, and ripened for less than six months.

Iberian ham is a dry-cured ham produced in the southwest of Spain with very distinctive sensory features as a consequence of the particular characteristics of both the pig breed and the processing conditions. These hams are produced with thighs from Iberian pigs, an autochthonous swine breed that has not been selected for productive performance or carcass leanness. Iberian pigs are usually free reared outdoors and fed on acorns and grass during the finishing fattening period during two or three months previous to slaughter. These pigs are slaughtered at very high weights (around 150 to 160 kg), which together with their inherent tendency to accumulate fat (Lopez-Bote 1998), lead to very fat carcasses, and meat with high levels of intramuscular fat, and hence, a very intense marbling (Ruiz-Carrascal et al. 2000). Dry-cured meats traditionally obtained from Iberian pigs have kept their peculiar appearance, juiciness, aroma and taste throughout several centuries. Several intrinsic (genetic) and extrinsic (environmental) factors explain the presence of high levels of fat and large concentrations of myoglobin and iron in Iberian pig muscles, which are appreciated quality traits (Ruiz et al. 2002a, Ventanas et al. 2007b). In addition, the intake of natural feeds, such as grass and acorns, during the last stage of the fattening period causes the deposition of high levels of oleic acid and tocopherols in their tissues, which contributes to obtain high quality dry-cured products (Ventanas et al. 2007c).

6.2.2 Quality characteristics in fresh meat

The marbling of Iberian pig meat is typically abundant and evident, much more intense than in the meat from pig commercial genotypes. This is a direct consequence of the high intramuscular fat (IMF) content of Iberian pig meat, with levels as high as 10% of the total fresh weight of the muscle (Mayoral et al. 1999), while common levels described for selected breeds are rarely above 3% (Martin et al. 2008c). This is a direct consequence of a higher total fat content of the carcass. For example, back fat thickness in Iberian pigs commonly reaches eight cm (Mayoral et al. 1999), while in commercial pigs is seldom above two cm (Martin et al. 2008c).

Such a high fat content in Iberian pigs is due to several factors. First, pigs are usually slaughtered at high weights (between 150 to 160 kg), which means that fat accumulation is much greater that in conventional selected lean pigs. Moreover, accumulation of IMF is more intense in older animals, in which there has been already a previous accumulation of subcutaneous fat. In addition to that, before the final finishing fattening stage, Iberian pigs follow a feed restriction phase. In the traditional rearing system this was a consequence of the scarce feeding resources during the summer season, which is hot and dry in those areas in which Iberian pigs are reared. Nowadays, restricted feeding is used on purpose before fattening in order to increase fat deposition during the finishing phase. This is most likely due to the fact that fat is one of the latest tissues to develop and grow. So, by the time animals are in the fattening phase, most of the energy from feeding is used for fat synthesis. Daza et al. (2008) observed a higher IMF in muscle of Iberian pigs which have been restricted before the fattening.

On top of that, free reared Iberian pigs are mainly fed on acorns and grass. This feeding regime shows a high caloric content due to the presence of acorns, and low protein content (Daza et al. 2007). It seems that a low ratio protein/calories in the diet leads to a higher fat deposition (Goerl et al. 1995). In addition to all this, Iberian pig is an anabolic breed (Lopez-Bote 1998) with a high tendency to accumulate fat, and it is also a fast maturing breed. Therefore, by the time animals are being fattened (12 to 14 months old), most of the ingested calories are directed to fat synthesis.

As will be discussed below, the high IMF content is one of the key reasons determining the quality of Iberian meat, upon which the productive system of Iberian pig is sustained. However, the enormous amount of subcutaneous fat constitutes an economical disadvantage, since fat synthesis is energetically very demanding, while the value of lard is very low. Nevertheless, such a costly and low performance productive system is necessary, since high IMF levels cannot be reached if there is not a previous enormous subcutaneaous fat accumulation (Lawrie and Ledward 2006). Also, this partially explains the higher price of Iberian meat products.

As compared to conventional pigs, the high IMF of Iberian meat has several consequences on the technological properties of the meat for dry curing processing, and it is also one of the main factors leading to the high sensory quality of the derived dry-cured products. Both, high levels of IMF and thick layer of subcutaneous fat, slow down moisture lost during the drying process, since fat shows a much lower diffusion rate for water than lean (Palumbo et al. 1977). In fact, a clear influence of fat content on weight loss during the processing of Iberian dry-cured hams has been evidenced (Fig. 6.1).

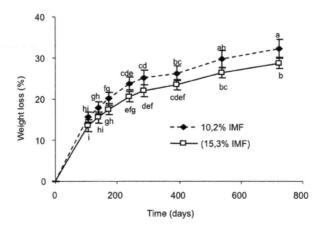

Figure 6.1 Weight loss in Iberian dry-cured hams with different IMF (adapted from Jurado 2005).

Moreover, due to the high IMF content, the initial stages in Iberian dry-cured ham processing, aimed to allow dehydration but also to permit salt distribution through the whole ham, need to be much longer than in hams from lean conventional pigs. This is due to the lower diffusion rate of sodium chloride in the fat than in the lean (Wood 1966): the higher the fat, the slower the reduction in water activity in the depth of the ham; hence, a longer time is needed for the hams to reach a water activity that allows to keep them at temperatures above refrigeration.

Such a long processing time could be also considered as an economical disadvantage: it implies high energetic costs and it also means that the potential revenues are delayed in time. However, from a sensory point of view it is not a drawback, since such a long ripening time allows the development of biochemical and chemical reactions (Ruiz et al. 1999b, Ruiz et al. 1999a), leading to the intense and particular flavour of Iberian dry-cured ham, which is highly appreciated by consumers. Lean hams from conventional pigs could also be processed for such long ripening times as Iberian hams are; however this would lead to an excessive dehydration, which in turn would make the ham very dry, tough and almost not chewable.

Another key consequence of the high IMF content of Iberian pigs as compared to conventional ones is the juiciness and mouth feeling of dry-cured products. In fact, it seems that, contrarily to other meat products, moisture content is not as crucial in determining juiciness, at least within the common levels of water content reported for this product (Andrés et al. 2005). The juiciness of meat products is considered to be a direct consequence of the moisture released during chewing and also due to the contribution from saliva (Winger and Hagyard 1994). IMF stimulates the saliva secretion and directly contributes to the juiciness perception by coating the tongue, teeth and other parts of the mouth. Since processing of dry-cured products involves a strong dehydration, both the moisture from saliva and the direct contribution of IMF play a very important role in juiciness. Given that juiciness has been pointed out as the main trait influencing overall quality of Iberian dry-cured ham (Ruiz et al. 2002a), it seems clear that in this type of dry-cured products, IMF content is directly linked to their acceptability by consumers.

Moreover, this association of IMF with the eating quality of dry-cured hams leads to an indirect effect on the appreciation of their appearance: while in other meat products, like Parma ham, marbling is not appreciated, consumers consider an intense marbling in Iberian meat products as an important appearance trait, and as a consequence, a major trait driving purchase intention (Ventanas et al. 2007b).

Together with juiciness, flavour intensity has been pointed out as the main determinant of consumer acceptability in Iberian dry-cured hams (Ruiz et al. 2002a), and in general in all dry-cured products (Morales et al. 2013, Font-i-Furnols and Guerrero 2014). The intense flavour of Iberian ham is due to the presence of high levels of low molecular weight non volatile compounds (mainly amino acids, peptides and sodium chloride) and a huge variety of low molecular weight volatile compounds (Ruiz et al. 1999a, Ruiz et al. 1999b). All these compounds arise from different biochemical (mainly lipolysis and proteolysis) and chemical reactions (mainly lipid oxidation, Maillard reactions and Strecker degradation of amino acids). IMF acts as a reservoir of compounds that subsequently undergo transformations leading to volatile compound formation. It also contains non polar compounds that are directly accumulated from the diet

(Tejeda et al. 1999), contributing to the distinctive flavour of hams from free reared animals. But in addition to that, fat influences the release of volatile compounds to the mouth, mainly retaining non polar compounds. The influence of fat content on the release of volatile flavour compounds in different foodstuffs has been extensively studied (Seuvre et al. 2002). However, little is known about its effect on whole meat products.

As far as the implication of the high IMF levels of Iberian pig meat on its nutritional quality, it is widespread knowledge that in developed countries, diet is one of the most important factors affecting wellbeing and health. In this sense, meat and meat products are important components in the diets of wealthy societies, where one of the aspects that most influence the "image" and hence the consumption of meat, is whether it is perceived as healthy (Jimenez-Colmenero et al. 2001). The association of certain meat constituents, such as fat content and cholesterol, with the risk of development of certain human diseases, has had a considerable impact on public opinion. In fact, meat is perceived as the major dietary source of fat and especially of saturated fatty acids (SFA) in developed countries (Wood et al. 2004). According to the recommendations of the World Health Organization, fat should provide between 15 and 30% of the calories in the diet and saturated fat should not provide more than 10% of these calories. Nowadays, consumers prefer low fat meat products and meat consumption is recommended in moderation. Due to that, during recent decades a great effort has been made aiming to reduce fat content of carcasses in order to fulfil consumer's demands for leaner meat. Though the high IMF content in meat and meat products from Iberian pigs is not in agreement with the current tendencies of meat consumption, it is essential in order to assure a correct processing and to obtain dry cured meat products with appropriate sensory attributes. However, the lipid content of Iberian cured products is in general lower than that of other meat products in which fat is added during the processing, such as sausages or burgers (Pérez-Palacios et al. 2008).

6.2.2.1 Fatty acid composition

The mean composition of fatty acids in pig meat approximately ranges from 35 to 40% SFA, 40 to 50% monounsaturated fatty acids (MUFA) and 8 to 12% polyunsaturated fatty acids (PUFA) (38). However, it is clear from the available scientific literature that this is highly variable and greatly influenced by diet. Fat deposition is a process that quantitative and qualitatively depends on the feed intake. Fatty acid composition in pig tissues depends on the proportion of those coming from the feed (direct deposition) and those produced endogenously (*de novo* synthesis) (Lawrie and Ledward 2006). The energy balance and the diet composition determine the relative importance of endogenous synthesis and direct accumulation of fatty acids from the feed, so that fat deposition in animal tissues depends upon the excess of energy consumed in relation to requirements. The priority for energy is to fulfil maintenance and protein synthesis. The excess of energy is accumulated as triglycerides in adipocytes. Carbohydrates are the preferential energy source for metabolic purpose, and only when dietary carbohydrates cannot fulfil the need for energy, would dietary fat be used for energy production. Therefore, when the intake of energy from carbohydrates is enough, almost all consumed fat would be accumulated in pig tissues with little modification. So, when pigs are fed a high fat content diet (i.e., 8 to 10% fat), the endogenous synthesis is considerably low and the lipid composition of pig tissues would reflect to a great extent dietary fatty acid composition. Traditional rearing system for Iberian pigs involved free rearing during the fattening, with free availability of acorns and grass. Acorn shows a high fat content (up to 7%) and high proportions of oleic acid (around 60 to 70%), while grass shows a high proportion of linolenic acid (Ruiz et al. 1998a), an omega-3 fatty acid. Consequently, fat from Iberian pig fed on acorns and grass is rich in oleic acid (up to 55 to 60%) (Andrés et al. 2000) and shows slightly higher proportions of n-3 fatty acids than conventional pigs, that are fed on concentrates (Muriel et al. 2002). Pigs fed on concentrates show a different fatty acid profile depending on the feeding composition (Daza et al. 2005). So, this explains the high proportion of oleic acid in Iberian pig tissues, and the slightly higher percentage of n-3 fatty acids in these pigs (Muriel et al. 2002). In fact, by feeding outdoors acorns and grass, the natural feeding resources of the surrounding ecosystem, Iberian pig farmers have been achieving for centuries what modern production systems for pigs are nowadays attempting: high levels of oleic acid and lower n-6/n-3 ratios.

Such a particular fatty acid profile of tissue lipids from Iberian pigs reared outdoors greatly affects technological, sensory and nutritional quality of meat, since it determines the physical state of the fat (liquid or solid), its prone to get oxidized, and the nutritional and metabolic effects on the consumer.

Fat consistency strongly influences the appearance, the feasibility for manipulation and the dehydration of meat products. The physical state of the fat mostly depends on the fatty acid melting point, which in turn depends upon the number of double bonds per fatty acid residue. In a saturated fatty acid chain, all carbon bonds are in cis configuration and orientated alternatively to produce a linear structure. When there is a double bond, two cis bonds are orientated consecutively in the same way to produce a bend structure, much more difficult to arrange in organised crystalline-like structures (hydrophobic bonds, etc.). Therefore, the energy needed to disorganise such fat structure (melting) is lower as the number of unsaturation rises. The melting point for tristearin (3 molecules of

C18:0 esterified to glycerol) is around 70°C, while for trilinolein (C18:2) it is below 0°C.

Among fatty acids present in pig tissues, stearic acid (C18:0) and linoleic acid (C18:2) show the highest correlations (positive and negative respectively) with fat consistency. However, from a practical point of view, linoleic acid is the most important fatty acid determining fat consistency, since its variability is much higher. Nevertheless, as far as Iberian pig fat is concerned, oleic acid rather than linoleic acid seems to be the main contributor to fat fluidity (Flores et al. 1988). Above 14 to 15% of linoleic acid, the incidence of "floppy meats" is high. For dry-cured sausages this value should not be higher than 12%, which could also be valid for dry-cured hams. These levels are set based on the effect of dietary fat unsaturation on water migration in dry-cured meat products (Girard et al. 1989). The levels of PUFA in fatty tissues of pigs used for dry-cured ham are critical, since an excessive accumulation in tissues could lead to an insufficient dehydration after the processing of dry-cured hams and loins. In fact, levels of linoleic acid over 12 to 15% of total fatty acids in pig fat lead to an extension of the processing time due to impaired water migration (Ruiz and Lopez-Bote 2002). For example, it has been estimated that in dry-cured hams from Iberian pigs fed diets containing high levels of linoleic acid, an additional year of drying is needed to achieve a moisture loss in the final ham product equivalent to that from pigs given diets enriched in SFA or MUFA. Due to this (and also to the effect on lipid oxidation), processing industries and Protected Designation of Origin (PDO) recommendations have established limits around 10% of linoleic acid in lard when meat is intended for Iberian ham production.

When the fat is mostly in liquid state at temperatures at which pig meat is normally commercialised and processed (4 to 7°C), consistency is low and appearance is oily. Fat consistency plays itself a significant role in consumer acceptability of dry-cured ham. However, this effect is not common for all types of ham. Consumers in Spain very much value a soft and oily fat in Iberian dry-cured ham (Ventanas et al. 2007b), while theses features negatively affect the quality of other types of dry-cured ham from conventional pig breeds (Ruiz and Lopez-Bote 2002). The positive effect of oiliness on Iberian dry-cured ham acceptability is most likely due to the positive link between higher oleic acid contents and a desirable flavour in dry-cured Iberian hams. Fat from Iberian pigs free reared shows a more fluid and oily aspect at room temperature than that from pigs reared in confinement. This different aspect is a direct consequence of higher levels of oleic acid in the fat of free reared pigs, while those in confinement usually shows a higher proportion of SFA. At the same time, dry-cured hams from free reared pigs shows a more intense and pleasing flavour, partially because of the fatty acid composition. Consequently, consumers and ham producers have used fat consistency to indirectly assess dry-cured ham quality for years, those hams dripping fat during the processing being considered as top quality ones (Ruiz et al. 2002a).

Lipid oxidation is one of the main causes of deterioration in the quality of meat products during storage and processing, due to the formation of compounds with rancid flavour, to discolouration, the formation of free radicals, the generation of cholesterol oxidation products, the insolubilization of proteins and so on (Morrissey et al. 1998). On the other hand, some of the volatile compounds with higher impact on the aroma of Iberian dry-cured meat products arise from oxidation of unsaturated fatty acids (Ruiz et al. 1998b). In fact, regardless of the conditions followed during the processing or the raw material composition, the volatile profile of cured products includes chemical compounds from autoxidation of unsaturated fatty acids. Among them, the most important aroma compounds are aldehydes and several unsaturated ketones and furan derivatives (Ruiz et al. 2002b). They include C3-C10 aldehydes, C5-C8 unsaturated ketones and pentyl or pentenyl furans. On the other hand, saturated and unsaturated aldehydes have been extensively used as markers for lipid oxidation in food (Andrés et al. 2004a). Some of them, such as 2,4-decadienal or hexanal impart rancid or pungent flavours, which are believed to be negative for the overall flavour of meat products (Flores et al. 1997).

Nevertheless, information about maximum acceptable levels of lipid oxidation in Iberian dry-cured products is scarce. Some volatile aldehydes from lipid oxidation, especially hexanal, have shown good correlations with rancid flavour in dry-cured ham (Flores et al. 1997). This is in agreement with previous findings in cooked meat, in which hexanal has been related to WOF (Shahidi et al. 1987). Moreover, the ratio between compounds from oxidation of PUFA (specially hexanal and 2,4-decadienal) and those from oxidation of oleic acid (mainly octanal and nonanal) has been pointed out as a good indicator of rancidity in Iberian ham (Antequera et al. 1992). This is sustained in the more pleasant aromatic notes of nonanal and octanal (Flores et al. 1997).

Some other compounds from fatty acid oxidation may positively affect the flavour of Iberian cured products. For example, 1-octen-3-ol, an unsaturated alcohol that may rise from autoxidation of either linoleic or arachidonic acids, shows an intense mushroom odour (Grosch 1987). Some of the furans found in these cured products show interesting aromatic notes (Flores et al. 1997). Lactones, which may also have a lipid oxidation origin, show very interesting aromatic notes and very low detection thresholds (Slaughter 1999). E-4-heptenal, which has been highlighted as a contributor to off-flavour in pâté, imparting metallic notes, is also a compound from the autoxidation of linoleic acid or arachidonic acids (Grosch 1987). Lower levels of this compound have been found in meat products containing Iberian pig fat than when using commercial pigs lard, due to the lower amounts of linoleic acid in that made with the lard from Iberian pigs (Ruiz et al. 1998a).

Furthermore, compounds from lipid oxidation may further react with amino compounds to give rise to Maillard reaction compounds (Mottram 1998) with very low odour thresholds, which have been highlighted as responsible for pleasant flavour notes in Iberian ham (Carrapiso et al. 2002). In fact, some of these volatiles have been also pointed out as main contributors to flavour of Iberian dry-cured loin (Ventanas et al. 2010). Formation of Strecker aldehydes in model systems at low temperatures, and the occurrence of Maillard reactions between reactive compounds from lipid oxidation and amino acids and proteins at room temperature have been evidenced (Ventanas et al. 2007a).

Therefore, although deep oxidation of unsaturated lipids may lead to rancid flavour in dry-cured ham, several compounds from lipid oxidation and from interaction between amino compounds and compounds from lipid oxidation play a key role in dry-cured ham flavour. Since the fatty acid profile of raw material highly influences the profile of volatile flavour compounds of processed meats, the increased levels of oleic acid in tissue lipids of Iberian pigs due to the outdoor feeding system, appears as a very interesting approach for reducing the amount of compounds showing rancid notes (those coming from PUFA), and at the same time, increasing those showing pleasant flavour notes, or at least not intense rancid aroma, such as octanal and nonanal, whose main origin is the autoxidation of oleic acid.

It seems that phospholipids (PLs) are the primary substrates of lipid oxidation in muscle foods, while triglycerides seem to play a minor role (Igene and Pearson 1979). Therefore, these lipids should be the target compounds in controlling fatty acid composition in fresh meat for Iberian pig cured products. Since PLs show a high turnover rate (Leray et al. 1995), they are more markedly affected by the feeding composition of the period previous to slaughter (Ruiz et al. 1998a). Therefore, the final fattening outdoors on acorns and grass fits the requirements for obtaining a pleasant flavour in the elaborated cured products, this having been developed for ages not on purpose, but due to the weather and ecological conditions of the environment.

Fatty acid composition plays an important role on the diet-health relationship, since each dietary fatty acid affects the plasmatic lipids levels differently and has different effects on atherogenic and thrombogenic processes. SFA increases blood total low density lipoprotein (LDL) cholesterol and the high density lipoprotein HDL/LDL ratio, which implies a risk factor for cardiovascular diseases. Myristic (C14:0) and palmitic (C16:0) fatty acids, commonly found in meat (around 1.4% myristic and 25% palmitic in pork), are the main SFA behind the cholesterol elevating effect. Stearic acid (C18:0) (around 10% in pork) is partially converted to oleic acid (C18:1) *in vivo* and has not shown an hypercholesterolemia effect since it does not raise cholesterol levels as the rest of SFA, though it decreases the HDL fraction (Lovegrove and Griffin 2013). Several studies have reported lower levels of SFA in tissues from free-range reared Iberian pigs than in those from intensively reared commercial pigs. The presence of MUFA and PUFA in human diet reduces the level of plasma LDL-cholesterol, although PUFA also depress the HDL-cholesterol, which has a positive effect in preventing from cardiovascular diseases. The intake of MUFA has been inversely associated with the risk of cardiovascular heart disease, although the association is weaker than for polyunsaturated fat (Hu et al. 1997). In accordance to recommendations ratio PUFA/SFA (P/S) should be above 0.4. Since some meat generally have a P/S ratio around 0.1, meat has been implicated in causing the imbalanced fatty acid intake of today's consumers (Wood et al. 2004). Meat and dry-cured products from Iberian pigs contain high levels of MUFA and particularly of oleic acid (C18:1). Traditionally, in Mediterranean countries, the sources of MUFA in the diet have been vegetable origin fats such as olive oil, which contains 78% of oleic acid. Meat and meat products from Iberian pigs might be an alternative source of MUFA. On the other hand, numerous strategies for achieving healthier meat and meat products involve replacing part of the animal fat with another more suited to humans needs, i.e., with less SFA and more MUFA (oleic acid) or PUFA. It has been proved that the substitution of saturated by unsaturated fat is more effective in the decrease of risk of cardiovascular disease than only reduction of total fat intake. Consequently, fat from Iberian pig might be a healthy fat source due to his optimal fatty acids composition especially taking into account the high levels of oleic acid.

6.2.2.2 Antioxidants

After microbial spoilage, lipid oxidation is the main factor reducing the quality of meat and meat products (Morrissey et al. 1998). The oxidation process involves the degradation of PUFA, vitamins and other tissue components, the formation of protein carbonyls, leading to the development of rancid odours and changes in colour and texture in foodstuffs (Kanner 1994). Oxidative damage of lipids and proteins in meat is accentuated after slaughter during handling, processing, cooking or storage. Thus, the oxidation of lipids and proteins in pig meat products has been evidenced during refrigerated storage of raw and cooked meat (Martin et al. 2008a), processing of dry-cured and fermented products (Muriel et al. 2007, Andrés et al. 2004a), and processing of cooked products (Roldán et al. 2014). Though the positive effect of lipid oxidation on the development of desirable aroma characteristics in cooked meat (Roldán et al. 2015, Sanchez del Pulgar et al. 2013), dry-cured (Flores et al. 1997) and cooked products (Martin et al. 2009) has been suggested, the development of oxidative reactions has been generally associated with loss of quality including aroma, colour and texture deterioration and generation of toxic compounds (Morrissey et al. 1998). As a consequence, several antioxidant strategies have been suggested for pork in order to enhance the oxidative stability of pig tissues, reduce the development of oxidative reactions during processing of meat products and minimise the adverse effects on meat quality. The improvement of the lipid oxidative stability of porcine meat products through dietary means involves the modification of the fatty acid composition of the tissues, commonly focussed in reducing the proportion of colour, and the supplementation with substances with proven antioxidant activity (Ruiz and Lopez-Bote 2002).

Some of the potential strategies that have been specifically used in the case of Iberian pigs include feeding on diets with high MUFA/PUFA ratios and supplementation with α-tocopherol at levels between 100 and 200 mg/kg (Rey et al. 2006). In fact, as explained above, free-range reared Iberian pigs have been traditionally fed on acorns and grass. The influence of this traditional feeding system on the chemical composition and oxidative stability of tissues from Iberian pigs has been profusely studied concluding that the high quality of Iberian pigs products can be mainly attributed to this feeding regime (Ruiz et al. 2005). Acorns provide high levels of MUFA (mainly oleic acid) and γ-tocopherol to Iberian pigs whereas grass is a well known source of ω-3 fatty acids (mainly linolenic acid) and α-tocopherol (Ventanas et al. 2007b) (Fig. 6.2).

The traditional procedures used for Iberian pig feeding is in absolute concordance with the current strategies carried out in order to enhance the

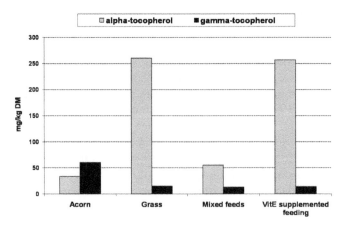

Figure 6.2 Levels of α- and γ-tocopherol in the main feeding sources for Iberian pigs (chart made with data from Daza et al. (2005)).

oxidative stability of meat from conventional pigs. Moreover, Iberian pigs reared in confinement are fed using MUFA-enriched diets with α-tocopherol supplementation up to 200 mg/kg in order to imitate the traditional free range feeding system. The use of such mixed diets for feeding Iberian pigs rapidly spread out as a result of the publication of papers in which MUFA and tocopherols were evidenced as the major contributors to the oxidative stability in tissues and meat products from Iberian pigs (Daza et al. 2005). It has been shown that the meat from Iberian pigs fed on natural resources (grass and acorns) contains similar or even higher tocopherol levels than that from animals fed with diets supplemented with α-tocopherol up to 200 mg/kg. Furthermore, the presence of γ-tocopherol is almost exclusive in tissues from pigs fed on acorns, not being detected in tissues from animals fed on α-tocopherol-supplemented diets (Rey et al. 1998). As a consequence, meat from Iberian pigs fed on acorns and pasture shows a lower susceptibility to lipid oxidation than meat from pigs fed on non supplemented mixed diets and shows a similar oxidative stability to those supplemented with 200 mg/kg α-tocopherol (Daza et al. 2005). The protective role of tocopherols against lipid oxidation has been reported also in elaborated products such as dry-cured Iberian ham, significantly reducing the generation of lipid oxidation products during ripening and improving some particular sensory characteristics such as flavour and odour intensity (Ventanas et al. 2007c). The persistence of relatively high tocopherols levels in cellular membranes after Iberian ham processing protects against lipid oxidation beyond the ripening stage: ham slices from supplement pigs showed lower discolouration and lower weight losses than those from pigs fed on control diets when stored imitating the conditions of marketing display (Ruiz and Lopez-Bote 2002).

Other substances accumulated in tissues from Iberian pigs as a consequence of the intake of natural resources could also contribute to enhance meat oxidative stability. Thus, a higher amount of phenolic compounds have been found in adipose tissue from Iberian pigs fed exclusively on natural resources (grass and acorns) than those fed with a mixed diet (9.11 vs. 6.74 mg/kg) (Gonzalez and Tejeda 2007). These authors suggest that the high oxidative stability attributed to Iberian pigs products could be explained not only by the presence of tocopherols but also by the likely protective role of phenolic compounds. Indeed, some studies have reported elevated polyphenol levels in acorns (Cantos et al. 2003).

6.2.2.3 Myoglobin and iron contents

Meat greatly contributes to the daily intakes of iron and other trace elements to the human diet (McAfee et al. 2010). Particularly interesting is the fact that meat is the food richest in heme iron, which seems to be more bioavailable than non heme iron (Carpenter and Mahoney 1992). However, the iron levels in meat vary considerably between species and breeds and it is affected by intrinsic (genetic) and extrinsic (environmental) factors (Lawrie and Ledward 2006). The variation in the iron levels is generally associated to different heme pigments concentrations due to the close relationship between both meat components.

Meat from Iberian pigs has been considered an excellent source of high available iron for humans. Several studies have reported higher concentrations of heme pigments and total iron in muscles from Iberian pigs compared to those from industrial genotype pigs (Ventanas et al. 2005). When the amounts of heme iron were compared, the differences between muscles from Iberian and industrial genotype pigs were consistent to results previously described. The levels of iron in meat from Iberian pigs are around twice higher than in meat from industrial genotype pigs (between 60 and 70 µg iron/g muscle for Iberian pig muscle vs. 20 to 30 µg iron/g muscle for conventional pigs) (Ventanas et al. 2005). In fact, muscles from Iberian pigs contain similar iron levels than those from other animal species such as beef or ostrich meat, habitually considered the highest iron content meats. The high levels of heme pigments and iron in muscles from Iberian pigs are mainly explained by the peculiar genetic characteristics of the non selected rustic pig breeds. It is known that muscles from commercial pig breeds selected for fast growth contain higher proportion of fast twitch fibres (glycolytic: IIB type) and lower of slow twitch ones (oxidative: types I and IIA) than muscles from rustic breeds (Andrés et al. 2000) which affects muscle heme pigments concentration and therefore, iron levels. In fact, Iberian pig muscle have shown high proportions of oxidative I and IIA fibres (around 48%), as compared to commercial genotype pigs, such as Large-White, Landrace or Yorkshire, in which the proportion of such oxidative fibres does not exceed 31% in the same muscles (Andrés et al. 2001, Abreu et al. 2006). Some other extrinsic factors associated to the traditional procedures of Iberian pig's rearing system affect myoblogin and iron contents in muscles. In order to obtain heavy and fatty carcasses, Iberian pigs are traditionally slaughtered with 12 to 14 months of age whereas considerably shorter times are used for industrial genotype pigs (around five months). The concentrations of myoglobin and iron in muscles are known to increase with age (Mayoral et al. 1999). In addition, there is a beneficial impact of physical exercise on myoglobin content in muscles through the increase of the proportion of oxidative fibres (Andrés et al. 2000). These authors reported higher amount of hemein muscles from free-range reared pigs than in those from intensively reared pigs.

From a nutritional point of view, the meat and meat products from Iberian pigs can be considered excellent sources of iron. Higher levels of iron in Iberian dry-cured hams (32 µg iron/g) than in those elaborated with raw material from selected pig breeds (22 μg iron/g) have been found (Ventanas et al. 2005).

On the other hand, iron is considered a potent promoter of oxidative reactions in muscle foods (Kanner 1994) and therefore, the presence of high iron levels in muscles from Iberian pigs could enhance their oxidative instability. However, the accurate knowledge of the proportion between the chemical forms of iron is of great interest since non heme iron is thought to have more ability to promote oxidative reactions than heme iron (Kanner 1994). Previous studies have reported similar non heme iron contents in muscles from Iberian and industrial genotype pigs, concluding that the higher amounts of iron in Iberian pig muscles are mainly due to differences in heme iron contents (Estévez et al. 2004). In conclusion, the high levels of iron in meat from Iberian pigs could be considered as an interesting nutritional benefit for human health that might not affect the oxidative stability of Iberian pig meat products.

From a sensory point of view, the high concentrations of myoglobin and iron in muscles from Iberian pigs have a direct impact on their colour traits due to the relationship established between those parameters (Andrés et al. 2000). The redness (a*-values) described in muscles from Iberian pigs (Muriel et al. 2004b) are considerably higher than those reported in the same muscles from industrial genotype pigs (Martin et al. 2008b). The colour standards displayed by meat from Iberian pigs are preferred by consumers who appreciate intense red colours in fresh pig meat (Ruiz et al. 2002a).

6.3 Fermented sausages

A wide variety of raw fermented and dried sausages are produced in Spain, reflecting the diversity of traditions and climate conditions in different regions and areas. The varieties of sausages differ in: raw materials (usually pork but also beef), ingredients (spices and seasoning), degree of chopping, type and diameter of casing and ripening conditions (Lorenzo et al. 2015). The name of the products depends of all these features and geographic origin; sometimes the names differ even between very close and small areas (Toldrá 2006).

The Mediterranean sausages can be classified according to a range of criteria (acidity, mincing size of the ingredients, addition or absence of starters' cultures, presence or absence of moulds on the surface, addition of same ingredients and of seasoning, diameter and type of casing used, etc.). Spanish fermented sausages can be classified as fermented meat products with a high acidity (because, usually, the fermentation and ripening temperature, concentration of curing salts used, pH, and addition of carbohydrates most commonly used in these products are similar to those described by Incze (1992) for that type of fermented sausages.

There are over 50 different varieties of these fermented sausages, although many are handmade on a small scale (MAPA 1997, Marcos 1991). Moreover, some of these products have protected their appellation (DOP: Denominaciones de Origen Protegidas or Protected Designation of Origin, IGP: Indicaciones Geográficas Protegidas or Protected Geographical Indication) (MAGRAMA 2015).

Chorizo and salchichon are the most important products (over 120 tons/year). Both are made from coarsely cut pork meat (or a mixture of pork and beef) with pork fat, together with spices and seasoning: garlic, oregano (*Origanum vulgare*), sodium chloride, Spanish paprika (which confers the typical colour) in chorizo, black pepper in salchichon, etc. (Lois et al. 1987).

Chorizo is the most typical, popular and widely consumed Spanish dry fermented sausage. The traditional manufacturing method includes pork lean (70 to 80%), pork back fat (20 to 30%) generally from heavy pigs (of live weight of 150 kg or more) (Lorenzo et al. 2015).

Chorizo are sausages made with meat and fat (usually pork, but can also be made with meat and fat of other animals), coarse or fine chopped, subject at a salting process. Paprika is added as the characterizing ingredient, although other spices, seasonings, ingredients and additives can be added. The sausages are kneaded and stuffed in natural or artificial and subjected to a curing-matured process, with or without fermentation, and optionally smoked, giving them typical aroma and flavour. They are generally characterized by its red colour and aroma and flavour, although the sausages made without paprika or its derivatives (oleoresins or extracts), called white sausage are also included in this section (Real Decreto 474/2014).

The traditional process takes place with ripening under natural environmental conditions during the coldest months of the year, eating (and sometimes smoking) at the beginning of the ripening process. Currently, in the industrial process, nitrate and/or nitrite, sugars, reducing agents and even colorants are added. Besides, the ripening phase takes place under controlled conditions.

Salchichon is similar to chorizo, except that the characteristic ingredient used is pepper (Real Decreto 474/2014). Salchichon is, after chorizo, the most produced and consumed dry-cured sausage in Spain. Black pepper (ground or grain) is typically the only spice used in its manufacture. There are slight differences between traditional and industrial process and also between different brands. The ingredients are lean pork and/or lean beef, pork back fat, salt, sugar, potassium nitrite, sodium nitrite, polyphosphates, sodium ascorbate and black pepper (Berian et al. 1993).

Fuet is one of the products included in this definition (Real Decreto 474/2014). This is a thin sausage with a characteristic mold bloom on the surface. It originates in Cataloni (northeast Spain) and is abundantly produced and consumed in Spain. The raw materials used are lean pork (65 to 75%) and pork back fat (25 to 35%), salt, sugars (glucose, lactose), spices (black pepper, white pepper, nutmeg) and additives (sodium glutamate, sodium ascorbate, sodium phosphates, sodium nitrate, potassium nitrite and colouring).

Although the composition of the raw cured sausages can be very different depending on the type, a generic composition is indicated by Ordoñez and De La Hoz (2001).

6.3.1 Processing

The manufacture of dry fermented sausages consists, of a number of sequential steps (Fig. 6.3).

1. The choice of type meat depends on eating habits, customs, and preferences availing in the geographical region where the fermented sausages is produced, but usually is pork. The meat also has a variable composition, and its fat content depends on the species used and the anatomical region of the animal from which is take. The meat used must be firm with buffer capacity and optimum capacity of water retention and pH values between 5.4 and 6. Generally, meat of adult animals y preferred owing to its higher myoglobin content, which favours a more appropriate colour (Lücke 1998). The fat should be firm, with a high melting point and a low content of polyunsaturated fatty acids, because this causes the fermented sausage to turn rancid more quickly and to exude fats (Frey 1985). But in Spain there is one exception because some regions manufacture sausages with adipogenic pork (Iberian pig) which contains a high level of unsaturated fats (Ordoñez et al. 1996) appreciated by the consumer.

2. Mincing of meat and fat is done at low temperature (between –5 and 0°C) to achieve a clean cut and to avoid the release of intramuscular fat from fatty meats, which could cause changes in the colour and the drying process during ripening (Frey 1985).

3. Once meat and fat have been combined ingredients, additives, curing agents and starter culture are added. The most used are: ingredients (sugars, glutamate, aromatic herbs and spices), additives (colorants, ascorbic acid, etc.), curing agents (salt, nitrates and nitrites), starters (lactic acid bacteria, belonging to the *Lactobacillus* genus in Mediterranean sausages, and the nitrate and nitrite reducer bacteria *Micrococcaceae*).

4. All the components must be mixed in a kneader with caution to avoid breaking the particles of fat. The meat mass with all the components must suffer a pre ripening during 24 to 48 hours at 1 to 5°C. This phase is not always performed.

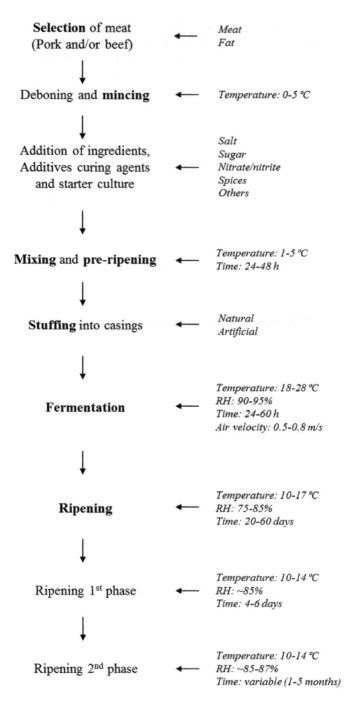

Figure 6.3 Steps in the manufacture of Spanish dry-fermented sausages.

5. After, the meat mass is stuffing into casings (natural or artificial). This operation is carried out in a vacuum to exclude, as far as possible, oxygen from the matrix and to prevent abnormal fermentations and the development of undesirable colours and flavours.

6. The next phase is the fermentation process. The sausages are placed in ripening rooms under conditions of controlled: temperature (18 to 28°C), relative humidity (90 to 95%) and air movement (0.5 to 0.8 m/s) during 24 to 48 hours. These conditions are important to improve the growth of microorganisms and to avoid external scab formation. This can appear when the relative humidity is low and would hinder the drying. During the fermentation important physic-chemical and microbiological change takes place; these will be described later.

7. After fermentation phase, the temperature and relative humidity are reduced (10 to 17°C and 85 to 87%). The air movement must be of 1 m/s and homogeneously distributed in the ripening room to insure a correct drying. This phase goes on for 20 to 60 days generally. During this stage different reactions take place. The dehydration of product is pronounced and this confers the characteristic texture. Water content decrease at 35% or less and aw arrives at values less than 0.90. In these conditions, growth of microorganisms is reduced if there is no water condensation in the surface of sausage. At the end of ripening, the pH increases slightly due to acid lactic production by superficial microorganisms and accumulation of degradation products.

Table 6.1 shows the characteristics (raw meat, ingredients and process) of different Spanish dry fermented sausages.

Table 6.1 Characteristics (raw meat, ingredients and process) of different Spanish dry fermented sausages.

Sausage variety	Origin	Meat raw	Other ingredients	Characteristics of process
Androlla	Galicia (NW Spain)	Lean pork (25%) Pork jowl (10%) Pork back fat (5%) (pork skin can be used)	Salt Spanish paprika (sweet and spicy) Marjoram Garlic	1. Meat ground 2. Ribs cut (3–4 cm) 3. Addition of ingredients 4. Mixture (15 min) 5. Resting (12 h at 20°C) 6. Stuffing (natural or artificial casings, 6–8 cm Ø and 25–25 cm of length) 7. Ripening (12°C, 75% RH, 20–30 days) * usually smoked with oak wood during the first 8 days of ripening

Table 6.1 cont....

Table 6.1 cont.

Sausage variety	Origin	Meat raw	Other ingredients	Characteristics of process
Botillo	Galicia CastillaLéon Asturias (NW Spain)	Ribs (70%, with their fleshy parts) Vertebrae (5%, with fleshy parts) Lean pork (10%) cooked pork skin (7%) Bacon (5%) Tail (2.5%) Tongue (2.5%)	Salt Spanish paprika (sweet and spicy) Garlic Marjoram	1. Meat ground 2. Ribs and the other bones cut (4–8 cm) 3. Addition of ingredients 4. Mixture (15 min) 5. Resting (48 h at 4–6°C) 6. Stuffing (by hand, into natural pig caps (cecum), stomach or bladder) 7. Ripening (10–12°C, 74–78 RH, 15–20 days) * usually smoked with oak or holm wood during the first 4 days of ripening ** always cooked before consumption
Chistorra	Navarra (N of Spain)	Lean pork (or pork and beef) Bacon Pig fat	Salt Spanish paprika Garlic	1. Stuffing (natural or artificial casings < 25 mm ø) 2. Short drying and ripening * with or without smoking
Chorizo Gallego	Variety of Chorizo of Galicia (NW Spain)	Lean pork (70–80%) Pork back fat (20–30%)	Salt Spanish paprika (sweet and spicy) Garlic	1. Mixing of lean meat and fat with salt 2. Resting (24 h under refrigeration) 3. stuffing (natural or artificial casings, 34–36 mm ø) 4. smoking (12–48 h, in heat) 5. drying and ripening (30–45 days) 6. making of strings of sausages, * characteristic: strings of sausages separated by small balls
Chorizo de Leon	León (N of Spain)		Not addition (in traditional process) of nitrifying salts and sugars	1. fermentation by intense smoking 2. Ripening under natural environmental conditions * Characteristic: high fat content

Table 6.1 cont....

Table 6.1 cont.

Sausage variety	Origin	Meat raw	Other ingredients	Characteristics of process
Chorizo of Pamplona	Navarra (N of Spain)	Lean pork (80%) Pork back fat (14%)	Salt Spanish paprika sugar	1. Mincing (with chopper disk of 3 mm) 2. Stuffing (casings of 55 mm ø and 20 cm length) 3. Fermentation (25°C, 60 h) 4. Drying and ripening (12°C, 65–85% RH, 6 weeks)
Fuet	Catalonia (NE of Spain)	Lean pork (65–75%) Pork back fat (25–35%)	Salt Sugar Black pepper White pepper nutmeg, etc.	1. Stuffing (casings < 40 mm ø) 2. Fermentation (28°C, 90% RH, 24 h) 3. Ripening (12–14°C, 5–85% RH, 30 days)
Longaniza of Aragón	Aragón (NE of Spain)	Pork Pork fat	Curing salts Sugars Ascorbic acid Pepper *Cloves and Nutmeg (sometimes) Garlic Onions, etc.	1. Stuffing (casings 25–40 mm ø and 20–70 cm of length into a horseshoe shape) 2. Ripening (12–18°C, 75–80% RH for less than 1 month)
salchichon				1. Stuffing (Casing 60 mm) 2. Fermentation (3 days) 3. Ripening (3 weeks)
sobrasada	Mallorca (Balearic Islands)	Lean pork Pork fat	Spanish paprika Salt pepper	1. Stuffing (pig guts) 2. Ripening (4–8 weeks)

Source: adapted from Lorenzo et al. (2015).

6.3.1.1 Fermentation

In fermentation phase, the sausages are placed in ripening rooms under conditions of controlled temperature (18 to 28°C), relative humidity (90%) and air movement (0.5 to 0.8 m/s). The sausages are kept here for one or two days (even three days in some products, e.g., Chorizo de Pamplona). In this phase, two basic microbiological reactions proceed simultaneously, formation of nitric oxide by nitrate and nitrite reducing bacteria (*Micrococcaceae*) and the generation of lactic acid by lactic acid bacteria (mainly *Lactobacillus* spp. in Mediterranean sausages). These two phenomena are well known, and several authoritative reviews have been published in which the mechanisms and interaction of the reactions taking place are thoroughly discussed (Palumbo and Smith 1977,

Liepe 1982, Lücke 1998, Nychas and Arkoudelos 1990, Demeyer et al. 1992, Ordoñez et al. 1999, Toldrá 2002). Although the nitrite reactions occurring during curing have been studied extensively, the actual chemical reactions are not conclusive due to the high reactivity of nitrite, the complexity of the meat substrate and the different types of the product (Cassens 1995). Nevertheless, several functions have been attributed to nitrites: it is involved in both colour (Eakes et al. 1975, Giddings 1977) and odour (Cross and Ziegler 1965, Bailey and Swain 1973) development; it prevents the growth of several undesirable microorganisms (Hauschild et al. 1982, Pierson and Smoot 1982) and it has an antioxidant activity (Sato and Hegarty 1971, MacDonald et al. 1980).

The sausage ingredients play an important role in the selection of the proper microbiota of the sausages. The initial microbial population of meat is always varied and is similar to that found in fresh meat. It includes gram negative (*Pseudomonas* spp., *Achromobacter* spp., *Moraxella* spp., *Enterobacteriaceae*, etc.) and gram positive (*Bacillus* spp., *Lactobacillus* spp., *Enterococcus* spp., *Brochotrixthermosphacta*, *Micrococcaceae*, etc.) bacteria (Gill 1986, Gill and Newton 1977). On the contrary, the water activity (a_w) has passed from 0.98 (in fresh meat) to 0.96 (in fresh sausage) due to the addition of curing salts and other solutes (sugars, etc.), which, together with the specific inhibitory effects of nitrate and nitrite and the low oxygen tension, favours the selection of both lactic acid and *Micrococcaceae* bacteria. These circumstances are normally brought about in naturally produced dry fermented sausages (without starters), which causes the inhibition of organisms responsible for the spoilage of fresh meat (gram negative bacteria, especially *Pseudomonas* spp.) and allows the rapid growth of the former lactic acid bacteria and *Micrococcaceae*, which are the dominant bacteria until the end of the ripening period. Similarly, these conditions stimulate the growth of the starter cultures when these are added. The sugar metabolism of lactic acid bacteria naturally present, or those included in the starter, in the Mediterranean dry fermented sausages is mainly homolactic, yielding (Gottschalk 1986) approximately 1.8 mol of lactic acid per mole of metabolized hexose and around 10% of other substances (formic and acetic acids and ethanol) ready to act as substrates for the synthesis of some flavour compounds. These phenomena produce a decrease in the pH from initial values of 5.8 to 6.2 close to 5.0 or below, which has several beneficial effects on both the manufacturing process and the shelf life of the product, such as the control of microbiota and enzymatic reactions, the reduction of water retention, capacity of proteins favouring the drying, acceleration of the gelation of myofibrillar proteins, and regulation of colour formation reactions.

The presence of molds in Mediterranean fermented sausages is usually attractive for the consumers and, in fact, spores now are often inoculated to enhance growth of a given species. Owing to its aerobic character, its growth is limited to the surface of the sausages (Geisen 1993). The count is very slow in the first few days (10^2–10^3 cfu/cm^2) but rapidly increases, reaching values of 10^6–10^7 cfu/cm^2 in around 25 days (Roncales et al. 1991). The molds most frequently isolated belong to the genera *Penicillium* and *Aspergillus*, although sometimes mouds of the genera *Clasdosporium*, *Scopularopsis*, *Alternaria* and *Rhizopus* (Leistner and Eckardt 1979, Pestka 1986) are identified.

The presence of molds on the sausage surface can lead to both desirable and detrimental effects. The main positive actions are the antioxidant effect by oxygen consumption, peroxide degradation and protection against light, which leads to the inhibition of oxidative phenomena and colour stabilization (Bruna et al. 2001, Sunensen and Stanhke 2003); the protection against spontaneous colonization with undesirable molds, including toxinogenic ones (Lücke and Hechelmann 1987); the favourable characteristic white or greyish appearance of the mycelium and conidia (Lücke 1998); and the contribution to the development of flavour compounds, which has been recently reviewed by Sunensen and Stahnke (2003). The adverse effects are usually related to the growth of unwanted species, and the main and antibiotics. The inoculation of nontoxinogenic strains is a very useful approach to avoid colonization by undesirable molds; *Penicilliumnalgiovense* is a much-used strain.

6.3.1.2 Ripening

The ripening can be divided in two phases. The first ripening with progressive reduction of temperature to 10 to 14°C and relative humidity to approximately 85%, and the second ripening that starts when these temperatures and RH are reached; its duration is very variable depending of the product (Ordoñez and De la Hoz 2007).

During this phase numerous reactions take place and the resulting products become the substrates of other ones. This is the reason that the composition of sausages changes constantly during this stage of process. A wide range of compounds are generated (peptides, free amino acids, free fatty acids, amines, ketones, aldehydes, etc.) and these contribute to the taste (amino acids and amines and long chain fatty acids, non-volatile sapid compounds) and/or odour (short chain fatty acids, etc.). All these products are responsible for the flavour and odour of the final product. The origin of these compounds can result of microbial phenomena, biochemical reactions of endogenous origin (lipases and proteases), chemical phenomena (nitric oxide reactions and lipid autoxidation) and the effect of the addition of certain ingredients (species and seasonings).

Edwards et al. (1999) analysed Spanish commercial dry sausages with different times of ripening, diameters (>3 cm versus Fuet <3 cm), and spices (versus Chorizo), which different mincing degrees (whole versus ground). Sulphurous compounds and 3-hexanol were detected only in Chorizo, derived respectively, from garlic and paprika. In sausages made with ground pepper, the terpenes were the major components and reached must higher levels than in dry fermented sausages that were manufactured with whole peppercorns. The narrower the diameter of sausages, the higher the levels of compounds from lipid oxidation, and finally, the longer the ripening times, the grater the ester contents.

For example, the casing diameter of salchichon versus Fuet, in the first case, the volatile fraction is dominated by compounds derived from microbial activities (esters, alcohols, etc.), in the second case, volatile compounds produces by autoxidative phenomena are more prevalent (aldehydes and ketones) due, in this case, to the mixture inside being more exposed to oxygen (Edwards et al. 1999).

6.3.2 Quality characteristics

Spanish law classifies the dry-cured sausages in the function of the ingredients used in the elaboration process. From this point of view the dry-cured sausages can be defined as "products made with sectioned

Table 6.2 Quality parameters of Spanish dry fermented sausages.

Product name	Commercial category	Fat (g/100 g DM)	Carbohydrates (g glucose/100 g DM)	Total protein (g/100 g DM)	Relationship collagen/ protein (%)	Protein added (g/100 g)
Chorizo.	extra	≤57	≤9	≥30	≤16	≤1
Salchichón.						
Chorizo sarta extra.		≤57	≤2	≥30	≤16	≤1
Chorizo of Pamplona.	extra	≤65	≤8	≥25	≤22	≤1
Salami.	extra	≤68	≤9	≥22	≤25	≤1
Salchichón of Málaga.	extra	≤50	≤5	≥37	≤14	≤1
Chorizo and Salchichónibérico.	extra	≤65	≤5	≥22	≤25	≤1
Chistorra.		≤80	–	≥14	–	≤3
Sobrasada.		≤85	≤5	≥8	≤35	≤3
Chorizo and Salchichón and other dry-cured sausages.		≤70	≤10	≥22	≤30	≤3

(Source: Real Decreto 474/2014).

meat (cut and chopped), fat and with or without edible meat offal, added of condiments, species and authorised additives, with a ripe drying process and smoking optionally". Many of the dry-cured sausages have been included on the "Catalogo de Embutidos y Jamones Curados de España" (Ministerio Agricultura, Pesca y Alimentación 1983) but Chorizo and Salchichon are the most characteristic.

The essential quality factors of the Spanish dry-cured sausages are mentioned in the quality standard of meat products (Real Decreto 474/2014). Table 6.2 shows the quality parameters of the Spanish dry-cured sausages. The proximate composition and counts of different microbial groups of the main Spanish fermented sausages were collected by Lorenzo et al. (2015).

References

Abreu, E., E. Quiroz-Rothe, A.I. Mayoral, J.M. Vivo, A. Robina, M.T. Guillén, E. Agüra and J.L. Rivero. 2006. Myosin heavy chain fibre types and fibre sizes in nuliparous and primiparous ovariectomized Iberian sows: Interaction with two alternative rearing systems during the fattening period. Meat Science 74: 359–372.

Andrés, A.I., R. Cava, A.I. Mayoral, J.F. Tejeda, D. Morcuende and J. Ruiz. 2001. Oxidative stability and fatty acid composition of pig muscles as affected by rearing system, crossbreeding and metabolic type of muscle fibre. Meat Science 59: 39–47.

Andrés, A.I., R. Cava, J. Ventanas, E. Muriel and J. Ruiz. 2004a. Lipid oxidative changes throughout the ripening of dry cured Iberian hams with different salt contents and processing conditions. Food Chemistry 84: 375–381.

Andrés, A.I., R. Cava, J. Ventanas, V. Thovar and J. Ruiz. 2004b. Sensory characteristics of Iberian ham: Influence of salt content and processing conditions. Meat Science 68: 45–51.

Andrés, A.I., J. Ruiz, A.I. Mayoral, J.F. Tejeda and R. Cava. 2000. Influence of rearing conditions and crossbreeding on muscle colour in Iberian pigs. Food Science and Technology International 6: 315–321.

Andrés, A.I., S. Ventanas, J. Ventanas, R. Cava and J. Ruiz. 2005. Physicochemical changes throughout the ripening of dry cured hams with different salt content and processing conditions. European Food Research and Technology 221: 30–35.

Antequera, T., C.J. López-Bote, J.J. Córdoba, C. García, M.A. Asensio, J. Ventanas, J.A. García-Regueiro and I. Díaz. 1992. Lipid oxidative changes in the processing of Iberian pig hams. Food Chemistry 45: 105–110.

Bailey, M.E. and J.W. Swain. 1973. Influence of nitrite in meat flavor. Proc. of Meat Industries and Research Conference, 29. Chicago, USA.

Berian, M.J., M.P. Peña and J. Bello. 1993. A study of the chemical components of Spanish saucisson. Food Chemistry 48: 31–37.

Bruna, J.M., M. Fernandez, J.A. Ordoñez and L. de la Hoz. 2001. Papel de la flora fúngica en la maduración de los embutidos crudos curados. Alim. Equip. Tecnol., Septiembre 79–84.

Cantos, E., J.C. Espín, C. López-Bote, L. de la Hoz, J.A. Ordóñez and F.A. Tomás-Barberán. 2003. Phenolic compounds and fatty acids from acorns (*Quercus* spp.), the main dietary constituent of freeranged Iberian pigs. Journal of Agricultural and Food Chemistry 51: 6248–6255.

Carpenter, C.E. and A.W. Mahoney. 1992. Contributions of heme and nonheme iron to human nutrition. Critical Reviews in Food Science and Nutrition 31: 333–367.

Carrapiso, A.I., J. Ventanas and C. Garcia. 2002. Characterization of the most odour active compounds of Iberian ham headspace. Journal of Agricultural and Food Chemistry 50: 1996–2000.

Cassens, R.G. 1995. Use of sodium nitrite in cured meats today. Food Technol. 49: 72–80.

Cava, R., J. Ventanas, J. Ruiz, A.I. Andrés and T. Antequera. 2000. Sensory characteristics of Iberian ham: Influence of rearing system and muscle location. Food Science and Technology International 6: 235–242.

Cross, C.K. and P. Ziegler. 1965. A comparison of the volatiles fractions of cured and uncured meats. J. Food Sci. 30: 610–614.

Daza, A., C.J. López-Bote, A. Olivares, D. Menoyo and J. Ruiz. 2008. Influence of a severe reduction of the feeding level during the period immediately prior to freerange fattening on performance and fat quality in Iberian pigs. Journal of the Science of Food and Agriculture 88: 449–454.

Daza, A., A. Mateos, A.I. Rey, I. Ovejero and C.J. López-Bote. 2007. Effect of duration of feeding under freerange conditions on production results and carcass and fat quality in Iberian pigs. Meat Science 76: 411–416.

Daza, A., A.I. Rey, J. Ruiz and J. López-Bote. 2005. Effects of feeding in free-range conditions or in confinement with different dietary MUFA/PUFA ratios and a-tocopheryl acetate, on antioxidants accumulation and oxidative stability in Iberian pigs. Meat Science 69: 151–163.

Demeyer, D., E.Y. Claeys, S. Ötles, L. Caron and A. Verpleaetse. 1992. Effect of meat species on proteolysis during dry sausage fermentation. Proc. 38th Inter. Cong. Meat Sci. Technol. Clermond Ferrand, 775–778.

Eakes, B.D., T.N. Blumer and R.J. Monroe. 1975. Effect of nitrate and nitrite on color and flavor of country style hams. J. Food Sci. 40: 973–976.

Edwards, R.A., J.A. Ordoñez, R.H. Dainty, E.M. Hierro and L. de la Hoz. 1999. Characterization of the headspace volatile compounds of selected Spanish dry fermented sausages. Food Chem. 64: 461–465.

Estévez, M., D. Morcuende, R. Ramírez, J. Ventanas and R. Cava. 2004. Extensively reared Iberian pigs versus intensively reared white pigs for the manufacture of liver pate. Meat Science 67: 453–461.

Flores, J., C. Biron, L. Izquierdo and P. Nieto. 1988. Characterization of green hams from Iberian pigs by fast analysis of subcutaneous fat. Meat Science 23: 253–262.

Flores, M., C.C. Grimm, F. Toldrá and A.M. Spanier. 1997. Correlations of sensory and volatile compounds of Spanish Serrano dry cured ham as a function of two processing times. Journal of Agricultural and Food Chemistry 45: 2178–2186.

Font-i-Furnols, M. and L. Guerrero. 2014. Consumer preference, behavior and perception about meat and meat products: an overview. Meat Science 98: 361–371.

Frey, W. 1985. Fabricación fiable de embutidos. Acribia, Zaragoza, Spain.

Geisen, R. 1993. Fungal starter cultures for fermented foods: Molecular aspects. Trend Food Sci. Technol. 4: 251–256.

Giddings, G.G. 1977. Basis of colour in muscle foods. CRC Crit. Rev. Food Sci. Nutr. 9: 81–114.

Gill, C.O. and K.G. Newton. 1977. The development of aerobic spoilage flora on meat stored at chill temperatures. J. Appl. Bacteriol. 43: 189–195.

Gill, C.O. 1986. The control of microbial spoilage in fresh meats. In: A.M. Pearson and T.R. Dutson (eds.). Advances in Meat Research. Meat and poultry Microbiology. AVI Publishing Co., Westport, Connecticut, USA.

Girard, J.P., C. Bucharles, J.L. Berdague and M. Ramihone. 1989. The influence of unsaturated fats on drying and fermentation processes in dry sausages. Fleischwirtschaft 69: 255–260.

Goerl, K.F., S.J. Eilert, R.W. Mandigo, H.Y. Chen and P.S. Miller. 1995. Pork characteristics as affected by two populations of swine and six crude protein levels. Journal of Animal Science 73: 3621–3626.

Gonzalez, E. and J.F. Tejeda. 2007. Effects of dietary incorporation of different antioxidant extracts and free-range rearing on fatty acid composition and lipid oxidation of Iberian pig meat. Animal 1: 1060–1067.

Gottschalk, G. 1986. Bacterial Metabolism. Springer-Verlag, pp. 214–224, New York, USA.

Grosch, W. 1987. Reactions of hydroperoxides: products of low molecular weight. *In*: H.W.S. Chan (ed.). Autoxidation of Unsaturated Lipids. Academic Press, London, UK.

Hauschild, A.H.W., R. Hilsheimer, G. Jarvis and D.P. Raymond. 1982. Contribution of nitrite to the control of *Clostridium botulinum* in liver sausages. J. Food Prot. 45: 500–506.

Hu, F.B., M.J. Stampfer, J.E. Manson, E. Rimm, G.A. Colditz, B.A. Rosner, C.H. Hennekens and W.C. Willett. 1997. Dietary fat intake and the risk of coronary heart disease in women. New England Journal of Medicine 337: 1491–1499.

Incze, K. 1992. Raw fermented and dried meat products. Fleischwirtschaft 72: 1–5.

Igene, J.O. and A.M. Pearson. 1979. Role of phospholipids and triglycerides in warmed-over flavor development in meat model systems. Journal of Food Science 44: 1285–1290.

Jimenez-Colmenero, F., J. Carballo and S. Cofrades. 2001. Healthier meat and meat products: their role as functional foods. Meat Science 59: 5–13.

Kanner, J. 1994. Oxidative processes in meat and meat products—quality implications. Meat Science 36: 169–189.

Lawrie, R.A. and D.A. Ledward. 2006. Chapter 2—Factors influencing the growth and development of meat animals. *In*: R.A. Lawrie and D.A. Ledward (eds.). Lawrie's Meat Science. Woodhead Publishing Limited. Cambridge, UK.

Leistner, L. and C. Eckardt. 1979. Occurrence of toxinogenic Penicilia in meat products. Fleischwirtschaft 59: 1892–1896.

Leray, C., M. Andriamampandry, G. Gutbier, T. Raclot and R. Groscolas. 1995. Incorporation of n-3 fatty acids into phospholipids of rat liver and white and brown adipose tissues—a time-course study during fish-oil feeding. Journal of Nutritional Biochemistry 6: 673–680.

Liepe, H.U. 1982. Starter cultures in meat production. *In*: H.J. Rhem and G. Reed (eds.). Biotechnology. Basilea: Verlag Chemie, Weinheim, Germany.

Lois, A.L., L.M. Gutiérrez, J.M. Zumalacárregui and A. López. 1987. Changes in several constituents during the ripening of chorizo. Meat Science 19: 169–177.

López-Bote, C.J. 1998. Sustained utilization of the Iberian pig breed. Meat Science 49: S17–S27.

Lorenzo, J.M., S. Martínez and J. Carballo. 2015. Microbiological and biochemical characteristics of Spanish fermented sausages. *In*: V.R. Rai and J.A. Bai (eds.). Beneficial Microbes in Fermented Food. CRC Press, Taylor & Francis Group, USA.

Lovegrove, J.A. and B.A. Griffin. 2013. The acute and long term effects of dietary fatty acids on vascular function in health and disease. Current Opinion in Clinical Nutrition and Metabolic Care 16: 162–167.

Lücke, F.K. and H. Hechelmann. 1987. Starter cultures for dry sausages and raw ham composition and effect. Fleischwirtschaft 67: 307–314.

Lücke, F.K. 1998. Fermented sausages. *In*: B.J.B. Wood (ed.). Microbiology of Food Fermentation. Appl. Sci. Publ., London, UK.

MacDonald, B., J.I. Gray and L.N. Gibbins. 1980. Role of nitrite in cured meat flavor. Antioxidant role of nitrite. J. Food Sci. 45: 893–897.

Magrama. 2015. Ministerio de Agricultura, Alimentación y Medio Ambiente. http://www.magrama.gob.es/es/alimentacion/temas/calidad-agroalimentaria/calidad-diferenciada/dop/default.aspx.

Mapa (Ministerio de Agricultura, Pesca y Alimentación). 1997. Inventario Español de productostradicionales. Madir, Servicio de Publicacionesdel MAPA.

Marcos, D. 1991. Embutidos crudos curados españoles. Ed. Ayala, Madrid, Spain.

Martin, D., T. Antequera, E. Muriel, A.I. Andrés and J. Ruiz. 2008a. Oxidative changes of fresh loin from pig, caused by dietary conjugated linoleic acid and monounsaturated fatty acids, during refrigerated storage. Food Chemistry 111: 730–737.

Martin, D., T. Antequera, E. Muriel, T. Pérez-Palacios and J. Ruiz. 2009. Liver pate from pigs fed conjugated linoleic acid and monounsaturated fatty acids. European Food Research and Technology 228: 749–758.

Martin, D., E. Muriel, T. Antequera, T. Pérez-Palacios and J. Ruiz. 2008b. Fatty acid composition and oxidative susceptibility of fresh loin and liver from pigs fed conjugated linoleic acid in combination with monounsaturated fatty acids. Food Chemistry 108: 86–96.

Martin, D., E. Muriel, E. Gonzalez, J. Viguera and J. Ruiz. 2008c. Effect of dietary conjugated linoleic acid and monounsaturated fatty acids on productive, carcass and meat quality traits of pigs. Livestock Science 117: 155–164.

Martin, M. 2001. Meat Curing Technology. *In*: O.A. Young, R.W. Rogers, Y.H. Hui and W.-K. Nip (eds.). Meat Science and Applications. CRC Press, Boca Raton, USA.

Mayoral, A.I., M. Dorado, M.T. Guillén, A. Robina, J.M. Vivo, C. Vázquez and J. Ruiz. 1999. Development of meat and carcass quality characteristics in Iberian pigs reared outdoors. Meat Science 52: 315–324.

McAfee, A.J., E.M. McSorley, G.J. Cuskelly, B.W. Moss, J.M. Wallace, M.P. Bonham and A.M. Fearon. 2010. Red meat consumption: an overview of the risks and benefits. Meat Science 84: 1–13.

Mercasa. 2013. Informe 2013 sobre Producción, Industria, Distribución y Consumo de Alimentación en España. http://www.mercasa-ediciones.es/alimentacion_2013/pdfs/pag_191-220_Carnes.pdf.

Morales, R., L. Guerrero, A.P.S. Aguiar, M.D. Guárdia and P. Gou. 2013. Factors affecting dry-cured ham consumer acceptability. Meat Science 95: 652–657.

Morrissey, P.A., P.J.A. Sheehy, K. Galvin, J.P. Kerry and D.J. Buckley. 1998. Lipid stability in meat and meat products. Meat Science 49: S73–S86.

Mottram, D.S. 1998. Flavour formation in meat and meat products: a review. Food Chemistry 62: 415–424.

Muriel, E., A.I. Andrés, M.J. Petron, T. Antequera and J. Ruiz. 2007. Lipolytic and oxidative changes in Iberian dry cured loin. Meat Science 75: 315–323.

Muriel, E., J. Ruiz, D. Martin, M.J. Petron and T. Antequera. 2004a. Physico-chemical and sensory characteristics of dry cured loin from different Iberian pig lines. Food Science and Technology International 10: 117–123.

Muriel, E., J. Ruiz, J. Ventanas and T. Antequera. 2002. Free-range rearing increases (n-3) polyunsaturated fatty acids of neutral and polar lipids in swine muscles. Food Chemistry 78: 219–225.

Muriel, E., J. Ruiz, J. Ventanas, M.J. Petrón and T. Antequera. 2004b. Meat quality characteristics in different lines of Iberian pigs. Meat Science 67: 299–307.

Nychas, G.J.E. and J.S. Arkoudelos. 1990. Staphylococci. Their role in fermented sausages. J. Appl. Bacteriol. 69: S167–S188.

Ordoñez, J.A. and L. de la Hoz. 2007. Mediterranean products. *In*: F. Toldrá (ed.). Handbook of Fermented Meat and Poultry. Blackwell Publishing, Ames, Iowa, USA.

Ordoñez, J.A., E.M. Hierro, J.M. Bruna and L. de la Hoz. 1999. Changes in the components of dry-fermented sausages during ripening. CRC Crit. Ver. Food Sci. Nutr. 39: 329–367.

Ordoñez, J.A., M.O. Lopez, E. Hierro, M.I. Cambero and L. de la Hoz. 1996. Efecto de la dieta de cerdos ibéricos en la composiciónen ácidos grasosdeltejido adiposo y muscular. Food Sci. Technol. Inter. 2: 383–390.

Ordoñez, J.A. and L. de la Hoz. 2001. Embutidos crudos curados. Tipos. Fenómenos madurativos. Alteraciones. *In*: S. MartínBejerano (coordinator). Encicloedia de la carne y de los productos cárnicos. Ediciones Martin & Macias, Plasencia, Caceres, Spain.

Palumbo, S.A., M. Komanowsky, V. Metzger and J.L. Smith. 1977. Kinetics of pepperoni drying. Journal of Food Science 42: 1029–1033.

Palumbo, S.A. and J.L. Smith. 1977. Chemical and microbiological changes during sausage fermentation and ripening. *In*: L. Orly and J. St. Angelo (eds.). ACS Symposium Series 47.

Pérez-Palacios, T., J. Ruiz, D. Martin, J.M. Barat and T. Antequera. 2011. Pre-cure freezing effect on physicochemical, texture and sensory characteristics of Iberian ham. Food Science and Technology International 17: 127–133.

Pérez-Palacios, T., J. Ruiz, D. Martin, E. Muriel and T. Antequera. 2008. Comparison of different methods for total lipid quantification in meat and meat products. Food Chemistry 110: 1025–1029.

Pestka, J.J. 1986. Fungi and mycotoxins in meats. *In*: A.M. Pearson and T.R. Dutson (eds.). Advances in Meat Research. Meat and Poultry Microbiology. AVI Publishing, Westport, Connecticut, USA.

Pierson, M.D. and L.A. Smoot. 1982. Nitrite, nitrite alteratives and the control of *Clostridium botulinum* in cured meats. CRC Crit. Rev. Food Sci. Nutr. 17: 141–187.

Sunensen, L.O. and L.H. Stahnke. 2003. Mold starter cultures for dry sauasge selection, application and effects. Meat Science 65: 935–948.

Real Decreto 474/2014, 2014. Norma de calidad de derivados cárnicos. Boletin Oficial del Estado, n° 147, sección I, 46058-46078.

Rey, A.I., A. Daza, C. López-Carrasco and C.J. López-Bote. 2006. Feeding Iberian pigs with acorns and grass in either freerange or confinement affects the carcass characteristics and fatty acids and tocopherols accumulation in *Longissimus dorsi* muscle and back fat. Meat Science 73: 66–74.

Rey, A.I., B. Isabel, R. Cava and C.J. López-Bote. 1998. Dietary acorns provide a source of gamma-tocopherol to pigs raised extensively. Canadian Journal of Animal Science 78: 441–443.

Roldán, M., T. Antequera, M. Armenteros and J. Ruiz. 2014. Effect of different temperature-time combinations on lipid and protein oxidation of sous-vide cooked lamb loins. Food Chemistry 149: 129–136.

Roldán, M., J. Ruiz, J.S. del Pulgar, T. Pérez-Palacios and T. Antequera. 2015. Volatile compound profile of sous-vide cooked lamb loins at different temperature-time combinations. Meat Science 100: 52–57.

Roncales, P., M. Aguilera, M. Beltrán, J.A. Jaime and J.M. Peiró. 1991. The effect of natural and artificial casing on the ripening and sensory quality of mold covered dry sausages. International Journal of Food Science and Technology 26: 83–89.

Ruiz, J., R. Cava, T. Antequera, L. Martin, J. Ventanas and C.J. López-Bote. 1998a. Prediction of the feeding background of Iberian pigs using the fatty acid profile of subcutaneous, muscle and hepatic fat. Meat Science 49: 155–163.

Ruiz, J., R. Cava, J. Ventanas and M.T. Jensen. 1998b. Headspace solid phase microextraction for the analysis of volatiles in a meat product: Dry cured Iberian ham. Journal of Agricultural and Food Chemistry 46: 4688–4694.

Ruiz, J., L. de la Hoz, B. Isabel, A.I. Rey, A. Daza and C.J. López-Bote. 2005. Improvement of dry-cured Iberian ham quality characteristics through modifications of dietary fat composition and supplementation with vitamin E. Food Science and Technology International 11: 327–335.

Ruiz, J., C. García, M.D. Diaz, R. Cava, J.F. Tejeda and J. Ventanas. 1999a. Dry cured Iberian ham non-volatile components as affected by the length of the curing process. Food Research International 32: 643–651.

Ruiz, J., C. García, E. Muriel, A.I. Andrés and J. Ventanas. 2002a. Influence of sensory characteristics on the acceptability of dry cured ham. Meat Science 61: 347–354.

Ruiz, J. and C. López-Bote. 2002. Improvement of dry cured ham quality by lipid modification through dietary means. *In*: F. Toldrá (ed.). Research Advances in the Quality of Meat and Meat Products. Research Signpost, Trivandrum, India.

Ruiz, J., E. Muriel and J. Ventanas. 2002b. The flavour of Iberian ham. *In*: F. Toldrá (ed.). Research Advances in the Quality of Meat and Meat Products. Research Signpost, Trivandrum, India.

Ruiz, J., J. Ventanas, R. Cava, A. Andrés and C. García. 1999b. Volatile compounds of dry-cured Iberian ham as affected by the length of the curing process. Meat Science 52: 19–27.

Ruiz-Carrascal, J., J. Ventanas, R. Cava, A.I. Andrés and C. García. 2000. Texture and appearance of dry-cured ham as affected by fat content and fatty acid composition. Food Research International 33: 91–95.

Sanchez del Pulgar, J., M. Roldán and J. Ruiz-Carrascal. 2013. Volatile compounds profile of sous-vide cooked pork cheeks as affected by cooking conditions (Vacuum Packaging, Temperature and Time). Molecules 18: 12538–12547.

Sato, K. and G.R. Hegarty. 1971. Warmed-over flavour in cooked meat. J. Food Sci. 36: 1098–1102.

Seuvre, A.M., M.A.E. Diaz and A. Voilley. 2002. Transfer of aroma compounds through the lipidic-aqueous interface in a complex system. Journal of Agricultural and Food Chemistry 50: 1106–1110.

Shahidi, F., J. Yun, L.J. Rubin and D.F. Wood. 1987. The hexanal content as an indicator of oxidative stability and flavor acceptability in cooked ground pork. Canadian Institute of Food Science and Technology Journal 20: 104–106.

Slaughter, J.C. 1999. The naturally occurring furanones: formation and function from pheromone to food. Biological Reviews of the Cambridge Philosophical Society 74: 259–276.

Sunensen, L.O. and L.H. Stahnke. 2003. Mold starter cultures for dry sausage—seletion, application and effects. Meat Science 65: 935–948.

Tejeda, J.F., T. Antequera, J. Ruiz, R. Cava, J. Ventanas and C. García. 1999. Unsaponifiable fraction and n-alkane profile of subcutaneous fat from Iberian ham. Food Science and Technology International 5: 229–233.

Toldrá, F. 2002. Dry cured Meat Products. Food & Nutr Press, Inc. Trumbull, USA.

Toldrá, F. and M. Reig. 2006. Sausages. In: Y.H. Hui (ed.). Handbook of Food Product Manufacturing. John Wiley & Sons???.

Ventanas, S., M. Estévez, C.L. Delgado and J. Ruiz. 2007a. Phospholipid oxidation, non enzymatic browning development and volatile compounds generation in model systems containing liposomes from porcine *Longissimus dorsi* and selected amino acids. European Food Research and Technology 225: 665–675.

Ventanas, S., J. Ruiz, C. Garcia and J. Ventanas. 2007b. Preference and juiciness of Iberian dry-cured loin as affected by intramuscular fat content, cross-breeding and rearing system. Meat Science 77: 324–330.

Ventanas, S., J. Ventanas, M. Estévez and J. Ruiz. 2010. Analysis of volatile molecules in Iberian dry-cured loins as affected by genetic, feeding systems and ingredients. European Food Research and Technology 231: 225–235.

Ventanas, S., J. Ventanas and J. Ruiz. 2007c. Sensory characteristics of Iberian dry-cured loins: Influence of cross-breeding and rearing system. Meat Science 75: 211–219.

Ventanas, S., J. Ventanas, J. Ruiz and M. Estévez. 2005. Iberian pigs for the development of high-quality cured products. Recent Research Developments in Agricultural & Food Chemistry 6: 1–27.

Winger, R.J. and C.J. Hagyard. 1994. Juiciness—its importance and some contributing factors. In: A.M. Pearson and T.R. Dutson (eds.). Quality Attributes and their Measurement in Meat, Poultry and Fish Products. Springer, USA.

Wood, F.W. 1966. Diffusion of salt in pork muscle and fat tissue. Journal of the Science of Food and Agriculture 17: 138–140.

Wood, J.D., R.I. Richardson, G.R. Nute, A.V. Fisher, M.M. Campo, E. Kasapidou, P.R. Sheard and M. Enser. 2004. Effects of fatty acids on meat quality: a review. Meat Science 66: 21–32.

CHAPTER 7

Slovenian Honey and Honey Based Products

Mojca Korošec,* Rajko Vidrih and *Jasna Bertoncelj*

7.1 Introduction

Honey has been a part of human nutrition since ancient times, as a sweetener and a home remedy. Prior to the rise in industrial sugar production at the start of the 19th century, honey was often the only sweetener available for humans (Allsop and Miller 1996, Crane 1975). Honey is a food product that honey bees (*Apis mellifera*) produce from flower nectar, secretions of living parts of plants, and excretions of plant-sucking insects that feed on the living parts of different plants. Bees collect this raw material, take it into the hive, transform it by adding their own substances, deposit it, and dehydrate it. Honey is then kept in honeycombs to ripen and mature. To preserve its original composition, bioactivity and characteristics, honey that is intended for human consumption should not be altered by man (European Council Directive 2002).

The largest amounts of honey are produced in China. This is estimated on average to be three-fold greater than for Turkey, Argentina, and the USA, which represent the following top honey-producing countries. The annual world honey production has increased in the last 20 years, from 1.11 to 1.59 million tons (FAOSTAT 2015). Nevertheless, this still only represents <1% of the annual world sugar production.

The contribution of the European Union to world honey production is significant. The total honey produced in the European Member States was 217,000 tons in 2011, which put Europe as the second-largest honey producer in the world. Spain is the main honey producer within the

Department of Food Science and Technology, Biotechnical Faculty, University of Ljubljana, Slovenia.
* Corresponding author: mojca.korosec@bf.uni-lj.si

European Union, with an annual production of over 30,000 tons (European Commission 2013). The quantities of honey produced in Slovenia, which average 2,000 tons per year, are in this respect negligible; however, these levels are important in terms of self-sufficiency. Indeed, in Slovenia, honey consumption is slightly higher than honey production (1.2 vs. 0.9 kg per capita) (Statistical Office of the Republic of Slovenia 2014). Due to the long tradition of beekeeping in Slovenia, mainly Slovenian honey is consumed in Slovenia, with imported honey having only a 14% share (Statistical Office of the Republic of Slovenia 2014). Along with Germany, Austria and Greece, Slovenia ranks among countries with the highest consumption of honey (1.0–1.6 kg per capita per year), while the opposite holds true for China and Argentina (0.1–0.2 kg per capita per year) (CBI Market Survey 2009).

7.2 Beekeeping technology

Beekeeping is deeply rooted in the Slovenian people. As people had no refined sugar in the past, nearly every farm in Slovenia had bee hives as well as domestic animals. In the past, two indispensable bee products were known: honey, as the only sweetener, and beeswax from which candles were produced. The most famous Slovenian beekeepers who were responsible for the progress of Slovenian beekeeping in the past were Peter Pavel Glavar (1721–1784) and Anton Janša (1734–1773). Peter Pavel Glavar proposed the establishment of beekeeping schools and published a manual in the Slovene language. Anton Janša extended beekeeping knowledge far beyond the frontier of Slovenia, as he was ordered by the Empress to become the first teacher at the Austrian Royal Court (Gregori et al. 2003).

Four species of the genus *Apis* prevail around the world. This species includes the giant or rock bee (*Apis dorsata*), the little bee (*Apis florea*), the Indian bee (*Apis indica*), and the honey bee (*Apis mellifera*). Twenty-five subspecies of honey bee are known, and are spread throughout Europe, Asia and Africa. Among these, five are particularly prevalent in Europe today: the Italian bee (*Apis mellifera ligustica*), the Caucasian bee (*Apis mellifera caucasica*), the black bee (*Apis mellifera mellifera*), the Macedonian bee (*Apis mellifera macedonica*), and the Carniolan bee (*Apis mellifera carnica*) (Fig. 7.1).

The Italian bee was originally spread over the Apennine peninsula, the Caucasian bee in the Caucasus territories, the black bee in central, northern and western Europe, the Macedonian bee in the southern Balkan peninsula, and the Carniolan bee in Slovenia, Austria and towards the east, to Romania and Bulgaria (Gregori et al. 2003). In the Slovenia territory of today, the Carniolan bee is the only bee subspecies used in apiculture. In the past, bees were kept in small wooden hives, known as

Figure 7.1 *Apis mellifera carnica* on acacia blossoms (Zlatko Kropf 2015).

Carniolan hives (syn. Kranjiči). Carniolan hives were set close together under wooden buildings (bee houses) to protect them from rain, cold and heat. Nowadays, beekeeping is still practised in bee houses (Fig. 7.2a) that have become one of the national features of Slovenia. The fronts of the beehives were painted to help the bees to find their own hives, as well as to help beekeepers to monitor the colonies. Hive painting is nowadays still practised, and it represents true Slovenian folk art. Indeed, together with hive paintings (Fig. 7.2b), bee houses represent one of the most characteristic national features of Slovenia.

Traditional Carniolan beehives can no longer compete with modern apiculture practised in Slovenia. Nowadays, Slovenian beekeepers use predominantly Slovenian style frame hives, called Žnideršič hives, or better known simply as the abbreviated: AŽ hives. Anton Žnideršič invented the AŽ hives, and he was a professional beekeeper from Ilirska Bistrica. The AŽ hive is still considered one of the best beehives for the transport of bees. Migratory beekeepers have trucks that have up to 100 beehives stacked close together. Traditional AŽ hives hold 10 frames in the brood section, and 10 frames in the honey section. During the season,

Figure 7.2 (a) Bee house and (b) a painted entrance to the beehive.

the frames with the brood are relocated to the honey section, to enable the queen to have free available combs in which to lay eggs. Relocation of frames is one of the most important interventions to prevent swarming (Lilek and Kandolf 2012, Božič 2015).

Recently, a few modifications to the AŽ hives were introduced. Some beekeepers have increased the number of frames to 11 in the brood and honey sections, and have modified the rear part of the hive to accommodate an additional six frames (i.e., three in each section), by removing the inner windows. Another modification led to the so-called three-story AŽ hive, which has one (or two) section(s) for the brood, and one or two for the honey (Božič 2015).

7.3 Slovenian honey

Natural geographical conditions and diversity of both climate and flora in Slovenia provide a rich selection of honey-bee pastures. Beekeepers take advantage of the variety of pastures by traditionally moving their bees across the country, to increase the quantity of honey produced and to ensure its varietal typicality. Nevertheless, the beekeeping technology used, as either stationary or moving bee hives, and the natural conditions, enable the production of different types of high quality honey, in terms of the botanical origins. The characteristics of a high quality honey are low water and hydroxymethylfurfural (5-hydroxymethylfuraldehyde; HMF) content, high enzymatic activity, and other physicochemical and sensory properties related to certain honey types. Traditionally, monofloral honey types where a certain botanical source prevails are produced in Slovenia, and Slovenian consumers are accustomed to having a wide honey choice.

7.3.1 Types of honey

The characteristics of honey reflect the diversity of the plants on which the bees forage. The availability of honey sources (i.e., nectar and/or honeydew) depends greatly on the plant characteristics and the weather conditions. On this basis, honeys differ from each other, although they can be classified according to their principal characteristics. For the honey source, two large groups of honey are recognized:

- nectar honey, which is produced from the nectar of different plants;
- honeydew honey, which is produced from the secretions of living parts of plants, or from excretions of plant-sucking insects (on the living parts of plants) (European Council Directive 2002).

According to the European Council Directive (2002) relating to honey, the botanical origins can be stated on the label when the honey is derived mainly, or on the whole, from the indicated botanical source, and thus has its typical physicochemical, sensory and microscopic characteristics. Indeed, different types of honey related to different botanical origins differ in their physicochemical and sensory characteristics and in the proportions and compositions of the pollen grains. As stated by Persano Oddo and Bogdanov (2004), honey with a botanical denomination, which represents almost half of the marketed honey in several European countries, is often regarded as more valuable and higher priced than blended nectar (multifloral) or honeydew (forest) honey.

The typical honey types traditionally produced in Slovenia include: the nectar types, as acacia (*Robinia pseudoacacia*) and multifloral honeys; the blended nectar and/or honeydew types, as linden (*Tilia* sp.) and chestnut (*Castanea sativa* Mill.) honeys; and three honeydew types, as spruce (*Picea abies* (L.) Karst.), fir (*Abies alba* Mill.), and forest honeys. Certain other honey types are produced in smaller quantities and in geographically limited areas (Table 7.1). The consumer choice usually depends on familiarity with or liking of the sensory characteristics of a certain honey type, while among the younger population, honeys with lighter colors are preferred over darker ones (Fig. 7.3).

Table 7.1 Types of honey produced in Slovenia (Korošec et al. 2016, reproduced by permission of Springer Science+Business Media).

Pasture source	Type of honey
Nectar	Multifloral, acacia (*Robinia pseudoacacia*), canola (*Brassica napus*), buckwheat (*Fagopyrum esculentum*), *Prunus mahaleb*, and dandelion (*Taraxacum officinale*) honeys
Nectar and/or honeydew	Linden (*Tilia* sp.), chestnut (*Castanea sativa* Mill.) and maple (*Acer pseudoplatanus* L., *A. platanoides* L.) honeys
Honeydew	Forest, spruce (*Picea abies* (L.) Karst.), fir (*Abies alba* Mill.) and *Metcalfa pruinosa* honeys

Figure 7.3 Colour of Slovenian honeys.

7.3.2 Composition

Honey is often referred to as a supersaturated sugar solution. Carbohydrates represent the majority of its components, predominantly as fructose and glucose, but the complexity of this food derives from its minor components. According to scientific data, honey contains more than 200 phytochemicals, the presence and content of which are related to the botanical as well as geographical origins of the honey. Amongst other minor components in honey, there are minerals, proteins, organic acids, phenolic acids, flavonoids, α-tocopherol, carotenoids, and products of the Maillard reaction (Bogdanov et al. 2008).

The content of the main and minor components, the number and origin of the pollen grains contained, the activity of the enzymes, and the sensory characteristics affect the quality and functional properties of a honey. As indicated above, honey types from different botanical sources differ in their pollen composition, physicochemical characteristics, color, odor, taste, and aroma. These properties can also be influenced by external factors that relate to the climate and the rock and soil and vegetation, as well as the beekeeping practices. The principal physicochemical parameters and their ranges in different types of Slovenian honey are given in Table 7.2. These were obtained from analyzing over 1000 honey samples in a ten-year period. The requirements for honey quality on the Slovenian market are set by the Rules relating to honey (2011), which are also in line with the European Council Directive (2002) relating to honey. These rules provide the criteria for the basic parameters of honey quality when sold on the market.

Table 7.2 Principal physicochemical parameters in Slovenian honey types (Korošec et al. 2016, reproduced by permission of Springer Science+Business Media).

Honey type	Water (g/100 g)	EC[1] (mS/cm)	pH	Free acids (meq/kg)	Diastase number	Protein (g/100 g)	Proline (mg/100 g)
Acacia	13.5–17.5	0.11–0.27	3.6–4.4	6.5–23.1	5.9–13.7	0.13–0.21	19.7–44.7
Multifloral	14.4–18.0	0.24–0.84	3.9–5.2	8.7–42.7	10.8–24.7	0.18–0.42	30.9–53.4
Linden	14.5–17.8	0.55–1.07	4.1–6.1	6.1–21.4	9.6–21.2	0.13–0.24	22.5–39.8
Chestnut	13.7–18.2	0.96–2.25	4.8–6.2	7.3–26.0	16.3–28.4	0.31–0.40	45.7–77.6
Forest	13.5–17.0	0.81–1.68	4.4–5.2	14.7–43.1	13.3–30.0	0.20–0.49	32.2–46.1
Fir	13.8–17.7	0.89–1.57	4.7–5.8	14.1–27.5	11.3–24.4	0.18–0.36	32.3–50.6
Spruce	14.3–18.5	0.92–1.63	4.3–5.5	17.7–45.1	12.7–24.2	0.18–0.38	23.1–49.5

[1]EC: electrical conductivity.

7.3.2.1 Water content

One of the most important criteria for honey quality is the water content. According to European Council Directive (2002), honey can contain up to 20% as water. Honeys with lower water content usually crystalize more quickly, are more viscous and thick, and are better preserved. A low water content in honey provides conditions that are unsuitable for the growth of osmophilic yeast, and thus prevents fermentation of the honey. The water content depends on the type and abundance of the pasture, the weather conditions during plant flowering and nectar secretion, the beekeeping practice, and the type of beehives. Slovenian honeys contain on average between 14% and 17% water, which is attributed also to beekeeping with the AŽ beehives, the most widespread type of beehives in Slovenia, which enable the production of honey with lower water content.

7.3.2.2 Electrical conductivity

The electric conductivity (EC) is a parameter that is used to distinguish between nectar and honeydew honeys, and can be useful for discrimination between certain types of honey. According to European Council Directive (2002) relating to honey, the EC of honey from nectar is ≤0.8 mS/cm, while that of honey from honeydew and chestnut honey is ≥0.8 mS/cm. Linden and chestnut honeys are excluded from this rule as they can be produced from a mixture of a nectar and honeydew of linden or chestnut, respectively. The EC of linden honey can be above or below this 0.8 mS/cm, while the EC of chestnut honey is significantly >0.8 mS/cm. Moreover, these EC values can provide useful information on honey quality and possible adulteration. The EC value depends on the content of mineral salts, organic acids, and proteins, and correlates with the amount of total inorganic matter in the honey.

The measurement of electrical conductivity is unsophisticated and can also be carried out in the field. It is applied as one of the basic analyses by beekeeping advisors in Slovenia, to provide additional help to the beekeeper and to monitor the accuracy of the honey labeling. Among Slovenian honey types, acacia honey has the lowest EC, while the highest ECs, even over 2 mS/cm in certain crop years, are typical of chestnut honey. The ranges of EC for individual honey types in Slovenia are given in Table 7.2, and these are similar to the ECs for certain honeys produced elsewhere in Europe.

7.3.2.3 Carbohydrates

Carbohydrates represent about three fourths of the honey mass, and they have significant contributions also from the nutritional aspect. The presentation and amount of different carbohydrates in honey depend on the type and botanical origin of the honey source, the physiological state of the bees, the strength of bee colonies, the presence of enzymes, and the climatic conditions. The carbohydrate composition affects the physicochemical properties of honey; i.e., viscosity, crystallization, and hygroscopicity (Kamal and Klein 2011).

Mixtures of fructose and glucose, as inverted sugars, represent some 65% to 90% of the carbohydrates in honey. The combined total for these two sugars is higher in nectar honeys (≥60%) than in honeydew honeys (≥45%). Fructose is very hygroscopic and quickly soluble in water, and it crystallizes very slowly. Glucose is less soluble in water and forms stable crystalline forms of α-D-glucose monohydrate at temperatures <50°C. The ratio between the fructose and glucose content (F/G) is calculated to estimate the rate of honey crystallization. The F/G is characteristic of each honey type, and it is strongly related to the botanical origin of the honey. In most honeys, the amount of fructose is slightly higher than the amount of glucose, and therefore the F/G is generally >1.0. The higher the F/G ratio, the slower the rate of crystallization, and thus the honey remains liquid for a longer time (Escuredo et al. 2014, da Silva et al. 2016). Acacia and chestnut honey are the most fructose-rich types among the Slovenian honeys, with F/Gs of 1.5 and 1.4, respectively.

Disaccharides, following by some oligosaccharides, are present in honey in much smaller amounts. Nevertheless, their distribution can be used as an index of honey authenticity and floral origin (Cotte et al. 2004, Kamal and Klein 2011, Korošec 2012, da Silva et al. 2016). The greatest differences in carbohydrate composition of nectar and honeydew honeys can be seen in the content of trisaccharides, while the amounts of disaccharides (e.g., sucrose, palatinose, turanose, maltose) usually do not differ significantly between these two groups of honey (Golob and Plestenjak 1999, Cotte et al. 2004, Korošec et al. 2009, de la Fuente et al. 2011). The trisaccharides erlose, panose, and maltotriose have been shown for the nectar and honeydew types of honey, while raffinose and melezitose are typical of honeydew honeys (Table 7.3). Also Escuredo et al. (2014) noted that melezitose in a nectar honey indicates the presence of honeydew in the honey.

7.3.2.4 Acidity and pH

Honey contains diverse organic and inorganic acids that impart distinctive tastes and aromas to honey, contribute to the preservation of honey, and have an impact on the antibacterial and antioxidant activities of honey. One of the important quality parameters of honey is the content of free acids, as higher values indicate the ongoing fermentation processes caused by osmophilic yeast. In this process, sugar is first converted into

Table 7.3 Sugar composition of different honey types from Slovenia.

Compound	Mean content ±SD (g/100 g) according to honey type						
	Acacia	Multifloral	Linden	Chestnut	Forest	Fir	Spruce
Fructose	41.1±0.84	36.6±2.27	35.4±1.49	37.5±0.59	33.2±2.20	30.0±1.06	35.2±0.19
Glucose	27.2±1.11	29.5±2.42	32.7±1.27	26.4±0.27	26.9±2.07	28.6±1.71	28.5±0.12
F/G ratio	1.31–1.67	0.95–1.32	0.91–1.19	1.35–1.56	0.93–1.29	0.97–1.16	1.13–1.38
Sucrose	1.78±2.20	1.96±1.37	0.39±0.07	0.48±0.36	1.44±1.10	1.75±1.63	1.28±0.18
Turanose	2.06±0.33	1.82±0.44	2.64±0.31	2.96±0.53	1.34±0.36	1.54±0.16	1.59±0.01
Palatinose	0.61±0.13	1.36±0.28	1.57±0.12	1.90±0.32	1.78±0.02	1.59±0.11	1.23±0.04
Maltose	4.09±0.56	5.85±0.73	2.95±0.60	5.81±0.63	4.82±0.67	5.25±0.68	2.99±0.01
Gentiobiose w. Melibiose[1]	0.17±0.01	0.17±0.04	0.23±0.04	0.26±0.05	1.38±0.08	1.30±0.05	1.20±0.04
Melezitose	0.01±0.03	0.21±0.05	0.90±0.25	2.65±0.54	2.44±0.40	3.25±0.61	3.54±0.61
Erlose	0.29±0.08	0.81±0.26	0.84±0.28	0.20±0.06	2.68±0.18	2.26±0.09	1.74±0.01
Raffinose	0.08±0.03	0.33±0.10	0.35±0.02	0.32±0.11	1.69±0.06	2.42±0.38	1.37±0.03
Maltotriose	0.16±0.03	0.27±0.03	0.23±0.05	1.09±0.13	0.84±0.01	0.62±0.02	1.50±0.03
Panose	0.29±0.06	0.30±0.04	0.19±0.03	0.23±0.04	0.59±0.01	0.62±0.09	0.58±0.03
Isomaltotriose	0.11±0.01	<LOQ	0.18±0.06	0.07±0.04	<LOQ	<LOQ	<LOQ

[1]chromatographic condition did not enable separation of the two carbohydrates; <LOQ, carbohydrate was detected, but the amount was too little to be quantified; SD, standard deviation.

alcohol, and then into acids and carbon dioxide. Honey on the market can contain up to 50 milliequivalents of free acids per kg honey (European Council Directive 2002). Unlike the total content of acids, the pH of honey is related to the botanical origin of the honey. Honeydew honeys are richer in minerals, which act as a buffer; consequently, they have higher pH and less acidic taste. The mean pHs of Slovenian honeys are in the range of 3.9 to 5.5, where acacia honey shows the lowest, and chestnut honey the highest (Table 7.2).

7.3.2.5 *Proteins and amino acids*

The amount of proteins in honey is low, and therefore of minor nutritional importance. The amount of protein found in honey is usually <0.5%, and this can show large standard deviations among the different honey types (White 1978, Anklam 1998). In different Slovenian honeys, the mean protein content was reported to be in the range from 0.16% in acacia honey up to 0.35% in chestnut honey (Kropf et al. 2010).

The nectar or honeydew, the pollen, and the honey bees are the sources of amino acids in honey. The total amino-acid content is low, and in Slovenian honeys ranges from 71.2 mg/100 g in acacia honey to 99.4 mg/100 g in forest honey (our unpublished data). As proline has the highest content in terms of amino acids, it is the only amino acid of note among the 27 that have been found in honey (Hermosin et al. 2003). The mean proline content ranges from 30.1 mg/100 g in acacia honey, to 61.7 mg/100 g in chestnut honey (Table 7.2), and this is related to the botanical source of the honey and the degree of processing of the raw material by the bees. Besides showing the maturity of honey, the proline content is related to the radical scavenging activity (Meda et al. 2005) and to the quality and authenticity of honey. The internationally recognized lowest admissible proline content in mature, unadulterated honey is 18.0 mg/100 g (International Honey Commission 2002).

7.3.2.6 Elements

Honey is not considered to be particularly important as a source of elements, with the total content in nectar honey usually <0.6%, and in honeydew honey, <1.0%. The variety of elements in an individual honey sample is greatly dependent on the composition of the nectar, honeydew, and soil, and the predominant pollen. From the group of macroelements in honey, only K is present in amounts above 200 mg/kg. A great variety of microelements can be found in honey, such as B, Ba, Br, Ca, Cl, Fe, Mg, Mn, Na, P, Rb, S, Sr and Zn, and their levels are above 1 mg/kg. Among the trace elements found in honey, there are Ag, As, Cd, Co, Cr, Cu, Li, Mo, Ni, Se, and Pb. Various studies have indicated that botanical factors have the greatest influence on the trace element content of honey (Golob et al. 2005, Bogdanov et al. 2007, Nečemer et al. 2009, Kropf et al. 2010).

A large number of Slovenian honey samples from three consecutive years was analyzed using total reflection X-ray spectroscopy (Nečemer et al. 2009, Kropf et al. 2010). This method enabled the determination of 14 elements in seven types of honey. For the analyzed samples, seven elements were found, as K, Cl, Ca, S, Rb, Mn, Br, which were then arranged from the most to the least abundant in terms of their mean values: K was the most abundant, and Br was the least abundant. The order of the remaining five of these elements was not the same in all of the honey types; namely, acacia and linden honey contained more Mn than Rb, and honeydew honeys contained more S than Ca (Table 7.4). The data obtained here has significantly contributed to the successful discrimination among honeys in terms of their botanical origin and geographical origin (one of four macroregions in Slovenia).

Table 7.4 Element content in different honey types of Slovenian origin (Korošec et al. 2016, reproduced by permission of Springer Science+Business Media).

Honey type	Mean content of elements ±SD (mg/kg)						
	K	Cl	Ca	S	Rb	Mn	Br
Acacia	278±78	95±52	17.3±7.7	47±19	0.72±0.32	1.68±1.27	0.60±0.26
Multifloral	1120±352	264±85	61±25	56±25	2.97±1.63	3.12±1.59	0.65±0.25
Linden	1800±349	379±139	69±23	50±27	5.5±2.9	3.55±1.56	1.02±0.43
Chestnut	3590±657	240±217	148±33	42±24	17.0±7.7	23.2±9.0	0.55±0.23
Forest	2940±561	310±79	59±19	57±21	13.7±7.8	6.74±2.51	0.59±0.25
Fir	3170±555	333±134	35±18	71±26	22.0±7.0	5.03±1.93	0.59±0.12
Spruce	2950±494	322±74	47±17	70±26	13.9±6.1	7.07±2.3	0.58±0.22

SD, standard deviation.

7.3.2.7 Antioxidants

Honey is often listed among the foods that can provide natural antioxidants in our diet. Flavonoids, phenolic acids, ascorbic acid, catalase, peroxidase, carotenoids, and Maillard reaction products are the components responsible for these antioxidative effects of honey. Their quantities vary widely, depending on the botanical and geographical origins of a honey. Moreover, beekeeping practices and processing, and the storage of honey might affect their contents (Turkmen et al. 2005, Küçük et al. 2007, Can et al. 2015).

The botanical origin of honey, its geographical provenance, and the climatic conditions in the area, influence the phenolic profiles of honey (Gómez-Caravaca et al. 2006). The flavonoids are one of the largest groups of naturally occurring phenolic compounds. They are derived from plants, and are transferred into honey via the nectar, honeydew and pollen collected. The phenolic profiles of honey are related to the plants foraged, the physiological state of the plant, the season, and the weather conditions. Flavonoids from three groups are present in honey: flavones, flavonols and flavanones (Viuda-Martos et al. 2008, Bertoncelj et al. 2011a, Gašić et al. 2014).

Slovenian honeys vary widely in total phenolic content, as determined by the modified Folin-Ciocalteu method, and as shown in Table 7.5. The lowest total phenolic contents are seen for the lightest, acacia honeys, and they increase in linden, multifloral, and chestnut honeys. Honeydew types of honey (i.e., fir, spruce, forest honeys), have the highest total phenolic contents, and particularly for the darkest in color. The mean total phenolic contents in Slovenian honeys are similar to those reported in other studies of European honeys (Beretta et al. 2005, Lachman et al. 2010).

Table 7.5 Total phenolic content (mg GAE/kg) and antioxidant activity in different types of Slovenian honey (Korošec et al. 2016, reproduced by permission of Springer Science+Business Media).

Honey type	Total phenolic content (mg GAE/kg)	Antioxidant activity (FRAP method) (μM Fe(II))
Acacia	25.7–67.9	56.8–86.0
Multifloral	126.8–194.6	181.1–262.9
Linden	63.4–109.0	94.6–155.1
Chestnut	146.8–272.3	238.3–469.5
Forest	192.3–270.1	371.6–494.1
Fir	163.4–285.7	320.8–582.2
Spruce	185.7–239.0	277.5–495.4

GAE, gallic acid equivalents; FRAP, ferric reducing antioxidant power.

Liquid chromatography with mass spectrometry has been applied for the determination of individual phenolic compounds in honey, after the removal of carbohydrates and other substances by solid-phase extraction (Bertoncelj 2008, Bertoncelj et al. 2011a). The results indicate that honeys have similar, but quantitatively different, phenolic profiles. Propolis-derived flavonoids, namely pinocembrin, pinobanksin, chrysin and galangin, are present in all honey types, but in variable amounts, as their quantities in honey arise from the presence and content of propolis. Kaempferol and apigenin are found in all honey types, and quercetin in most honey types. The contents of the other three flavonoids are very low, with myricetin not detected in acacia and linden honey, and naringenin not found in chestnut honey. The phenolic acids *p*-coumaric and caffeic acid have been reported for all analyzed honey types, while chlorogenic and ellagic acids have been found only in single samples of Slovenian honey types, except in acacia honey. Due to the different extraction and detection techniques that have been used, it is not possible to directly compare the phenolic profiles of Slovenian honeys with the literature data.

The antioxidant capacity of Slovenian honeys has been determined with the widely used FRAP (ferric reducing antioxidant power) assay (Benzie and Strain 1996, Taormina et al. 2001, Beretta et al. 2005, Blasa et al. 2006, Küçük et al. 2007). The antioxidant activities of different types decreased in the following order of honeys: fir > forest > spruce > chestnut > multifloral > linden > acacia honey, as presented in Table 7.5. The phenolics are believed to be one of the main components in honey to be responsible for its antioxidant effects, basing on the significant correlation between the total antioxidant activity (determined by the FRAP method) and the phenolic content. As reported by Gheldof et al. (2002) and Gašić et al. (2014), phenolic compounds have a significant contribution to the antioxidant activity of honey, although the antioxidant activity appears to be a result of the mutual activity of the phenolic compounds, peptides, organic acids, enzymes, and Maillard reaction products in honey.

The antioxidant activity and total phenolic content are related to the color of a honey. These two parameters reflect the bioactive properties of a honey, and they vary greatly among different types of honey. The highest values are found in darker honey types, namely, chestnut, fir, spruce, and forest, while the lightest honey types of acacia and linden show low total phenolic contents and lower antioxidant activities (Bertoncelj et al. 2007, Bertoncelj et al. 2011b).

7.3.2.8 Hydroxymethylfurfural

This furanic compound is formed through the decomposition of carbohydrates, and especially of fructose and glucose, in the acidic environment of honey (Molan 1996). The rate of HMF formation depends on the temperature and storage conditions (da Silva et al. 2016). On this basis, it presents an important criterion in the determination of the quality and freshness of honey. The content of HMF in fresh honey is as low as 0.0 mg/kg to 0.2 mg/kg. Increased HMF indicates inappropriate heating or incorrect storage of honey. The legally set limit for HMF content in honey from moderate climates is 40 mg/kg (European Council Directive 2002). As crystallization of honey is a natural process and liquefaction is needed prior to further processing of honey, appropriate heating conditions need to be applied (tempering at up to 40°C, or short-term higher temperature) to prevent a significant rise in the HMF formation.

The majority of honey in Slovenia is sold directly from the beekeepers. It is important therefore that the producers, who are also the sellers, are informed about appropriate heating techniques for honey. They are educated and trained through seminars and field work organized by the Slovenian Beekeepers' Association within the framework of their beekeeping advisor activities. The implementation of education into beekeeping practices showed positive results in a 5-year monitoring of HMF content in craft honeys, where over 600 samples were collected and analyzed. The mean HMF was ≤5 mg/kg HMF in 75% of the analyzed samples, and none of the samples in the monitoring contained HMF above the legal limit (40 mg/kg) (Kandolf et al. 2010).

7.3.3 Antibacterial activity

Honey is predominantly a sweetener; however, in many cultures it has also been used as a remedy in traditional medicine (Gomez-Caravaca et al. 2006). Bogdanov (1997) classifies the substances in honey with antimicrobial activity into two systems:

- The action of the hydrogen peroxide in honey that is produced by glucose oxidase in the presence of heat and light. Although hydrogen peroxide is poorly stable and prone to decay, a certain quantity is always present in honey, as it is continuously produced by this enzyme. The peroxide content depends on the type of honey as well as on the freshness of a certain sample. As reported in other studies, honey with a naturally low enzyme activity also has a minimum content of peroxide, as found for Slovenian types of honey. The highest content of peroxide was determined in Slovenian forest honey (77.8 µg H_2O_2/g honey/h) and fir honey (76.6 µg H_2O_2/g honey/h), and the lowest in acacia honey (19.7 µg H_2O_2/g honey/h) (our unpublished data).

- The stable non-peroxide activity that is independent of light and heat. In terms of antibacterial effects, non-peroxide antibacterial activity appears to be more important than peroxide activity. However, this activity diminishes with time, and therefore, for optimum antibacterial activity, honey should be stored in a cool, dark place, and consumed when fresh (Alvarez-Suarez et al. 2009).

Typical antibacterial properties of honey contribute to its stability during storage. Taormina et al. (2001) indicated that darker honeys have greater antibacterial effects than lighter ones. Darker honeys also show higher antioxidant activity, which contributes to the control of the growth of some foodborne pathogens. They studied the inhibitory effectiveness of honey on microorganisms responsible for food spoilage (i.e., *Escherichia coli, Salmonella typhimurium, Shigella sonnei, Listeria monocytogenes, Staphylococcus aureus, Bacillus cereus*).

Kralj Kunčič et al. (2012) studied antibacterial and antifungal properties of Slovenian chestnut, forest, fir, linden, acacia, and multifloral honey against bacterial species (i.e., *E. coli, Enterococcus faecalis, Pseudomonas aeruginosa* and *S. aureus*) and fungal species (i.e., *Aspergillus niger, Aureobasidium pullulans, Candida albicans, Candida parapsilosis, Candida tripicalis, Cladosporium cladosporioides, Penicillium chrysogenum, Rhodotorula mucilaginosa*). Highly significant and positive antimicrobial effects were found for chestnut honey, which appeared to be mainly due to its high peroxide activity. On the contrary, in Manuka honey (from *Leptospermum scoparium*), the high antimicrobial activity results from non-peroxide components (Weston et al. 1999). Kralj Kunčič et al. (2012) proposed chestnut honey to be applied for medicinal use. The antimicrobial effects of other types of Slovenian honey varied considerably, and were less important.

7.3.4 Protected honeys

Slovenia is positioned at the intersection of four major European geographical zones: the Alps, the Panonnian Basin, the Dinaric Highlands, and the Mediterranean. The pedogeographical diversity, varied terrain, and different, often locally restricted, climatic conditions give rise to a great diversity of flora. Consequently, individual types of honey can be produced only in certain natural geographical regions, while amongst those produced in two or more regions, differences in some parameters can be found.

European Union legislation enables the protection of agricultural and food products produced or processed in certain geographical areas, or made according to traditional recipes. Such products must be distinguished from similar ones, in terms of their characteristics, the production process, and the composition, which in the end result in their quality. Three European Union schemes are used to promote and protect names of quality agricultural and food products: Protected Designation of Origin (PDO), Protected Geographical Indication (PGI) and Traditional Speciality Guaranteed (TSG) (Regulation (EU) 1151/2012). These schemes provide consumers with information on the specific character of a product, while providing a warranty against misuse and imitations.

7.3.4.1 "Slovenski Med" PGI

The "Slovenski Med" ("Slovenian honey") name was PGI registered in November 2013, and can be used for honey produced on the territory of the Republic of Slovenia. However the use of the name is limited to acacia, linden, chestnut, fir, spruce, multifloral, and forest honeys, which are the most common types in Slovenia. In the production of honey, stationary or mobile hives moving within the geographical area can be used. As the Carniolan honey bee (*Apis melifera carnica*) originates from this region, the beekeeping is based exclusively on the use of this indigenous honey bee.

The specificities of "Slovenski Med" PGI were recorded through practising beekeeping and several years of honey analyses. The pollen spectrum reflects the characteristics of flora in the region where the honey is produced. Due to widespread chestnut trees, chestnut pollen can be found in most samples of "Slovenski Med" and the moving of bee hives even increases this possibility. Climatic conditions in Slovenia enable the production of honey with low water and HMF contents. "Slovenski Med" should not contain more than 18.6% water and 15 mg/kg HMF. Additional parameters that certain types of "Slovenski Med" need to comply with are the EC, pH and sucrose content, and the proportions of specific pollen varieties for acacia, lime, chestnut, and multifloral honeys (Slovenian Beekeepers' Association 2014).

A system of internal honey control is basing on a network of beekeeping inspectors, who are educated and managed by the Slovenian Beekeepers' Association. These inspectors monitor the beekeeping practices and provide field analysis of water content and EC, as well as assisting the beekeepers who produce the honey under the label "Slovenski Med" PGI. Honeys that meet the requirements are bottled in glass jars of a distinctive shape and are labeled with a uniform label that seals the lid and the jar. The type of honey should be indicated on the jar.

7.3.4.2 *"Kočevski Gozdni Med" PDO*

The group of "Kočevski Gozdni Med" ("Kočevje forest honey") Slovenian honeys was registered in 2011. "Kočevski Gozdni Med" PDO comprises linden, fir, spruce, and forest honeys, produced and bottled in the wider area of Kočevska, a part of the Dinaric Highlands. All of the beehives need to be located within this area of production, which overlaps with the Kočevska-Kolpa area, a Natura 2000 nature conservation area. As the area is predominantly covered with forests, all four types of honey are characterized by the presence of honeydew, and consequently by a higher EC. Indeed, the EC is one of the main parameters for differentiation among "Kočevski Gozdni Med" honeys and similar honeys from other regions. Notable differences can also be observed in the sensory properties, which again reflect the presence of honeydew in this honey.

All four types of "Kočevski Gozdni Med" PDO should contain ≤18.6% water and ≤10 mg/kg HMF, while the EC needs to be ≥0.8 mS/cm in linden honey, ≥0.85 mS/cm in forest honey, and ≥0.95 mS/cm in fir and spruce honey. For the scope of the compliance analysis, sensory analysis is also carried out, with scoring of appearance, smell, taste, and aroma, with sensory compliance monitored according to a score of up to 12 points. It is required that a sample has at least 9.5 points out of 12 to qualify for the label with the protected name (Kočevski med Association 2011). The internal honey control is carried out in the same manner as described for "Slovenski Med" PGI.

7.3.4.3 *"Kraški Med" PDO*

Honey with the "Kraški Med" ("Karst Honey") PDO indication is produced from nectar or honeydew that honey bees collect from vegetation within the broader Karst (Kras) region. The specific geoclimatic conditions and phytogeographical characteristics, together with the beekeeping traditions, affect the varieties of honeys produced in this area, and their quality and properties. St. Lucie cherry honey (from *Prunus mahaleb*) and winter savoury honey (from *Satureja montana*) are produced exclusively in this area. Other honey types from this region that can qualify as "Kraški

Med" PDO are acacia, linden, multifloral, wild cherry, chestnut, and forest honeys.

The water content in all of these honeys should not be >18.0% and HMF content should not be >15 mg/kg. The total fructose and glucose content in honeys labeled as "Kraški med" PDO must be ≥45 g per 100 g honey. Individual types of honey must meet the required EC and proportion of pollen grains of the plant that the honey is named after. Furthermore, the sensory properties have to comply with the description of certain honey types, where 9.5 points out of 12 is the lowest acceptable score. Beekeeping practice, records on locations of bee hives, and quantities of honeys produced are inspected under the internal control system, by the beekeeping inspectors. These inspectors also carry out field analysis for water content and EC, while the honey type must be determined according to the producer (Beekeepers' Society Sežana 2012). Honeys with the protected name "Kraški Med" PDO are bottled in glass jars of different volumes, and they are labeled with a uniform label indicating the protected name and the type of honey.

7.3.5 Traditional products with honey

7.3.5.1 Mead

Mead is thought to be the oldest alcoholic beverage known to man (Allsop and Miller 1996). In ancient Greece, this was called mead, Ambrosia, or Nectar. The ancient Greeks believed it to be the drink of the Gods, and it came down from the heavens as dew and was gathered in by the bees. As for many great discoveries, mead was probably also discovered accidentally. Although many theories exist for this, three of them appear quite credible.

Pottery was a well-known utensil and a lot of historical evidence has confirmed that pottery was used to process mead in the early times. Before the introduction of pottery, some accidentally driven events probably happened. One theory states that hunters might have found an upturned tree with the remnants of a bee colony or any type of ancient beehive that had filled with water. The honey was dissolved in the water, and fermentation took place. The hunters enjoyed the sweet taste of mead, unaware of what might happen. They were encouraged to drink further when the alcohol provoked great amazement. The first consumers of mead probably experienced intoxication by alcohol. Due to these aspects, mead will have provoked feelings that the fermentation took on mystical and religious qualities. While mystical for many centuries, fermentation was in the end understood in the times of Pasteur in the middle 1800s.

Some historians hint towards animal skins and organs, like bladders providing the first water-tight vessels (Schramm 2003). According to

Crane (1983), African honey hunters were using animal skin and pumpkin containers (both as types of vessels) to carry the water supplies during their expeditions. Schramm (2003) proposed the hypothesis that the first fermentation was carried out in skin or pumpkin containers after water and mead were accidentally put together.

The mythology of mead was well known in ancient times and still exists in Slovenian culture. In many ancient cultures, bees were thought to be messengers of the Gods. The Scandinavian word for honeymoon derives from an ancient northern European custom when newlyweds drank a daily cup of honey wine, called mead, in the first month of their married life. Another explanation is that the term "honeymoon" comes from the ancient tradition when bridal couples were given a 'moon's' worth of honey-wine. This habit should ensure a fruitful union between the newlywed couple, and was believed to promote the male-gender of the first-born child.

In early history, mead was the alcoholic beverage of choice throughout ancient Europe, Asia and Africa (Gayre and Papazian 1983). Mead was thus popular in ancient China, Egypt, Greece, the Roman Empire, and Medieval Europe. The earliest archeological evidence of mead production dates back to 7,000 B.C. Remnants of 9,000-year-old pottery jars were found in northern China in Henan province. This evidence revealed that the beverage was probably not pure mead made solely from honey, but with ingredients like wild grapes and rice also incorporated during its processing. Schramm (2003) states that there is good agreement among anthropologists that the beginnings of mead date back to approximately 8,000 years B.C. Also, as honey was widely available in prehistoric times, several cultures appear to have invented mead processing spontaneously. Bees were already domesticated in Egypt, with hives made from unbaked clay or wooden baskets. Records of mead production date back to 2,000 years B.C. (Schram 2003). Schram (2003) reported that honey was an important trading good that was also used for fermentation. Many procedures regarding beekeeping and mead processing have been found depicted in ancient tombs.

As in other ancient cultures, honey was the only sweetener also in ancient Greece. Ancient Greeks introduced removable bars in their hives to separate the brood from the honey compartment. While producing mead, ancient Greeks mastered the application of herbs to improve its taste. It was probably Greek settlers who brought the crucial knowledge to the Roman Empire. Romans further advanced their apiculture, with dozens of different hives made of different materials. As had the ancient Greeks, the Romans also practised the addition of different spices to enrich the taste of mead. According to Schram (2003), this beverage was named metheglin, which refers to mead with herbs, from the Latin medicus, meaning medicinal, and the Irish Ilyn, meaning liquor. They used honey to sweeten grape wines, which was known as hydromel. Etymological research has shown that mead is such an ancient beverage that the linguistic root for mead, medhu, is the same in all of the Indo-European languages. Examples are presented in Table 7.6.

Today, mead is considered as a rather exotic alcoholic drink that is made by fermenting a mixture of honey and water. Nowadays mead is not as popular as it was in the past. The reason appears to be the great availability of relatively cheap raw materials for beer and wine production. Wine became the drink of choice among the upper classes, leaving mead, along with beer and cider, as the drink of the poor.

Table 7.6 Word origin of mead in other languages from the etymology dictionary (Vidrih and Hribar 2016, reproduced by permission of Springer Science+Business Media).

Language	"Mead" root	Meaning
Old English	medu	mead
Proto-Germanic	meduz	mead
cf. Old Norse	mjöðr	mead
Old Norse	mjöð	mead
Danish	mjød	mead
Old Frisian and Middle Dutch	mede	mead
German	Met	mead
Old High German	metu	mead
Proto-Indo-European	medhu-	honey, sweet drink
Sanskrit	madhu	sweet, sweet drink, wine, honey
Greek	methy	wine
Old Church Slavonic	medu	mead
Lithuanian	medus	honey
Old Irish	mid	mead
Welsh	medd	mead
Breton	mez	mead
Slovenian	medica	mead
Polish	miód	mead
Baltic	midus	mead

Modern mead production

Contemporary mead production exploits all of the knowledge available. Instead of wild microflora (Dumont 1992), which were used to carry out fermentation from ancient times until the 18th century, selected yeast strains (Pereira 2009) are used to provide standardised quality. Selected yeast strains can start fermentation much more quickly, and can withstand higher concentrations of ethanol. Wild microflora provoked uncontrolled fermentation that spoiled mead, giving unacceptable quality (Mendes-Ferreria et al. 2010). It is of note that wild microflora seldom complete fermentation, leaving some residual sugar.

Another technique to prevent wild microflora actions is pasteurisation of the wort before fermentation (Stong 1972). Pasteurisation successfully prevents cloudiness of mead, which is otherwise very difficult to remove and can persist for months. Some studies have reported changes in the colour and aroma profile (Kime 1991a) and the antioxidative potential (Wintersteen et al. 2005). Due to the drawbacks caused by high temperature pasteurisation, the above mentioned studies proposed instead pasteurisation at higher temperature, but for a shorter time, to minimise the heat load. Kime et al. (1991b) proposed the application of ultrafiltration, which selectively removes the particles responsible for cloudiness. Mead produced by means of ultrafiltration is often described as fresher and cleaner in taste.

Completely dry mead with no remaining sugar left might be short in taste, with little or no after taste, or 'body'. To overcome this drawback, the simplest method is to leave some sugar unfermented or to add honey after the completion of fermentation. The optimal residual sugar content depends on the type of honey used, with the widely accepted range from 30 g/L to approximately 100 g/L. The other methods include the addition of plant extracts that are rich in phenols, which can improve the body of a mead, as well as its antioxidative potential.

Mead production in Slovenia

Although very popular in the past, mead (known as medica) in Slovenia has now been left behind for more popular beverages, like beer, wines, and spirits. Mead is produced mostly by some enthusiastic beekeepers. They produce small batches that are often put into well-decorated bottles and sold on special occasions or in gift boutiques. The producers are well aware that only quality can keep the mead business alive, so they put in every effort to ensure sufficient quality.

Another product similar to mead is "Sparkling Mead" which is actually made from mead under secondary fermentation. The secondary fermentation takes place in the bottle following the addition of honey and yeast used for the production of sparkling wines. The technology used is similar to the classic technology for the production of sparkling wines. Sparkling mead is fresher in taste than mead due to the CO_2 entrapped that slowly escapes after the opening of the bottle.

7.3.5.2 Other alcoholic products

Honey liquor is produced from different types of honey, and usually this fruit brandy is popular in Slovenia during the winter, and especially around Christmas. It is made by mixing honey and a fruit spirit that preferably should be more neutral in taste. Honey liquor contains from 20 vol% to 30 vol% alcohol and 100 g/L to 300 g/L of unfermented sugar. The taste depends on the type of honey, and this can vary from very mild when acacia honey is used, to relatively astringent and full bodied when the honey liquor is made from chestnut, fir or spruce honeys.

Honey brandy: Mead can be distilled to produce honey brandy. For this the mead should ferment completely, because any remaining sugar would be left in the distillation equipment and thus lost. The quality depends on the mead quality attributes as well as the distillation. It is worth mentioning that the mead quality is of great importance, because its compounds are distilled and collected in the receiving tank. Care is taken to remove the first flow, which is rich in low boiling point volatiles that include acetaldehyde as the main compound. Acetaldehyde provokes a rather pungent flavour and smell that resembles green apple. Compounds with higher boiling points, at approximately 120°C, include the higher alcohols (e.g., 1-propanol, 1-butanol, iso-butanol, isoamyl alcohol) that are distilled toward the end of the distillation process. In order not to have a lot of these compounds in the final product (because of their green grass like aroma), the distillation should be brought to an end before the higher alcohols start to boil. Honey brandy has a specific honey-like odour, and it is usually not as intense as fruit brandies.

References

Allsop, K.A. and J.B. Miller. 1996. Honey revisited: A reappraisal of honey in pre-industrial diets. British Journal of Nutrition 75(4): 513–520.

Alvarez-Suarez, J.M., S. Tulipani, S. Romandini, E. Bertoli and M. Battino. 2009. Contribution of honey in nutrition and human health: a review. Mediterranean Journal of Nutrition and Metabolism 3(1): 15–23.

Anklam, E. 1998. A review of the analytical methods to determine the geographical and botanical origin of honey. Food Chemistry 63(4): 549–562.

Beekeepers' Society Sežana. 2012. Specification for "Kraški (Karst) honey" with Protected Designation of Origin, pp. 1–15.

Benzie, I.F.F. and J.J. Strain. 1996. The ferric reducing ability of plasma (FRAP) as a measure of "Antioxidant Power": The FRAP assay. Analytical Biochemistry 239(1): 70–76.

Beretta, G., P. Granata, M. Ferrero, M. Orioli and R. Maffei Facino. 2005. Standardization of antioxidant properties of honey by a combination of spectrophotometric/fluorimetric assays and chemometrics. Analytica Chimica Acta 533(2): 185–191.

Bertoncelj, J. 2008. Identification and quantification of some antioxidants in Slovenian honey. PhD Thesis, University of Ljubljana, Biotechnical Faculty, Ljubljana. http://www.digitalna-knjiznica.bf.uni-lj.si/dd_bertoncelj_jasna.pdf.

Bertoncelj, J., T. Polak, U. Kropf, M. Korošec and T. Golob. 2011a. LC-DAD-ESI/MS analysis of flavonoids and abscisic acid with chemometric approach for the classification of Slovenian honey. Food Chemistry 127(1): 296–302.

Bertoncelj, J., T. Golob, U. Kropf and M. Korošec. 2011b. Characterisation of Slovenian honeys on the basis of sensory and physicochemical analysis with a chemometric approach. International Journal of Food Science & Technology 46: 1661–1671.

Bertoncelj, J., U. Doberšek, M. Jamnik and T. Golob. 2007. Evaluation of the phenolic content, antioxidant activity and colour of Slovenian honey. Food Chemistry 105(2): 822–828.

Blasa, M., M. Candiracci, A. Accorsi, M. Piera Piacentini, M.C. Albertini and E. Piatti. 2006. Raw Millefiori honey is packed full of antioxidants. Food Chemistry 97(2): 217–222.

Bogdanov, S. 1997. Nature and origin of the antibacterial substances in honey. Lebensmittel-Wissenschaft und-Technologie 30(7): 748–753.

Bogdanov, S., T. Jurendic, R. Sieber and P. Gallmann. 2008. Honey for nutrition and health: a review. Journal of the American College of Nutrition 27(6): 677–689.

Bogdanov, S., M. Haldimann, W. Luginbühl and P. Gallmann. 2007. Minerals in honey: environmental, geographical and botanical aspects. Journal of Apicultural Research and Bee World 46(4): 269–275.

Božič, J. 2015. A-Ž beekeeping with the Slovenian hive. Blomsted, W. (ed.). Mija, Ljubljana, Slovenia.

Can, Z., O. Yildiz, H. Sahin, E.A. Turumtay, S. Silici and S. Kolayli. 2015. An investigation of Turkish honeys: Their physico-chemical properties, antioxidant capacities and phenolic profiles. Food Chemistry 180: 133–141

CBI Market Survey. 2009: The honey and other bee products market in the EU, pp. 1–32. (www.cbi.eu; http://www.fepat.org.ar/files/eventos/759630.pdf).

Cotte, J.F., H. Casabianca, S. Chardon, J. Lheritier and M.F. Grenier-Loustalot. 2004. Chromatographic analysis of sugars applied to the characterisation of monofloral honey: Analytical and Bioanalytical Chemistry 380(4): 698–705.

Crane, E. 1975. History of honey. pp. 439–488. In: E. Crane (ed.). Honey, a Comprehensive Survey. William Heinemann, London. UK.

Crane, E. 1983. The archaeology of beekeeping. Duckworth, London. UK.

da Silva, P.M., C. Gauche, L. Valdemiro Gonzaga, A.C. Oliveira Costa and R. Fett. 2016. Honey: Chemical composition, stability and authenticity. Food Chemistry 196: 309–323.

de la Fuente, E., A.I. Ruiz-Matute, R.M. Valencia-Barrera, J. Sanz and I. Martinez Castro. 2011. Carbohydrate composition of Spanish unifloral honeys. Food Chemistry 129(4): 1483–1489.

Dumont, D. J. 1992. The ash tree in Indo-European culture. Mankind Quarterly 32(4): 323–336.

Escuredo, O., I. Dobre, M. Fernandez-Gonzalez and M. Carmen Sijo. 2014. Contribution of botanical origin and sugar composition of honeys on the crystallization phenomenon. Food Chemistry 149: 84–90.

European Commission. 2013. Report from the Commission to the European Parliament and the Council on the implementation of the measures concerning the apiculture sector of Council Regulation (EC) No 1234/2007, pp. 1–11. http://eur-lex.europa.eu/legal-content/EN/TXT/PDF/?uri=CELEX:52013DC0593&from=en.

European Council Directive 2002. Directive 2001/110/EC of 20 December 2001 relating to honey. Official Journal of the European Community, L10, pp. 47–52.

FAOSTAT. 2015. Food and Agriculture Organization of the United Nations Statistics Division. Production quantities by country. Average 1992–2012. http://faostat3.fao.org/browse/Q/QL/E.

Gašić, U., S. Kečkeš, D. Dabić, J. Trifković, D. Milojković-Opsenica, M. Natić and Ž. Tešić. 2014. Phenolic profile and antioxidant activity of Serbian polyfloral honeys. Food Chemistry 145: 599–607.

Gayre, R. and C. Papazian. 1983. Brewing Mead: Wassail! In Mazers of Mead. Boulder, Brewers Publications.

Gheldof, N., X.-H. Wang and N.J. Engeseth. 2002. Identification and quantification of antioxidant components of honeys from various floral sources. Journal of Agricultural and Food Chemistry 50(21): 5870–5877.

Golob, T., U. Dobersek, P. Kump and M. Nečemer. 2005. Determination of trace and and minor elements in Slovenian honey by total reflection X-ray fluorescence. Food Chemistry 91(2): 313–317.

Golob, T. and A. Plestenjak. 1999. Quality of Slovene honey. Food Technology and Biotechnology 37: 195–201.

Gómez-Caravaca, A.M., M. Gómez-Romero, D. Arráez-Román, A. Segura-Carretero and A. Fernández-Gutiérrez. 2006. Advances in the analysis of phenolic compounds in products derived from bees. Journal of Pharmaceutical and Biomedical Analysis 41(4): 1220–1234.

Gregori, J., J. Poklukar and J. Mihelič. 2003. The Carniolan honey bee (*Apis mellifera carnica*) in Slovenia. Brdo pri Lukovici: The Beekeepers' Association of Slovenia.

Hermosin, I., R.M. Chicon and M.D. Cabezudo. 2003. Free amino acid composition and botanical origin of honey. Food Chemistry 83(2): 263–268.

International Honey Commission. 2002. Harmonised methods of the International Honey Commission, 63 p. http://www.bee-hexagon.net/files/file/fileE/IHCPapers/IHC-methods_2009.pdf.

Kamal, M.A. and P. Klein. 2011. Determination of sugars in honey by liquid chromatography. Saudi Journal of Biological Sciences 18(1): 17–21.

Kandolf, A., A. Pereyra Gonzales and N. Lilek. 2010. Slovene honey. Apimedica & Apiquality Forum. https://docs.google.com/leaf?id=0B1XegMck_4sXNTJlNjZhYTktZjE1NC00O WNkLTkxYzQtNmU2YzhmMTY3NTcy&sort=name&layout=list&pid=0B1XegMck_4 sXMDJiZWRhMGYtODlmOC00NTU0LTg2YjAtYjc1NWEzMGMwYzZm&cindex=27.

Kime, R.W., M.R. McLellan and C.Y. Lee. 1991a. An improved method of mead production. American Bee Journal 131(6): 394–395.

Kime, R.W., M.R. McLellan and C.Y. Lee. 1991b. Ultra-filtration of honey for mead production. American Bee Journal 131(8): 517–521.

Kočevski med Association. 2011. Specification "Kočevje forest honey" with Protected Designation of Origin, 19 p. http://www.mkgp.gov.si/fileadmin/mkgp.gov.si/pageuploads/podrocja/Kmetijstvo/zascita_kmetijskih_pridelkov_zivil/specifikacija_kocevski_gozdni_med.pdf.

Korošec, M., J. Bertoncelj, A. Pereyra Gonzales, U. Kropf, U. Golob and T. Golob. 2009. Monosaccharides and oligosaccharides in four types of Slovenian honey. Acta Alimentaria 38(4): 459–469.

Korošec, M. 2012. Determination of physical and chemical parameters for verifying the honey authenticity. PhD Thesis, University of Ljubljana, Biotechnical Faculty, Ljubljana. http://www.digitalna-knjiznica.bf.uni-lj.si/zivilstvo/dd_korosec_mojca.pdf.

Korošec, M., U. Kropf, U. Golob and J. Bertoncelj. 2016. Functional and nutritional properties of different types of slovenian honey. pp. 323–336. In: K. Kristbergsson and S. Ötles (eds.). Functional Properties of Traditional Foods, Springer Science+Business Media, New York. USA.

Kralj Kunčič, M., D. Jaklič, A. Lapanje and N. Gunde-Cimerman. 2012. Antibacterial and antimycotic activities of Slovenian honeys. British Journal of Biomedical Science 69(4): 154–158.

Kropf, U., T. Golob, M. Nečemer, P. Kump, M. Korošec, J. Bertoncelj and N. Ogrinc. 2010. Carbon and nitrogen natural stable isotopes in Slovene honey: Adulteration and botanical and geographical aspects. Journal of Agricultural and Food Chemistry 58(24): 12794–12803.

Küçük, M., S. Kolaylı, Ş. Karaoğlu, E. Ulusoy, C. Baltacı and F. Candan. 2007. Biological activities and chemical composition of three honeys of different types from Anatolia. Food Chemistry 100(2): 526–534.

Lachman, J., M. Orsák, A. Hejtmánková and E. Kovářová. 2010. Evaluation of antioxidant activity and total phenolics of selected Czech honeys. LWT - Food Science and Technology 43(1): 52–58.

Lilek, N. and A. Kandolf. 2012. Čebelarji se predstavimo. Brdo pri Lukovici: The Beekeepers' Association of Slovenia.

McGovern, P.E., J.H. Zhang, J.G. Tang, Z.Q. Zhang, G.R. Hall, R.A. Moreau, A. Nunez, E.D. Butrym, M.P. Richards, C.S. Wang, G.S. Cheng and Z.J. Zhao. 2004. Fermented beverages of pre- and proto-historic China. Proceedings of the National Academy of Sciences of the United States of America. 101(51): 7593–17598.

Meda, A., C.E. Lamien, M. Romito, J. Millogo and O.G. Nacoulma. 2005. Determination of the total phenolic, flavonoid and proline contents in Burkina Fasan honey, as well as their radical scavenging activity. Food Chemistry 91(3): 571–577.

Mendes-Ferreira, A., F. Cosme, C. Barbosa, V. Falco, A. Ines and A. Mendes-Faia. 2010. Optimization of honey-must preparation and alcoholic fermentation by *Saccharomyces cerevisiae* for mead production. International Journal of Food Microbiology 144(1): 193–198.

Molan, P.C. 1996. Authenticity of honey. pp. 259–303. *In*: P.R. Ashurst and M.J. Dennis (eds.). Food Authentication, Blackie Academic & Professional, London. UK.

Morse, R.A. and K.H. Steinkraus. 1975. Wines from the fermentation of honey. pp. 392–407. *In*: E. Crane (ed.). Honey: a Comprehensive Survey. Heinemann, London. UK.

Nečemer, M., I. Košir, P. Kump, U. Kropf, M. Jamnik, J. Bertoncelj, N. Ogrinc and T. Golob. 2009. Application of total reflection X-ray spectrometry in combination with chemometric methods for determination of the botanical origin of Slovenian honey. Journal of Agricultural and Food Chemistry 57(10): 4409–4414.

Pereira, A.P., T. Dias, J. Andrade, E. Ramalhosa and L.M. Estevinho. 2009. Mead production: Selection and characterization assays of *Saccharomyces cerevisiae* strains. Food and Chemical Toxicology 47(8): 2057–2063.

Persano Oddo, L. and S. Bogdanov. 2004. Determination of honey botanical origin: problems and issues. Apidologie 35: S2–S3.

Regulation (EU) No. 1151/2012 of the European Parliament and of the Council of 21 November 2012 on quality schemes for agricultural products and foodstuffs. Official Journal of the European Community, L 343, pp. 1–29.

Riches, H.R.C. 2009. Mead: making, exhibiting & judging. Hebden Bridge, Northerm Bee Books.

Rosch, M. 2005. Pollen analysis of the contents of excavated vessels—direct archaeobotanical evidence of beverages. Vegetation History and Archaeobotany 14(3): 179–188.

Rose, A. (ed.). 1977. Alcoholic Beverages. Academic Press, Michigan, USA.

Rules relating to honey. 2011. Official gazette of Republic of Slovenia, 4/2011, pp. 345–347.

Schramm, K. 2003. The compleat meadmaker: Home production of honey wine from your first batch to award-winning fruit and herb variations. Brewers Publications, Boulder, Colorado, USA.

Slovenian Beekeepers' Association. 2014. Specification for Slovenian honey with Protected Geographical Indication, pp. 1–17. http://www.mkgp.gov.si/fileadmin/mkgp.gov.si/pageuploads/podrocja/Kmetijstvo/zascita_kmetijskih_pridelkov_zivil/SMGO_potr_sp_7_1_15.pdf.

Statistical Office of the Republic of Slovenia. 2011. SI-Stat Data portal. The balance of production and consumption of sugar (1000T), the marketing year, Slovenia, annually. http://www.stat.si/pxweb/Dialog/varval.asp?ma=1563403S&ti=Bilanca+proizvodnje+in+porabe+sladkorja+%281000t%29%2C+tr%9Eno+leto%2C+Slovenija%2C+letno&path=../Database/Okolje/15_kmetijstvo_ribistvo/12_prehranske_bilance/01_15634_trzne_bilance/&lang=20.

Stong, C.L. 1972. Mead, drink of Vikings, can be made (legally) by fermenting honey a home. Scientific American 227(3): 185.

Swallow, K.W. and N.H. Low. 1994. Determination of honey authenticity by anion-exchange liquid chromatography. Journal of AOAC International 77(3): 695–702.

Taormina, P.J., B.A. Niemira and L.R. Beuchat. 2001. Inhibitory activity of honey against foodborne pathogens as influenced by the presence of hydrogen peroxide and level of antioxidant power. International Journal of Food Microbiology 69(3): 217–225.

Turkmen, N., F. Sari, E.S. Poyrazoglu and Y.S. Velioglu. 2005. Effects of prolonged heating on antioxidant activity and colour of honey. Food Chemistry 95(4): 653–657.

Vidrih, R. and J. Hribar. 2016. Mead: The oldest alcoholic beverage. pp. 325–338. *In*: K. Kristbergsson and J. Oliveira (eds.). Traditional Foods: General and Consumer Aspects, Springer Science+Business Media, New York. USA.

Vidrih, R. and J. Hribar. 2007. Studies on the sensory properties of mead and the formation of aroma compounds related to the type of honey. Acta Alimentaria 36(2): 151–162.

Viuda-Martos, M., Y. Ruiz-Navajas, J. Fernández-López and J.A. Pérez-Alvarez. 2008. Functional properties of honey, propolis, and royal jelly. Journal of Food Science 73(9): 117–124.

Weston, R.J., K.R Mitchell and K.L. Allen. 1999. Antibacterial phenolic components of New Zealand manuka honey. Food Chemistry 64(3): 295–301.

White, J.W. 1978. Honey. Advances in Food Research 24: 287–374.

Wintersteen, C.L., L.M. Andrae and N.J. Engeseth. 2005. Effect of heat treatment on antioxidant capacity and flavor volatiles of mead. Journal of Food Science 70(2): C119–C126.

CHAPTER 8

Portuguese Sheep Cheeses

Gil Fraqueza

8.1 Introduction

Sheep milk differs from the milk of other animals due to its rich constituents. It has favorable features for the preparation of dairy products, as it is rich in almost all its components. This milk is more concentrated and has more than twice the fat content of cow and goat milk, making it ideal for producing cheese. Sheep milk cheese is a product in itself with a high market price (Assenat 1991).

Sheep milk is favourable for processing due to its high fat content that totals about 6.5% of its composition, and protein of about 6% (Boyazogl and Morand-Fehr 2001).

The white and creamy sheep milk is much more concentrated than cow or goat milk. It has much higher levels of fat and casein (an important substance for the formation of curd), which is essential for the development of hard and soft cheeses. These cheeses are known for their quality and are also in demand to allergie to cow or goat milk (Timperley and Norman 1997).

8.2 Characteristics of sheep milk

Information about the composition and physicochemical characteristics of sheep milk is essential for the development of industry and the market for specific products (Park et al. 2007). There are marked differences between sheep, goat, cow and human milk (Table 8.1).

Sheep milk is considered the richest type of milk and its physical and chemical composition differs greatly from cow milk. Sheep milk is much richer in total solids. This explains the high yield in manufacturing

Department of Food Engineering, Institute of Engineering and CCMAR – Centre of Marine Sciences, University of Algarve, Faro, Portugal.
Email: gfraque@ualg.pt

Table 8.1 Chemical composition of milk from different species (Bencini and Purvis 1990, Jandal 1996, Zamiri et al. 2001, Nudda et al. 2002, Assis et al. 2004, Sevi et al. 2004, Park et al. 2007).

	Sheep	Goat	Cow	Human
Solids, total %	19.3	12.97	12.01	12.5
Energy, kcal	108	69	61	70
Energy, kJ	451	288	257	291
Protein, total %	5.98	3.56	3.29	1.03
Casein, total %	4.2	2.4	2.6	0.4
Lipids, total %	7	4.14	3.34	4.38
Carbohydrates, total %	5.36	4.45	4.66	6.89
Ash, total %	0.96	0.82	0.72	0.2
Ca, mg	193	134	119	32
Fe, mg	0.1	0.05	0.05	0.03
Mg, mg	18	14	13	3
P, mg	158	111	93	14
K, mg	136	204	152	51
Na, mg	44	50	49	17
Zn, mg	0.57	0.3	0.38	0.17
Ascorbic acid, mg	4.16	1.29	0.94	5
Thiamine, µg	80	40	40	20
Riboflavin, mg	0.355	0.138	0.162	0.036
Niacin, mg	0.417	0.277	0.084	0.177
Pantothenic acid, mg	0.407	0.31	0.314	0.223
Vitamin B6, µg	80	60	60	10
Folate, total, µg	5	1	6	5
Vitamin B12, µg	0.711	0.065	0.357	0.045
Vitamin A, µg	83	44	52	58
Vitamin D, µg	0.18	0.11	0.03	0.04
Vitamin E, mg	0.11	0.03	0.09	0.34
Vitamin C, mg	5	1	1	4

cheeses—100 litres of sheep milk can produce up to 22 kg of Roquefort cheese. There is no carotene present in the milk fat, which makes the milk very white in colour, whereas cow milk has higher amounts of fatty acids such as caproic (hexanoic), caprylic (octanoic) and capric (decanoic) (Furtado 2003).

The differences in the composition of cow and sheep milk are related to their physicochemical properties. Some of these physico-chemical properties are shown in Table 8.2.

Table 8.2 Some physico-chemical properties of sheep, goat and cow milk (Park et al. 2007).

Properties	Sheep milk	Goat milk	Cow milk
Specific gravity (density)	1.0347–1.0384	1.029–1.039	1.0231–1.0398
Viscosity, Cp	2.86–3.93	2.12	2
Refractive index	1.3492–1.3497	1.450 ± 0.39	1.451 ± 0.35
Freezing point (°C)	0.570	0.540–0.573	0.530–0.570
Acidity (lactic acid %)	0.22–0.25	0.14–0.23	0.15–0.18
pH	6.51–6.85	6.50–6.80	6.65–6.71

Visually, sheep milk is pearly white in colour, similar to porcelain and with a greater opacity than cow or goat milk. The viscosity of sheep milk is also higher than that of cow milk (Assenat 1991, Jandal 1996).

While the sheep milk has a higher viscosity and acidity, it has a similar freezing point and lower refractive index compared to cow milk (Park and Haenlein 2006, Park et al. 2007). The difference in density and freezing points between cow milk and small ruminant milk are explained primarily by high dry matter content that is fat free (Alichanidis and Polychroniadou 1997). The physicochemical properties of milk vary, depending on production conditions and individual characteristics of each animal (Pavić et al. 2002).

Compared to cow, goat and human milk, sheep milk has higher levels of total solids, protein, casein, fat, calcium, iron, magnesium, zinc, thiamin, riboflavin, vitamin B6, vitamin B12, vitamin D, medium fatty acids and short chain monounsaturated fatty acids, linolenic acid and all essential amino acids (Haenlein 2001, Mayer and Fiechter 2012). These differences, particularly in casein, results in lower clotting time, and curd firmness increases when compared to the production of cheese from cow milk (Wendorff 2002, Jooyandeh and Aberoum 2010). Lipids of goat and sheep milk are similar. The most significant difference is the presence of short chain fatty acids, like caproic (hexanoic), caprylic (octanoic) and capric (decanoic) in higher proportions in goat milk. The carbohydrate fraction of goat and sheep milk is composed by lactose. The lactose present in goat milk is slightly greater than in sheep milk, however the total ash content is lower.

Sheep milk contains approximately 160 mg of calcium and 145 mg of phosphorus, while goat milk contains 194 mg of calcium and 270 mg of phosphorus per 100 g of milk.

The level of mineralization of the sheep milk micelles is superior to those found in cow milk and have similar characteristics to those of goat milk (Park et al. 2007, Raynal-Ljutovac et al. 2008). Raynal-Ljutovac et al. (2008) reported that the sheep milk casein micelles differ from bovine in the content of caseins and in its organization and mineralization and this determines the specific technological behavior, without, however, the nutritional impact of these characteristics being known.

The caseins are composed of the following fractions: κ-casein, β-casein, $α_{s1}$-casein and $α_{s2}$-casein and their percentages in milk vary with the species, since these are genetic traits due to different levels of phosphorylation, amino acid substitutions, differences in glycosylation, changes in electrical charge, hydrophobicity and molecular weight of the proteins (Haenlein 2001, Jooyandeh and Aberoum 2010). $α_{s2}$-casein in sheep milk is higher than in goat milk, but $α_{s1}$-casein and $α_{s2}$-casein concentrations are lower than in cow milk. β-casein represents approximately a percentage of 50% from the total of milk caseins, and represents 2/3 and 1/3 in goat milk and cow milk, respectively. The differences in amounts of casein present in sheep milk explain the differences in the micelle structure, which determine variations in stability and coagulation, reducing the occurrence of the bitter taste of cheeses obtained from such milk, compared to goat milk (Jooyandeh and Aberoum 2010). Another important feature of sheep milk is the low concentration of calcium and inorganic phosphorus in the soluble fraction, leading to a more concentrated colloidal milk portion (Raynal-Ljutovac et al. 2008). β-casein and colloidal calcium phosphate present in milk, may negatively affect the industrial performance of cheese (Lawrence Neto 1998). The storage of milk at low temperature leads to a partial solubility and colloidal concentration of these salts in the sheep milk favor a lower stability. Cooling may also affect the yield of cheese manufacture. Thus, the largest β-casein and milk lipids present in the form of globules are characteristically abundant in smaller sizes than three micrometers in the case of milk from goat and sheep. Some studies suggest that the average size of the globules is smaller in the sheep milk, followed by goat milk. This feature is interesting because it is associated with better digestibility and more efficient metabolism of these lipids, compared to that of cow milk (Park et al. 2007).

Another difference in sheep milk is the presence of significant higher level (about 16% of total fatty acids) of short and medium chain (C6–C12) compared to cow milk. These are associated with the flavour of cheese, and can be also seen as fraud indicators by mixing milk of different species (Park et al. 2007, Raynal-Ljutovac et al. 2008). This is an interesting feature, because these fatty acids are metabolized so differently from those of long chain, and can be released by hydrolysis by the enzymes in the gut and pancreas, as well as be directly absorbed and transported to the liver where they will be oxidized. They are, therefore, a rapid energy source, ideal for elderly or malnourished people. Moreover, being quickly metabolized, they reduce levels of circulating cholesterol and promote deposition of the lower fat in adipocytes (Raynal-Ljutovac et al. 2008). On the other hand, palmitic acid (C16) has hypercholesterolemic properties, and because of the health problems associated with the intake of unsaturated fatty acids, sheep milk consumption may not be desirable (Zhang et al. 2006a). Conjugated linoleic acid content (CLA) showed higher value in sheep milk (1.08%), followed by cow milk (1.01%) and goat milk (0.65%) (Jarheis et al. 1999, Park et al. 2007). Considering the various nutritional benefits associated with these fatty acids, sheep milk is desirable from a nutritional point of view. In addition, research has demonstrated the existence of a certain case of manipulation of the diet in order to increase the CLA in sheep milk and to make it even more suitable for the diet of humans. In turn, sheep milk has higher trans fatty acid content than cow and goat milks, but the amount observed in these milks is not sufficient to pose risks of coronary heart disease associated with ingestion (Park et al. 2007).

Lactose is the principal solid component of sheep milk, as previously referred, and the main carbohydrate in milk. Other carbohydrates are many oligosaccharides, which may give rise to monosaccharides, which may have acidic characteristics (sialic acid), or may be neutral. These elements are important substrates for bifidobacteria and can play an important role in protecting the intestinal mucosa against pathogens and are important for neonatal brain development. Lactose content in sheep milk, compared to cow milk, is practically the same level as fat, and presents a high content of protein. Lactose content in these species represents a smaller proportion in the total solids of 22 to 27%, while cow milk presents around 33 to 40% (Raynal-Ljutovac et al. 2008).

Goat milk and sheep milk have adequate amounts of vitamin A, thiamine, riboflavin and pantothenic acid but are deficient in vitamins C and D, folic acid and cyanocobalamin and may also be deficient in pyridoxine.

The quality of sheep milk is associated to its ability to be processed as dairy products and high yields of such products per litre of milk, because the majority of sheep milk produced worldwide is processed into cheese and, to a lesser extent, into yogurt and ice cream (Bencini and Pulina 1997). The high total solids content (sum of the macro nutrients, protein, fat, lactose and salts) in sheep milk contributes to better yields in cheese production compared to cow milk (Gutierrez 1991). The total solids content is almost 2-fold, having higher protein content, especially the fraction of casein and fat. This yield is between 18 to 25%, that is, it only takes 4 to 5 kg sheep milk to produce 1 kg of cheese.

8.2.1 Factors affecting the production and composition of sheep milk

There are many factors that contribute to the variation in the production of sheep milk: race, age, stage of lactation, number of suckling lambs, nutritional level during pregnancy and lactation, environment, milking techniques, health status and infections of the udder (Peeters et al. 1992, Gonzalo et al. 1994, Bencini 2001). Since virtually all produced sheep milk is processed into cheese and other derivatives such as yogurt and ice cream, as previously referred, milk quality is directly related to industrial income. The quantity and quality of the cheese produced per litre of milk depends mainly on milk coagulation properties, that is, the coagulation time of clot formation rate and consistency of the clot. These properties are affected by the milk composition, the somatic cell count and the industrial processing. Therefore, any factor that affects milk composition, will also affect the production and quality of dairy products (Bencini 2001).

8.3 Types of Cheese

Sheep milk is rarely eaten fresh. Most of the milk produced is processed into cheeses known worldwide for its characteristics and countries of origin (Table 8.3) (Campos 2011).

Table 8.3 Different types of cheeses made with milk from sheep and their countries of origin.

Country	Cheese
Portugal	Serra da Estrela, Serpa, Azeitão, Castelo Branco
Bulgaria	Katschkawalj
Czech Republic	Abertam
England	Friesla, Olde York
France	Roquefort, Abbaye de Belloc, Perail
Greece	Kefalotiri, Myzithra, Feta*
Hungary	Liptoi
Ireland	Orla
Italy	Canestrato Pugliese, Fiore Sardo, Pecorino, Roman, Sardo, Toscano
Libya	Al Zahra, Jibnet Grus, Al Naseem
Romania	Brinza
Spain	Castellano, Idiazabal, Manchego, Roncal, Zamorano
Turkey	Beyaz Peynir, Mihalic Peynir

* It can also be made from goat and cow milk. (Source: Adapted from Harbutt 1999 (cited by Meunier-Goddik and Nashnush 2006)).

The cheeses can be classified in various ways, according to the raw material used, the consistency, the treatment of the mass and the method for obtaining the forms of clotting the fat in dry matter and humidity. There are different types of cheeses made with sheep milk, which can be classified according to the moisture content. Scholz (1995) classifies:

1) Fresh cheese, which should be thin and delicate in mass, with mild flavour, slightly acid and slightly resembling the typical sheep milk flavour (e.g., traditional Austrian cheese, cheese and quark cheese acid and slow coagulation in bags);

2) Soft cheeses, with relatively high moisture content (55 to 65%) can be ripened or not, and are relatively acid and core reminiscent of the curd (e.g., Serra da Estrela, Azeitão); and

3) Fungal ripened cheeses (such as Roquefort).

8.3.1 Composition

Considering the literature since 1990, most research about cheeses made with milk from small ruminants, mainly from Spain, Italy and Greece, evaluated the chemical composition of hard and semi hard cheeses (fat, lactose and protein). The chemical composition of the cheese depends mainly on the type of cheese (hard, soft, whey cheeses, among others). Table 8.4 shows the composition of a variety of different cheeses made with sheep milk. Since the milk of small ruminants is rarely standardized for the manufacture of cheese, fat and protein content of these derivatives vary depending on the breed of sheep and feeding systems (Raynal-Ljutovac et al. 2008). In turn, these authors considered that most of the research that reveals the composition of Spanish and Italian cheeses cannot be applied to the French cheeses with goat and sheep milk, as these are mostly fresh cheeses with average moisture, while the former usually has longer periods of ripening, with stiff dough and less moisture content.

8.3.2 Traditional Portuguese sheep cheeses

The Mediterranean countries, in particular Portugal, are characterized by the existence of various types of artisan or traditional cheeses made from raw sheep milk, which represent a very important role in rural activity and the local economy. The popularity of these cheeses has increased among consumers, especially the Serra da Estrela, Serpa and Azeitão cheeses due to specific characteristics, such as texture and flavour (Freitas et al. 2000), contributing for an increase in consumption and production intensification (Roseiro et al. 2003a, Roseiro et al. 2003b, Zhang et al. 2006b).

Table 8.4 Percentage composition of sheep milk cheeses of different breeds (adapted from Raynal-Ljutovac et al. 2008).

Cheese	Breed	Ripening time*	Total solids	Fat	Protein	Ash
Canestrano		1–56 d	39	31	25	
Feta		3–240 d	45	22	18	
Fiore sardo			70	29	28	
Halloumi		fresh	65	32	23	
Los Pedroches	Merino	1–200 d	35	31/33	26	8
Manchego	Manchega	90 d	70	30/42/37	23	
Acidic curd		0–33 d	53	26		
Manchego	Manchega	1–9 mon	37	31	25	7
Ossau-Iraty			61	32	24	4
Pecorino	Sarda	1 and 60 d	70	37 and 36	26	
Pecorino roman			65	30	27	
Pugliese		10–12 mon	67	30	27	
Ricota	Sarda		30	18		
Roquefort			57	33	19	6
Serena		58 d				9
Serra da Estrela		1, 7, 21, 35 d				8
Terrincho		0–60 d	46	25	21	8

These Portuguese cheeses are an important part of Portuguese culture and industry, not so known to the rest of the world. Portugal is home of many high quality farmhouse cheeses, which are produced mainly in the mountainous regions and the central highlands using sheep and goat milk. Of these, perhaps the best known are the Serra da Estrela, Serpa and Azeitão cheeses. In Table 8.5, is possible to verify that the composition of such cheeses does not show great qualitative differences but that quantitative differences are a result of different curing times and consequently different water contents.

These cheeses contain a high concentration of essential nutrients, particularly proteins, high-quality calcium, as well as other nutrients such as phosphorus, zinc, vitamin A, riboflavin and vitamin B12. Many cheeses, particularly ripened cheeses such as Katschkawalj and Pecorino, containing little or no lactose, can be digested by many who are lactose intolerant. During the process of cheese production, most of the lactose in the whey is drained. The remaining amount in the curd is converted into lactic acid during cheese ripening. Only trace amounts of lactose remain

Table 8.5 Composition of the main Portuguese sheep cheeses (Source: INSA 2007).

	Milk, sheeps, raw	Cheese, "Azeitão"	Cheese, "Serra", fresh	Cheese, "Serra", cured	Cheese, "Serpa"
	(by 100 g*)	(by 100 g*)	(by 100 g*)	(by 100 g*)	(by 100 g*)
Energy, kcal	93	313	333	389	335
Energy, kJ	388	1308	1391	1629	1400
Macrocomponents					
Water, g	86.9	48.3	45.5	36.5	43
Protein, g	5.1	21	21	25.5	24.8
Fat, total, g	6.2	25	27	31.5	25.7
Carbohydrate, total available, g	4.2	0.1	0.2	0.2	0.3
Carbohydrate, total available (expressed in monosaccharides), g	4.4	0.1	0.2	0.2	0.3
Mono+disaccharides, g	4.2	0.1	0.2	0.2	0.3
Organic acids, g	0	1.1	1.6	1	1
Fatty Acids					
Fatty acids, total saturated, g	3.3	13.1	14.2	16.6	13.5
Fatty acids, total monounsaturated, g	1.5	6	6.5	7.6	6.2
Fatty acids, total polyunsaturated, g	0.2	0.9	0.9	1.1	0.9
Fatty acids, total trans, g		1	1.1	1.3	1.1
Linoleic acid, g	0.2	0.8	0.8	1	0.8
Vitamins					
Vitamin A (retinol equivalent), µg	50	210	240	285	195
Carotene, total, mg	0	70	80	10	6
Vitamin D, µg	0.17	0.2	0.21	0.24	0.22
Alpha-tocopherol, mg	0.11	0.4	0.45	0.53	0.43
Thiamin, mg	0.06	0.04	0.05	0.05	0.05
Riboflavin, mg	0.15	0.66	0.6	0.65	0.7
Niacin equivalents, total, mg	1.1	5.5	5.6	6.8	4.7
Niacin, preformed, mg	0.2	0.6	0.7	0.8	0.8

Table 8.5 cont....

Table 8.5 cont.

	Milk, sheeps, raw	Cheese, "Azeitão"	Cheese, "Serra", fresh	Cheese, "Serra", cured	Cheese, "Serpa"
	(by 100 g*)	(by 100 g*)	(by 100 g*)	(by 100 g*)	(by 100 g*)
Vitamins (contd.)					
Niacin equivalents from tryptophan, mg	0.9	4.9	4.9	6	3.9
Vitamin B6, total, mg	0.074	0.082	0.09	0.11	0.09
Vitamin B12, μg	0.15	0.37	1.4	1.4	1.1
Vitamin C, mg	5	0	0	0	0
Folate, total, μg	5	31	31	38	37
Minerals					
Ash, g	0.93	4.4	4.72	5.3	5.2
Sodium, mg	37	896	706	792	991
Potassium, mg	116	76	80	90	85
Calcium, mg	190	540	700	815	710
Phosphorus, mg	140	420	475	585	600
Magnesium, mg	15	50	53	59	56
Iron, total, mg	0.2	0.7	0.7	0.8	1.3
Zinc, mg	0.7	2.6	2.8	3.1	3

* of edible part.

in the cheese. Cheeses with low content in lactose are also good sources of calcium. For this reason, cheese is a concentrated source of many important nutrients found in breast milk.

Most Portuguese cheeses have been given the designation of Protected Designation of Origin (PDO). With the same objective of the PDO, the Designation of Origin (DO) or the Controlled Designation of Origin (CDO), the PDO guarantees that the cheese is produced within the demarcated region using traditional methods and ingredients. These cheeses are named and labeled with their city or town of origin and the PDO label of approval.

8.3.2.1 Serra da Estrela

The Serra da Estrela cheese is handmade with sheep milk from Bordaleira Serra da Estrela and/or Churra Mondegueira, in Portugal, in the winter months, being December to April the best time for their production. It is coagulated by enzymes (cardosins) from the thistle flower (*Cynara cardunculus* L.), a native plant of the region and its maturation can last for 30 to 45 days (buttery cheese) or up to six months (old cheese) (Barbosa 1990).

Serra da Estrela cheese (Fig. 8.1), probably the most famous among all Portuguese traditional cheeses, is manufactured at the farm level only from raw sheep milk. The cheese is manufactured by following artisanal

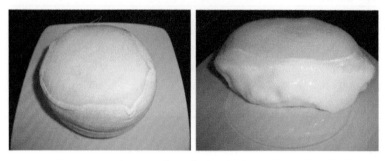

Figure 8.1 Serra da Estrela cheese.

protocols which includes coagulation with aqueous infusions of flowers of wild thistle (*Cynara cardunculus* L.), without the deliberate addition of any starter or non starter cultures.

Great variability is usually encountered between distinct dairies, and even within the same manufacturing batch. This variability obviously occurs owing to the lack of consistency of the microbiological, biochemical and physicochemical characteristics of the cheese milk and the protocols of manufacture.

Flavour profile of a given food product is generally the most important criterion for consumer preference. The Serra da Estrela cheese is legally defined by NP-1922 as a cured sheep cheese, semi soft, buttery, white or slightly yellowish, with little or no eyes, obtained by slow draining of the curd after coagulation of raw sheep milk, due to the action of the thistle (*Cynara cardunculus* L.) and weighing between 0.7 kg and 1.7 kg. The geographical area of production includes the municipalities of Carregal do Sal, Celorico da Beira, Fornos de Algodres, Gouveia, Mangualde, Manteigas, Oliveira do Hospital, Penalva do Castelo and Seia and even some areas from the municipalities of Aguiar da Beira, Arganil, Covilhã, Guarda, Viseu and Trancoso.

8.3.2.2 Serpa

Serpa cheese (Baixo-Alentejo): Serpa cheese is produced in the Alentejo, a well-defined and protected region with its own regulation (DR 1987). Serpa cheese is a ripened cheese, semi-soft also with Protected Designation of Origin (PDO), obtained by slow draining or syneresis of curd after coagulation of the raw milk of sheep, by the action of an infusion of *Cynara cardunculus* L. at a minimum of 30 days of ripening (DR 1987, Roseiro et al. 2003a).

Serpa cheese (Fig. 8.2) was traditionally made with sheep milk of local sheep breeds, Merino and Campaniça, according to the extensive system

Figure 8.2 Serpa cheese.

of production, characteristic of the Alentejo. Currently, local breeds have been replaced with more productive dairy sheep breeds, such as the Serra da Estrela and Lacaune (Pinheiro et al. 2003).

This award-winning cheese is made from raw sheep milk and ripened for a minimum of one month to two years. The consistency can range from very soft and creamy to hard with small holes. The rind has a very distinctive brick orange colour, resulting from its regularly brushing with olive oil mixed with paprika and is produced in small to medium sized rounds. The flavour is a unique mix of strong, spicy and slightly sweet-tart, as a result of the paprika, and has gained distinction as one of the most extraordinary products in the world.

8.3.2.3 Azeitão

Azeitão cheese (Fig. 8.3), is also produced with sheep milk. It is sold after about 20 days of curing, usually wrapped in paper. The peel is thin and soft with straw-yellow colour. It is a soft paste cheese with some darker

Figure 8.3 Azeitão cheese.

yellow eyes, very buttery, with flavour and aroma similar to the Serra da Estrela cheese, although a little more acidic. It is a handmade cheese, recognized by experts from around the world (Franco 1981).

8.3.3 Processing

The traditional handmade manufacturing method of Portuguese sheep cheeses is very simple (Fig. 8.4).

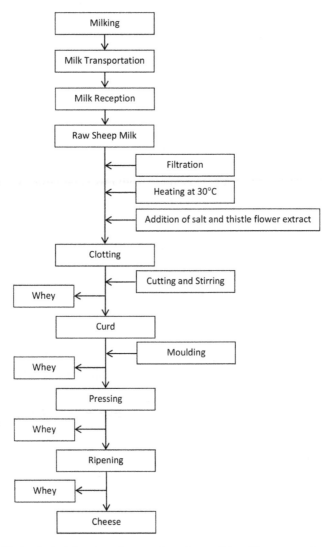

Figure 8.4 Technological flow chart for manufacturing traditional Portuguese sheep cheeses.

Milk reception

Sheep milk is the raw material for production, with milking (manual or mechanical) done twice a day. Raw milk is delivered to the producer (if the manufacturer and producer are not the same). This milk is sent directly to an isothermal tank, through a stainless steel tube and a milk pump.

Milk storage

The milk received is stored in refrigerated tanks at 4°C so that it retains all of its properties and to prevent microbiological contamination. If the temperature is kept below 4°C, the milk may be maintained for 48 hours until laboring; at a temperature between 4°C and 6°C, it can be kept for 36 hours.

Milk transportation

The milk is transported in the morning in cylinders at a temperature between 0 to 10°C.

Filtration

The milk is sieved in order to remove larger impurities (feces of the animal, etc.).

Heating

After the removal of impurities, the milk is placed in vats at a temperature of 30°C. This allows optimal conditions for the action of enzymes from the thistle flower on the casein micelles.

Addition of salt and thistle flower extract

Salt addition enables the long life of the cheese, and determines the taste of the paste; the thistle flower extract will promote milk clotting. For each litre of milk, there is an addition of 20 g of salt and 0.15 g thistle; and 5.5 litres of milk are required to produce 1 kg of cheese.

Clotting

This is the initial stage of processing milk into cheese, in which it unfolds in two phases: solid phase (curd) and liquid phase (whey), and it only ends when the "curd" is formed, e.g., when the milk solidifies in resistant form. This process takes about 60 minutes due to the action of enzymes of thistle flower and from this point on, it is ready to be worked. Precipitation or coagulation of the milk casein takes place, with a white clot formation and homogeneous texture through the action of enzymes from thistle flowers. It is necessary to take into account and control factors such as temperature, pH and the milk clotting time.

The micelle estabilization and milk clotting

To understand the dynamics of clotting, as well as the processes that occur during ripening, it is essential to know an acceptable model for the behavior of protein molecules. The protein molecules are associated in particles that are known as casein micelles and contain about 80% of milk proteins. It consists of several sub-micelles bound together by colloidal calcium phosphate (Fig. 8.5) containing α_{S1}-casein, α_{S2}-casein, β-casein and κ-casein in different proportions (Schmidt 1982, Alais 1985, Dalgleish 1999, Horne 2002, Farrell Jr. et al. 2006).

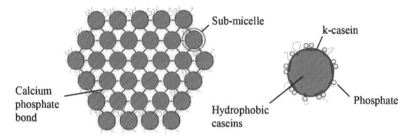

Figure 8.5 Casein micelle structure (adapted from Amiot et al. 2002).

In the central part the sub-micelles are associated by hydrophobic and electrostatic interactions, with hydrophobic regions pointed out inside and the hydrophilic zones to the surface. Therefore, the serine phosphate groups of α_{S1}-casein, α_{S2}-casein and β-casein are in the micelle surface due to interaction with calcium phosphate, and phosphate-calcium-phosphate interaction promotes the bonding of the sub-micelles. In models of structural organization of the known micelles, the κ-casein appears on the micelle surface with the terminal amine more hydrophobic, interacting with the central part of the micelles and the carboxylic terminal segment, more hydrophilic, interacting with the surrounding aqueous medium— this plays a fundamental role in the stability of the colloid (Heth and Swaisgood 1982, Schmidt 1982, Dalgleish 1999, Walstra 1999, Horne 2006). The sub-micelles with less amount of κ-casein take up more internal positions in the micelle, while having those with higher amounts of κ-casein are positioned at the periphery. So these proteins have surface properties (Chakraborty and Basak 2008).

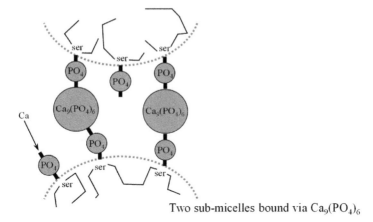

Two sub-micelles bound via Ca$_9$(PO$_4$)$_6$

Figure 8.6 Calcium phosphate bound (adapted from Alais 1985).

The α_{S1}-casein, α_{S2}-casein, β-casein and κ-casein are phosphorylated at specific serine (Ser) residues (Farrell Jr. et al. 2004). The phosphate groups are esterified in form of serine mono-ester and, this association, has the ability to strongly bind to polyvalent cations (Fig. 8.6). Due to the protein and calcium concentrations existing in milk, the α_{S1}-casein, α_{S2}-casein and β-casein have tendency to precipitate from this binding as result of charge neutralization (Farrell Jr. et al. 2004).

Calcium phosphate bound in micellar stability is important since it is this that connects the various calcium-sensitive caseins (Horne 2006) and in the sub-micellar model, this bound has a responsibility to unite the different aggregates (sub-micelles) between them. Due to its fundamental role in the stability of the micelle structure, this connection has been the subject of studies to clarify the bound mechanism (Horne 2003, 2006).

The structure of the casein micelle is sensitive to external factors, such as the presence of enzymes, pH and temperature, factors which contribute to its destabilization. The destabilization of the micelles can be promoted by physicochemical factors, such as calcium removal by chelating agents, pH decrease and increase of the temperature or by enzymes. The enzymatic process of casein micelles destabilization is of the most technological interest, being used in the first phase of manufacture of the cheese. This process is caused by specific proteolysis of the protein stabilizer micelle, κ-casein.

Clotting phases

Clotting is the initial stage of the technological process of cheesemaking and is entirely dependent on the composition of milk used and the conditions, such as the clotting agent and the temperature. Technological suitability of milk or milk technological properties essentially depends on their behavior in the clotting process because this will determine the curd behavior in subsequent phases.

Clotting, has a determining role in the characteristics of the finished product. Enzymatic clotting occurs due to physicochemical modifications in casein micelles promoted by the action of proteolytic enzymes which enable the coagulation of casein. The micellar phosphate binds with calcium in solution and calcium phosphate contributes to the aggregation of the casein micelles, leading to the formation of a protein network called clot, gel or curd (Eck 1987).

There are three distinct types of clotting, which are related to the type of coagulant used: enzymatic, acid or mixed. In traditional Portuguese sheep cheeses, the clotting of milk is due to an enzyme action.

The milk clotting by proteolytic enzymes is divided into three phases (Fig. 8.7), but the essential steps for clotting are the first two (Alais 1985, Dalgleish 1999) and may be carried out using enzymes of animal, plant and microbial origin (Alais 1985, Macedo et al. 1993, Macedo et al. 1996, Esteves et al. 2003, Gunasekaren and Ak 2003).

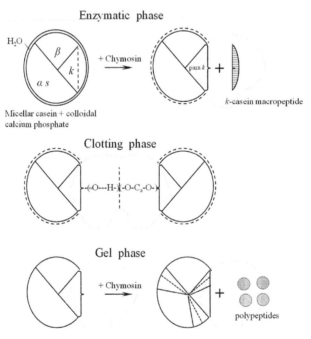

Figure 8.7 Three phases of enzymatic milk clotting (adapted from Alais 1985, St-Gelais and Tirard-Collet 2002).

Primary phase or enzymatic phase

The first phase is designated by the enzymatic phase. It occurs when a milk clotting enzyme, that promotes the hydrolysis of the peptide bond existent between position 105 and 106, consisting of amino acids phenylalanine and methionine in κ-casein (Phe105-Met106), is added. As a result of this action, κ-casein gives rise to two peptides: the κ-casein macropeptide (CMP) or the κ-casein glycomacropeptide (GCMP) which is soluble, hydrophilic, consisting of amino acids 106–169, and present in the whey and the para-κ-casein is insoluble, hydrophobic, consisting of amino acids 1–105, remains in modified micelles bound to the α_{S1}-casein, α_{S2}-casein and β-casein.

This process leads to destabilization of the micelles due to a decrease in the surface potential of the micelles with consequent weakening of the electrostatic repulsion of the protein aggregates. In addition there is also a reduction in the hydration coating of the micelles and loss of the protective effect of carbohydrates (Alais 1985, Dalgleish 1999, Gunasekaren and Ak 2003, Awad 2007).

Secondary phase or clotting phase

This phase follows the proteolysis occurred in the primary phase, starting even before the entire κ-casein can be proteolyzed. Aggregation only begins when 85 to 90% of κ-casein is hydrolyzed. Calcium ionic and the micellar calcium phosphate play a decisive role in the aggregation phenomenon (Eck 1987). In the second phase of clotting, the modified micelles tend to aggregate and form a network initially weak that then traps the fat globules, water and water soluble materials (El-Omar and Zayat 1986, Lawrence et al. 1987).

The aggregation corresponds to a gel formation by the association of residues of modified micelles, and is highly dependent on pH, temperature, content of calcium and casein, and is accelerated by the presence of cations (Van Hooydonk and Walstra 1987, Horne and Banks 2004).

The calcium phosphate linkages in the aggregation of modified casein micelles play an important role in this phase (Mellema et al. 2002, Pandey et al. 2003, Karlsson et al. 2007). The number of these links increases with the concentration of Ca^{2+} in the medium (Najera et al. 2003, Awad 2007) but on the other hand, decreases with low pH, as colloidal calcium phosphate level decreases and begins to solubilize at pH less than 5.8 (Le Graët and Gaucheron 1999, Lucey et al. 2003, Karlsson et al. 2007).

For clotting to occur, free calcium in ionic form must be available for casein (which is in the form of calcium phosphocaseinate) to bind to free calcium, and form calcium phosphoparacaseinate (Sá and Barbosa 1990).

The aggregation of casein micelles assumed by the protein network is known as curd (Walstra et al. 1999). This type of clotting leads to mineralized calcium products (Rajbhandari et al. 2013).

Tertiary phase or gel phase

This is not an essential step in clotting, but it involves successive phenomena for the action of clotting enzymes. At this stage, the general proteolysis process occurs (non specific proteolysis) affecting all caseins. In fact, it turns out that the mass per unit surface hydrophobicity keeps increasing after clotting, which indicates the existence of conformational changes, which are a result of the enzymatic hydrolysis of α_s-caseins, β-casein and κ-casein (Ono et al. 1987, Ostoa-Saloma 1990).

In parallel with this non specific proteolysis, due to the nature of the pores and the dynamics of aggregation of the micelles, the curds have a tendency to contract causing exudation of whey and retraction of the new structural rearrangement of caseins. This phenomenon is known as syneresis (Alais 1985, Walstra et al. 1999, Mellema et al. 2002, Gunasekaren and Ak 2003). The control of syneresis allows the cheese producer to control the final moisture content and water activity (a_w) of fresh curds and thus exert the decisive influence on the dynamic ripening process.

Milk clotting enzymes

There are many endopeptidases from animal, vegetable, or microbial origin, capable of hydrolyzing the linkage phenylalanine105-methionine106 of κ-casein which promotes the clotting of milk. However, this property is not sufficient for all these enzymes being suitable for promoting the clotting of milk in the cheesemaking industry. In fact, a good enzyme for cheese can be characterized by a high clotting power, and a low general proteolytic activity, that is, the enzyme will rapidly hydrolyze the linkage phenylalanine105-methionine106 of κ-casein but must have low and non specific proteolytic activity compared to other caseins. This proteolytic activity, if it is too high, has several disadvantages, such as low yield, over blandness, friable curd texture at the time of molding and the appearance of bitter flavours (Martins 1999).

Thus, the clotting may be performed by proteases from different sources. Some are not specific and have fast and intense proteolysis in any type of protein. In contrast, there are enzymes that are classified as aspartic proteases, and have as their primary action the destabilization of the colloidal casein structure, exploiting fragility of a particular peptide bond of the k-casein protein with great responsibility in the stability of casein micelles of milk. The most traditional coagulant is called rennet and consists of chymosin and pepsin animal extract from the fourth stomach of young ruminants which have not yet been weaned (Martins 1999).

There is another clotting agent also widely used in most traditional cheese factories in Mediterranean countries, especially Portugal and Spain, which is an extract from a plant, thistle (*Cynara cardunculus* L.), as previously referred. The use of thistle flower extracts in traditional Portuguese cheese factories (in sheep cheese manufacturing) dates back to the Roman occupation. This plant extract contains a mixture of enzymes (cardosin or ciprosin) with high non specific proteolytic activity, but with coagulant behavior that is very effective in sheep milk, providing a manufacture of very characteristic cheeses (Vasconcelos 1990, Martins 1999).

Enzymes from clotting agents of plant origin, especially flower extracts of *Cynara cardunculus* L., have been the subject of several studies (Faro 1991, Sousa and Malcata 1998, Fernandez-Salguero and Sanjuan 1999, Martins 1999, Sousa et al. 2001, Barros et al. 2003, Esteves et al. 2003, Roseiro et al. 2005, Silva and Malcata 2005, Tejada et al. 2007, Pereira et al. 2008). It is well established by these authors that the existing system enzyme in the flower extracts of *Cynara cardunculus* L. consists in three acid proteases forms (I, II and III).

Since forms I and II are similar, it is common to group them together and designate this group as cardosin A, and the form III as cardosin B. The enzyme complex cardosin A presents similarities with chymosin, while cardosin B is similar to pepsin. That is, the proteolytic action of cardosin B is not specific (Silva and Malcata 2005). On the other hand, the plant enzymes demonstrate a higher proteolytic ability than enzymes of animals and are specific for a larger number of peptide bonds (Faro 1991, Férnandez-Salguero and Sanjuán 1999, Sousa et al. 2001, Esteves et al. 2003, Roseiro et al. 2003c, Tavaria et al. 2003, Prados et al. 2006).

The type of clotting enzyme that is used has a great relevance due to the ratio of the specific clotting activity/not-specific proteolytic activity (C/P) that characterizes the clotting enzymes (Martins and Vasconcelos 2004). The ratio of chymosin/pepsin is important, since the power of the clotting influences the amount of enzyme to add in milk (Eck 1987). The clotting time is inversely proportional to the amount of added enzyme, so the larger amount of clotting enzyme is added, the faster is the clotting time (Eck 1987, Behmer 1995).

Milks from different species have different behaviors in relation to the clotting enzymes, and may have different abilities to form a gel. This gel requires rheological characteristics that allow, without losses, the mechanical operations needed for draining. Some clot slowly, other gels clot faster—in the latter case, the gel is firm and drains the whey well, giving the curd a good quality moisture and texture for cheeses (Eck 1987).

Sheep milk is very sensitive to the rennet since it has a high proportion of caseins β/α_s, and so clotting is faster than in cow milk (Park and Haenlein 2006).

Effect of temperature and pH

In the coagulation of milk, temperature control must be considered, because below 10°C there is no micellar aggregation and, at temperatures above 65°C there may be no clotting by denaturing of the enzyme (Eck 1987). Optimal clotting temperatures are around 28 to 37°C (Sá and Barbosa 1990).

In this process, pH should be maintained and controlled as pH change may influence the clotting time and the firmness of the gel which are the parameters most sensitive to pH changes. The ideal pH would be 5.5; with low pH, the clotting time is shorter and stronger gel is formed. With values above pH 7, clotting does not occur because the enzyme is inactivated (Eck 1987).

The most important factors in the enzymatic phase are the effect of enzyme concentration, pH and temperature. In the micellar aggregation or clotting phase there is no enzymatic influence, but it is extremely sensitive to calcium ion concentration and temperature, with small influence by the pH (Martins and Vasconcelos 2004).

The technological ability of milk is related to milk behavior associated with cheese production techniques in order to process the milk into cheese. However, for manufacture of quality cheeses, the type of technological process is not the only factor to be taken into account. Other factors influence the milk behaviour—its physical and chemical characteristics and its hygienic/sanitary conditions.

Cutting the Curd—This provides a larger free surface, facilitating the whey output. The cutting and stirring of curd, done with lire and agitators, increases the surface for draining of the clot due to its reduction into smaller fragments. This is to remove the natural whey, which can be used later to make cottage cheese; when more whey is removed, the cheese produced is harder and drier. The temperature of the curd should be between 0°C and 10°C.

Draining can also be defined as the removal of whey; at this stage the mass will be squeezed in "francela". Draining is very important for the quality and level of preservation of the cheese. The whey should be removed fast and smoothly. The work on the mass will remove sugar and moisture excess.

Molding—in this process, the curd mass is placed in molds called "cincho" and pressed manually until the whey is removed. Then, the mass is squeezed slowly to give the cheese mass form—the form/format depends on the cheese mold.

Pressing—This allows to improve the consistency, texture and shape of the cheese, completely eliminating the whey excess. The cheese is pressed for approximately two hours and then removed from the forms. In same industrial processes the pressing is done by a pneumatic press for four hours at three bar. At this point it will have the shape of the used mold, and it is then placed in a cloth strap, which is changed every time it is washed, over a period of 40 days. Processes with higher pressure will promote harder cheese mass.

Ripening/Curing—Ripening is carried out in chambers with adequate ventilation and control of humidity and temperature after pressing. In the first stage of ripening, the cheeses are placed in a chamber for about 15 to 20 days at 6 to 7°C at 95% of humidity and are washed daily. This is the first curing step. The low temperature and high humidity are important for the control of the action of microorganisms and the formation of peel in the mass.

In the second ripening phase, the cheeses are placed through a second climatic chamber in an atmosphere with a lower humidity (85%), where they remain for 15 to 20 days at a temperature of 11 to 12°C. Then, the cheese is washed in warm water and lightly salted. This completes the ripening stage. The slight increase of temperature and aeration and the slight decrease in moisture help the cheese to form its buttery folder and its typical yellowish crust. The cheese is now ready for consumption. Since Serra da Estrela cheese is not a fresh product, it needs an extra 45 days to become perfectly cured. In the case of old Serra da Estrela cheese, the manufacturing process is exactly the same, varying only at the ripening time which takes about 120 days. Often some cheeses have an orange colour, which is due to paprika and olive oil, however this is not a mandatory practice (Reis and Malcata 2007).

This type of cheese has a soft paste, and is easy to spread due to intense proteolysis in the ripening phase. The production technology of these cheeses is described in legal regulations and there have been a few studies to establish artisanal technology and compare it with a slightly mechanized technology in the context of Serpa cheese (Canada 2001, Roseiro et al. 2003b) and the Serra da Estrela cheese technologies (Reis and Malcata 2007).

In the production of this type of cheese, the used raw milk may contain pathogens. However, in cheeses made from raw milk, it is not possible to establish this important step in terms of food safety. This is because there are natural mechanisms to promote the safety of the final products. In the ripening phase, the beneficial microorganisms in the cheese will ferment lactose under the action of air oxygen and produce antimicrobials such as lactic acid, carbon dioxide, hydrogen peroxide, ethanol, and promote the reduction of toxic compounds such as aflatoxins (Giraffa 2004), diacetyl (Veljovic et al. 2007) and bactericidal (Nuñez et al. 1997), that facilitate inhibition or the elimination of pathogenic microorganisms. The production of lactic acid during this period decreases the pH, promotes the expulsion of whey from curd, inhibits the growth of pathogenic bacteria, ensures the predominance of lactic acid bacteria and automatically decreases the risk of microbial contamination (Canada 2001).

Since the fermentation process is not instantaneous, in terms of food safety, these cheeses have a minimum ripening time, which in most Portuguese cheeses obtained from raw milk is 30 days. Ripening is a critical period, in which the production of antimicrobial compounds occurs, and thus, it contributes for food safety.

Prepackaging and packaging—This step should be carried out by the manufacturer in accordance with good hygienic practices and be commercially viable without being called into question its features during the storage and sale. Cheeses are not always pre packed, however, this practice is mandatory if the cheeses are cut. After ripening, the cheeses are labeled and packaged in cardboard boxes for further distribution.

Labelling, storage, distribution and marketing—The Serra da Estrela, Serpa and Azeitão cheeses are PDO cheeses, as previously referred and so, are strictly labeled.

Long storage periods can be carried out at a temperature below 1°C, if authorized and controlled by the private organization for control and certification. In a normal situation, the cheese must be kept between 0 and 5°C avoiding the spreading of undesirable microorganisms.

The transportation of the product to the point of sale must be carried out at temperatures below 10°C. Sometimes it may be sold by the producer at fairs or exhibitions.

8.3.4 Milk composition and cheese yield

Since milk costs approximately 85% of the cost of producing cheese, it is of extreme importance to know the milk composition and how many cheeses is possible to obtain from a specific milk source. Cheese is a product in which the protein and fat of the milk are concentrated, so it is clear that cheese yield is related to casein and fat content of the milk (Van Boekel 1994).

The majority of the whey proteins, lactose and water are separated out in the form of whey. Since the moisture portion of cheese is actually whey, small amounts of whey proteins and lactose are retained in the final cheese, in the same proportion of the moisture content. However, it is obvious that the fat and casein content of the milk will be the key constituents of milk that will contribute to the yield of the cheese. Several studies on the stability of frozen sheep milk (Wendorff 2001) revealed problems with destabilized casein in sheep milk that had been frozen and stored in home freezers. The destabilized casein did not impair the coagulation process but there was a significant loss in cheese yield from that frozen milk.

Fat (F) and casein (C) are the two primary milk components that are recovered in the cheesemaking process and are directly related to cheese yield. Since casein is the key component in making up the curd matrix that entraps the fat globules, observing casein relationships with other milk constituents allows predicting the potential cheese quality and cheese yield. The C/F ratio is critical in controlling the final fat in the dry matter (FDM) of the finished cheese. Minimum FDM specifications are established for many of the cheeses with standards of identity. The casein/total protein ratio will supply some potential information on the amount of intact casein that is present in the milk to give a good gel structure during curd formation (Economides et al. 1987, Alichanidis and Polychroniadou 1996, Pirisi et al. 2000, Jaeggi et al. 2001).

8.3.4.1 Sheep cheese yields

Cheese yields for sheep milk have been reported in literature by various researchers in the following manner:

1) Gross cheese yield after 1 day, lb/100 lb (Barron et al. 2001).
2) Adjusted cheese yield to x percent of moisture (Pirisi et al. 1999).
3) Quantity of milk (kg) necessary to make 1 kg of full fat cheese (Kukovics et al. 1999).
4) $Y = -0.20 + 0.011$ fat $+ 0.025$ protein (Economides and Mavrogenis 1987).
5) Yield $= (0.93F + C-0.1)1.09/100$-W (Van Slyke and Price 1979).
   where:
   > F = fat concentration in the milk, percentage
   > C = casein concentration in the milk, percentage
   > W = moisture, expressed as kg water per kg of cheese
6) Yield $= (0.85$ percentage of fat$) + ($percentage of casein $- 0.1)1.13/1 - ($cheese moisture$/100)$ (Barbano 1996).
7) Yield $= [(RFxF) + (RCxC)]RS/1.0 - W$ (Kerrigan and Norback 1986)
   where:
   > RF = fat retention factor
   > F = percentage of fat in the milk
   > RC = casein retention factor
   > C = percentage of casein in the milk
   > RS = solids retention factor
   > W = moisture in the final cheese (percentage)

Cheese yield formulas are very useful in pricing milk and providing an estimate of cheese yield under defined conditions. Calculated yields are only theoretical cheese yields. The actual cheese yields obtained may also be influenced by processing conditions in the factory. Some of the factors that the cheesemaker has control over are as follows:

1. Storage of milk
2. Milk standardization
3. Heat treatment of milk
4. Homogenization
5. Type of coagulant
6. Curd firmness at cut
7. Salt addition
8. Moisture loss during ripening.

Good process control, record keeping and consistent cheesemaking operations allow the cheesemaker to be able to have uniform composition and cheese yields throughout the season. This should then result in an efficient and profitable business for the dairy sheep industry (Wendorff 2002).

References

Alichanidis, E. and A. Polychroniadou. 1996. Special features of dairy products from ewe and goat milk from the physicochemical and organoleptic point of view. pp. 21–43. *In*: Proceedings of IDF/CIVRAL, Seminar on Production and Utilization of Ewe and Goat Milk, Crete, Greece, International Dairy Federation, Brussels, Belgium.

Alichanidis, E. and A. Polychroniadou. 1997. Special features of dairy products from ewe and goat milk from the physicochemical and organoleptic point of view. Sheep Dairy News 14: 11–18.

Amiot, J., F. Fournier, Y. Lebeuf, P. Paquin and R. Simpson. 2002. Composition, propriétés physicochimiques, valeur nutritive, qualité technologique et technique d'analyse du lait. pp. 1–73. *In*: Science et Technologie du lait : Transformation du lait. Presses internationales Polytechnique, Montréal, Canada.

Assenat, L. 1991. Leche de oveja: composición y propiedades. *In*: F.M. Luquet, J. Keilling and R. de Wilde (eds.). Leche y productos lácteos: vaca–oveja–cabra. Editorial Acribia, Zaragoza, Spain.

Assis, A.J., J.M.S. Campos, S.C. Valadares Filho, A.C. Queiroz, R.P. Lana, R.F. Euclydes, J.M. Neto, A.L. Rodrigues Magalhães and S.S. Mendonça. 2004. Polpa cítrica em dietas de vacas em lactação. 1. Consumo de nutrientes, produção e composição do leite. Revista Brasileira de Zootecnia 33(1): 242–250.

Awad, S. 2007. Effect of sodium chloride and pH on the rennet coagulation and gel firmness. Food Science and Technology 40: 220–224.

Barbano, D.M. 1996. Mozzarella cheese yield: Factors to consider. pp. 28–38. *In*: Proceedings Seminar on Maximizing Cheese Yield. Center for Dairy Research, Madison, Wisconsin, USA.

Barbosa, M. 1990. The production and processing of sheep milk in Portugal: Serra da Estrela cheese. *In*: J. Bougler and J.L. Tisser (eds.). Les petits ruminants etleurs productions laitières dans la region méditerranéenne. Montpellier: CIHEAM (Options Méditerranéennes: Série A. SéminairesMéditerranéens, 12).

Barron, L.J.R., E.F. de Labastida, S. Perea, F. Chavarri, C. de Veja, M.S. Vicente, M.I. Torres, A.I. Najera, M. Virto, A. Santisteban, F.J. Perez-Elortaondo, M. Albisu, J. Salmeron, C. Mendia, P. Torre, F.C. Ibanez and M. de Renobales. 2001. Seasonal changes in the composition of bulk raw ewe milk used for Idiazabal cheese manufacture. International Dairy Journal 11: 771–778.

Barros, R.M., C.I. Extremina, I.C. Goncalves, B.O. Braga, V.M. Balcao and F.X. Malcata. 2003. Hydrolysis of α-lactalbumin by cardosin A immobilized on highly activated supports. Enzyme and Microbial Technology 33: 908–916.

Bencini, R. and G. Pulina. 1997. The quality of sheep milk: a review. Australian Journal of Experimental Agriculture 37(5): 485–504.

Bencini, R. and I.W. Purvis. 1990. The yield and composition of milk from Merino sheep. Proceedings of the Australian Society of Animal Production, Melbourne, Autralia 18: 144–147.

Bencini, R. 2001. Factors affecting the quality of ewe milk. *In*: Proceedings of 7th Great Lakes dairy sheep symposium, Wisconsin Sheep Breeders Cooperative. EauClaire. Wisconsin, USA.

Behmer, M.L. 1995. Tecnologia do Leite: leite, queijo, manteiga, caseína, iogurte, sorvete e instalações, produção industrialização e análise 15ª edição. Editora Nobel, São Paulo, Brazil. 320 pp.

Boyazoglu, J. and P. Morand-Fehr. 2001. Mediterranean dairy sheep and goat products and their quality: a critical review. Small Ruminant Research 40: 1–11.

Canada, J.S.B. 2001. Caracterización Sensorial y Físico-Química del Queijo Serpa. Ph.D. Thesis, Universidad de Extremadura, Cáceres, Spain.

Campos, L. 2011. Aspectos benéficos do leite de ovelha e seus derivados. Informativo Casa da Ovelha. Available in: http://www.casadaovelha.com.br/files/pesquisa_tecno_cientifica.pdf. Accessed on: June 16, 2015.

Chakraborty, A. and S. Basak. 2008. Effect of surfactants on casein structure: a spectroscopic study. Colloids and Surfaces B: Biointerfaces 63: 83–90.

Dalgleish, D.G. 1999. The enzymatic coagulation of milk. *In*: P.F. Fox (ed.). Cheese: Chemistry, Physics and Microbiology. Aspen Publishers, INC, Gaithersburg, USA.

DR. 1987. Decreto Regulamentar Nº 39/87—Cria a Região Demarcada de Queijo Serpa. Diário da República—I Série—No. 146 de 29 de Junho, 2499–2500.

Eck, A. 1987. O Queijo-Vol.1. Publicações Europa-América Lda., Mem-Martins, Portugal. 337 pp.

Esteves, C.L.C., J.A. Lucey, D.B. Hyslop and E.M.V. Pires. 2003. Effect of gelation temperature on the properties of skim milk gels made from plant coagulants and chymosin. International Dairy Journal 13: 877–885.

Economides, S.E.G. and A.P. Mavrogenis. 1987. The effect of different milks on the yield and chemical composition of Halloumi cheese. Technical Bulletin—Cyprus Agricultural Research Institute 90: 2–7.

Faro, C.C. 1991. Purificação e Caracterização Físico-Química da Protease de *Cynara cardunculus* L. Ph.D. Thesis, Universidade de Coimbra, Coimbra, Portugal.

Farrell, Jr., H.M., E.L. Malin, E.M. Brown and P.X. Qi. 2004. Nomenclature of the proteins of cow milk-sixth revision. Journal of Dairy Science 87: 1641–1674.

Farrell, Jr., H.M., E.L. Malin, E.M. Brown and P.X. Qi. 2006. Casein micelle structure: What can be learned from milk synthesis and structural biology? Current Opinion in Colloid & Interface Science 11: 135–147.

Férnandez-Salguero, J. and E. Sanjuan. 1999. Influence of vegetable and animal rennet on proteolysis during ripening in ewe milk cheese. Food Chemistry 64: 177–183.

Franco, F.P.S. 1981. O queijo Azeitão. Available in: http://www.azeitao.net/azeitao/queijo/. Accessed on: April 27, 2015.

Freitas, A.C., A.C. Macedo and F.X. Malcata. 2000. Review: technological and organoleptic issues pertaining to cheeses with denomination of origin manufactured in the Iberian Peninsula from ovine and caprine milks. Food Science and Technology International 6: 351–370.

Furtado, M.M. 2003. Queijos finos maturados por fungos. Milkbizz, São Paulo, Brazil.

Giraffa, G. 2004. Studying the dynamics of microbial populations during food fermentation. Microbiology Reviews 28: 251–260.

Gonzalo, C., J.A. Carriedo, J.A. Baro and F. San Primitivo. 1994. Factors influencing variation of test day milk yield, somatic cell count, fat, and protein in dairy sheep. Journal of Dairy Science 77(6): 1537–1542.

Gunasekaren, S. and M.M. Ak. 2003. Cheese Rheology and Texture, CRC Press, London, UK.

Gutiérrez, R.B. 1991. Elaboración artesanal de quesos de oveja. Comunidad del Sur, Montevideo, Uruguay.

Haenlein, G.F.W. 2001. The nutritional value of sheep milk. International Journal of Animal Science 16: 253–268.

Harbutt, J.A. 1999. Complete illustrated guide to the cheeses of the world. Annes Publishing Inc. New York, USA.

Heth, A.A. and H.E. Swaisgood. 1982. Examination of casein micelle structure by a method for reversible covalent immobilization. Journal of Dairy Science 65: 2047–2054.

Horne, D.S. 2002. Casein structure, self assembly and gelation. Current Opinion in Colloid & Interface Science 7: 456–461.

Horne, D.S. 2003. Casein micelles as hard spheres: limitations of the model in acidified gel formation. Colloids and Surfaces A: Physicochemical and Engineering Aspects 213: 255–263.

Horne, D.S. 2006. Casein micelle structure: Models and muddles. Current Opinion in Colloid & Interface Science 11: 148–153.

Horne, D.S. and J.M. Banks. 2004. Rennet-induced coagulation of milk. *In*: P.F. Fox, P.L.H. McSweeney, T.M. Cogan and T.P. Guinee (eds.). Cheese: Chemistry, Physics and Microbiology. 3rd Edition. Elsevier Academic Press, San Diego, USA.

INSA—Instituto Nacional de Saúde. Dr. Ricardo Jorge 2007. Tabela de Composição de Alimentos. Centro de Segurança Alimentar e Nutrição do Instituto Nacional de Saúde Dr. Ricardo Jorge. Lisboa, Portugal.

Jandal, J.M. 1996. Comparative aspects of goat and sheep milk. Small Ruminant Research, Amsterdam 22(2): 177–185.

Jarheis, G., J. Fritsche and J. Kraft. 1999. Species dependant, seasonal, and dietary variation of conjugated linoleic acid in milk. *In*: M.P. Yurawecz, M.M. Mossoba, L.K.G. Kramer, M.W. Pariza and G.J. Nelson (eds.). Advances in Conjugated Linoleic Acid. Champaign: American Oil Chemistry Society.

Jaeggi, J.J., Y.M. Berger, M.E. Johnson, R. Govindasamy-Lucey, B.C. McKusik, D.L. Thomas and W.L. Wendorff. 2001. Evaluation of sensory and chemical properties of Manchego cheese manufactured from ovine milk of different somatic cell levels. pp. 84–93. *In*: Proceedings of the 7th Great Lakes Dairy Sheep Symposium, Eau Claire, WI, University of Wisconsin-Madison, USA.

Jooyandeh, H. and A. Aberoum. 2010. Physicochemical, nutritional, heat treatment effects and dairy product aspects of goat and sheep milks. World Applied Science Journal 11: 1316–1322.

Karlsson, A.O., R. Ipsen and Y. Ardo. 2007. Influence of pH and NaCl on rheological properties of rennet-induced casein gels made from UF concentrated skim milk. International Dairy Journal 17: 1053–1062.

Kerrigan, G.L. and J.P. Norback. 1986. Linear programming in the allocation of milk resources for cheese making. Journal of Dairy Science 69: 1432–1440.

Kukovics, S., L. Daroczi, P. Kovacs, A. Molnar, I. Anton, A. Zsolnai, L. Fesus, M. Abraham and F. Barrillet. 1999. The effect of β-lactoglobulin genotype on cheese yield. pp. 524–527. *In*: Proceedings of the 6th International Symposium on the Milking of Small Ruminants, Athens, Greece. EAAP Publ. N° 95, WageningenPers, Wageningen, Netherlands.

Lawrence, R.C., L.K. Creamer and J. Gilles. 1987. Texture development during cheese ripening. Journal of Dairy Science 70: 1748–1760.

Le Graët, Y. and F. Gaucheron. 1999. pH-induced solubilization of minerals from casein micelles: Influence of casein concentration and ionic strength. Journal of Dairy Research 66: 215–224.

Lucey, J.A., M.E. Johnson and D.S. Horne. 2003. Invited review: Perspectives on the basis of the rheology and texture properties of cheese. Journal of Dairy Science 86: 2725–2743.

Lourenço Neto, J.P.M. 1998. Leite resfriado: matéria-prima da queijaria moderna. Revista Leite e Derivados 41: 18–34.

Macedo, I.M.Q., C.J.F. Faro and E.M.V. Pires. 1993. Specificity and kinetics of the milk-clotting enzyme from cardoon (*Cynara cardunculus* L.) toward bovine κ-casein. Journal of Agricultural and Food Chemistry 41: 1537–1540.

Macedo, A.C., M.L. Costa and F.X. Malcata. 1996. Assessment of proteolysis and lipolysis in Serra cheese: effects of axial cheese location, ripening time and lactation season. Lait 76: 363–370.

Martins, A.P.L. 1999. A flor de cardo (*Cynara cardunculus* L.) como agente coagulante no fabrico do queijo. Caracterização e influência dos processos de conservação na actividade coagulante. Ph.D. Thesis. Universidade Técnica de Lisboa. Instituto Superior de Agronomia. Lisboa, Portugal.

Martins, A.P.L. and M.M. Vasconcelos. 2004. A qualidade do queijo fabricado com leite cru. Efeito dos principais factores tecnológicos. Pastagens e Forragens 25: 15–33.

Mayer, H.K. and G. Fiechter. 2012. Physical and chemical characteristics of sheep and goat milk in Austria. International Dairy Journal 24(2): 57–63.

Mellema, M., P. Walstra, J.H.J. van Opheusden and van T. Vliet. 2002. Effects of structural rearrangements on the rheology of rennet-induced casein particle gels. Advances in Colloid and Interface Science 98: 25–50.

Meunier-Goddik, L. and H. Nashnush. 2006. Producing Sheep Milk Cheese. OSU Extension Publication. EM 8908.

Najera, A.I., M. de Renobales and L.J.R. Barron. 2003. Effects of pH, temperature, CaCl$_2$ and enzyme concentrations on the rennet-clotting properties of milk: a multifactorial study. Food Chemistry 80: 345–352.

NP 1922 (1985). Queijo Serra da Estrela. Definição, Características e Marcação. Instituto Portugues de Qualidade, Lisboa.

Nudda, A., R. Bencini, S. Mijatovic and G. Pulina. 2002. The yield and composition of milk in Sarda, Awassi, and Merino sheep milked unilaterally at different frequencies. Journal of Dairy Science 85(11): 2879–2884.

Nuñez, M., J.L. Rodriguez, E. Garcia, P. Gaya and M. Medina. 1997. Inhibition of *Listeria monocytogenes* by enterocin4 during the manufacture and ripening of Manchego cheese. Journal of Applied Microbiology 83: 671–677.

Omar, M.M. and A.I. El-Zayat. 1986. Ripening changes of Kashkaval cheese made from cow milk. Food Chemistry 22: 83–94.

Ono, T., R. Yada, K. Yutani and S. Nakai. 1987. Comparison of conformations of κ-casein, para-κ-casein and glycomacropeptide. Biochimica et Biophysica Acta 911: 318–325.

Ostoa-Saloma, P., J. Ramirez and R. Perez-Montfort. 1990. Causes of the decrease in fluorescence due to proteolysis of α-casein. Biochimica et Biophysica Acta 1041: 146–152.

Pandey, P.K., H.S. Ramaswamy and D. St-Gelais. 2003. Effect of high pressure processing on rennet coagulation properties of milk. Innovative Food Science and Emerging Technologies 4: 245–256.

Park, Y.W. and G.F.W. Haenlein. 2006. Overview of milk of non-bovine mammals. *In*: Y.W. Park and G.F.W. Haenlein (eds.). Handbook of Milk of Non-bovine Mammals. Blackwell Publishing Professional, Ames, Iowa, USA.

Park, Y.W., M. Juárez, M. Ramos and G.F.W. Haenlein. 2007. Physicochemical characteristics of goat and sheep milk. Small Ruminant Research 68(1): 88–113.

Pavić, V., N. Antunac, B. Mioč, A. Ivanković and J.L. Havranek. 2002. Influence of stage of lactation on the chemical composition and physical properties of sheep milk. Czech Journal of Animal Science 47: 80–84.

Peeters, R., N. Buys, L. Robijns, D. Vanmontfort and J. van Isterdael. 1992. Milk yield and milk composition of Flemish milksheep, Suffolk and Texel ewes and their crossbreeds. Small Ruminant Research 7(4): 279–288.

Pereira, C.I., E.O. Gomes, A.M.P. Gomes and F.X. Malcata. 2008. Proteolysis in model Portuguese cheeses: effects of rennet and starter culture. Food Chemistry 108: 862–868.

Pinheiro, C., C. Bettencourt, C. Matos and G. Machado. 2003. Efeito da raça na composição do leite de ovelha utilizado no queijo serpa. XIII Congresso de Zootecnia, published on CD, without page numbers. Universidade de Évora/APEZ. Évora, Portugal.

Pirisi, A., G. Piredda, C.M. Papoff, R. di Salvo, S. Pintus, G. Garro, P. Ferranti and L. Chianese. 1999. Effects of sheep αs1-casein CC, CD and DD geneotypes on milk composition and cheesemaking properties. Journal of Dairy Research 66: 409–419.

Pirisi, A., G. Piredda, M. Corona, M. Pes, S. Pintus and A. Ledda. 2000. Influence of somatic cell count on ewe milk composition, cheese yield and cheese quality. pp. 47–59. *In*: Proceedings of the 6th Great Lakes Dairy Sheep Symposium, Guelph, Ontario, University of Wisconsin-Madison, USA.

Prados, F., A. Pino, F. Rincon, M. Vioque and J. Fernandez-Salguero. 2006. Influence of the frozen storage on some characteristics of ripened Manchego type cheese manufactured with a powdered vegetable coagulant and rennet. Food Chemistry 95: 677–682.

Rajbhandari, P., J. Patel, E. Valentine and P.S. Kindstedt. 2013. Effect of storage temperature on crystal formation rate and growth rate of calcium lactate crystals on smoked Cheddar cheeses. Journal of Dairy Science 96(6): 3442–3448.

Raynal-Ljutovac, K., G. Lagriffoul, P. Paccard, I. Guillet and Y. Chilliard. 2008. Composition of goat and sheep milk products: an update. Small Ruminant Research 79: 57–72.

Reis, P.J.M. and F.X. Malcata. 2007. Improvements in small scale artisanal cheesemaking via a novel mechanized apparatus. Journal of Food Engineering 82: 11–16.

Roseiro, L.B., M. Garcia-Risco, M. Barbosa, J.M. Ames and R.A. Wilbey. 2003a. Evaluation of Serpa cheese proteolysis by nitrogen content and capillary zone electrophoresis. International Journal of Dairy Technology 56: 99–104.

Roseiro, L.B., R.A. Wilbey and M. Barbosa. 2003b. Serpa cheese: technological, biochemical and microbiological characterisation of a PDO ewe milk cheese coagulated with *Cynara cardunculus* L. Lait 83: 469–481.

Roseiro, L.B., J.A. Gomez-Ruiz, M. Garcia-Risco and E. Molina. 2003c. Vegetable coagulant (*Cynara cardunculus*) use evidenced by capillary electrophoresis permits PDO Serpa cheese authentication. Lait 83: 343–350.

Roseiro, L.B., D. Viala, J.M. Besle, A. Carnat, D. Fraisse, J.M. Chezal and J.L. Lamaison. 2005. Preliminary observations of flavonoid glycosides from the vegetable coagulant *Cynara* L. in PDO cheeses. International Dairy Journal. The Fourth IDF Symposium on Cheese: Ripening, Characterization and Technology 15: 579–584.

Sá, F.V. and M. Barbosa. 1990. Leite e os seus produtos. Clássica Editora, Nova Coleção Técnica Agrária. Lisboa, Portugal.

Schmidt, D.G. 1982. Association of caseins and casein micelle structure. *In*: P.F. Fox (eds.). Developments in Dairy Chemistry-1-proteins. Applied Science Publishers, London, UK.

Sevi, A., M. Albenzio, R. Marino, A. Santillo and A. Muscio. 2004. Effects of lambing season and stage of lactation on ewe milk quality. Small Ruminant Research 51(3): 251–259.

Silva, S.V. and F.X. Malcata. 2005. Studies pertaining to coagulant and proteolytic activities of plant proteases from *Cynara cardunculus*. Food Chemistry 89: 19–26.

Sousa, M.J., Y. Ardo and P.L.H. McSweeney. 2001. Advances in the study of proteolysis during cheese ripening. International Dairy Journal 11: 327–345.

Sousa, M.J. and F.X. Malcata. 1998. Proteolysis of ovine and caprine caseins in solution by enzymatic extracts from flowers of *Cynara cardunculus*. Enzyme and Microbial Technology 22: 305–314.

St-Gelais, D. and P. Tirard-Collet. 2002. Fromage. Dans : Science et technologie du lait : transformation du lait. Presses internationales Polytechnique, Montréal, Canada. pp. 349–415.

Tavaria, F.K., I. Franco, F.J. Carballo and F.X. Malcata. 2003. Amino acid and soluble nitrogen evolution throughout ripening of Serra da Estrela cheese. International Dairy Journal 13: 537–545.

Tejada, L., R. Gomez and J. Fernandez-Salguero. 2007. Sensory characteristics of ewe milk cheese made with three types of coagulant: calf rennet, powdered vegetable coagulant and crude aqueous extract from *Cyanara cardunculus*. Journal of Food Quality 30: 91–103.

Timperley, C. and C. Norman. 1997. The Cheese Book, Salamander Press. Atlanta, Georgia, USA. 119 pp.

Veljovic, K., A. Terzic-Vidojevic, M. Vukasinovic, I. Strahinic, J. Begovic, J. Lozo, M. Ostojic and L. Topisirovic. 2007. Preliminary characterization of lactic acid bacteria isolated from Zlatar cheese. Journal of Applied Microbiology 103: 2142–2152.

Van Boekel, M.A.J.S. 1994. Transfer of milk components to cheese: Scientific considerations, 19–28. In Proceedings of IDF Seminar on Cheese Yield and Factors Affecting its Control, Cork, Ireland, International Dairy Federation, Brussels, Belgium.

Van Hooydonk, A.C.M. and P. Walstra. 1987. Interpretation of the kinetics of the renneting reaction in milk. Netherlands Milk and Dairy Journal 41: 19–47.

Van Slyke, L.L. and W.V. Price. 1979. Cheese Ridgeview Publishing Company, Atascadero, California, USA. 522 pp.

Vasconcelos, M.M. 1990. Estudo do queijo de Azeitão. Melhoramento da tecnologia tradicional e sua influência nas características do queijo. Access proof to Auxiliary Researcher. NTLD, ENTPA, Instituto Nacional de Investigação Agrária. Lisboa, Portugal.

Walstra, P. 1999. Casein sub-micelles: do they exist? International Dairy Journal 9: 189–192.

Walstra, P., T.J. Geurts, A. Nooman, A. Jellema and M.A.J.S. van Boekel. 1999. Dairy Technology—Principles of Milk Properties and Processes, Marcel Dekker, Inc., New York, USA. 794 pp.

Wendorff, W.L. 2001. Freezing qualities of raw ovine milk for further processing. Journal of Dairy Science 84(E Suppl.): E74–E78.

Wendorff, B. 2002. Milk composition and cheese yield. pp. 104–117. In: Proceedings of the 8th Great Lakes Dairy Sheep Symposium, Ithaca, New York, USA.

Zamiri, M.J., A. Qotbi and J. Izadifard. 2001. Effect of daily oxytocin injection on milk yield and lactation length in sheep. Small Ruminant Research 40(2): 179–185.

Zhang, R., A.F. Mustafa and X. Zhao. 2006a. Effects of feeding oilseeds rich in linoleic and linolenic fatty acids to lactating ewes on cheese yield and on fatty acid composition of milk and cheese. Animal Feed Science and Technology 127: 220–233.

Zhang, R., A.F. Mustafa, K.F. Ng-Kwai-Hang and X. Zhao. 2006b. Effects of freezing on composition and fatty acid profiles of sheep milk and cheese. Small Ruminant Research 64: 203–210.

CHAPTER 9

Portuguese Galega Kale

Teresa R.S. Brandão,[1,*] *Fátima A. Miller,*[1] *Elisabete M.C. Alexandre,*[2]
Cristina L.M. Silva[1] and *Sara M. Oliveira*[1]

9.1 Introduction

Cabbage vegetables are considered valuable food products, characterized by good nutritional and healthy properties. Botanically, those vegetables are included in the family *Brassicaceae*, which contains well known species such as *Brassica oleracea* (e.g., kale, cabbage, broccoli and cauliflower). One of these vegetables that plays an important role in the Portuguese diet is the Galega kale (*Brassica oleracea* L. var. *acephala*) (Martínez et al. 2010). It is consumed cooked and used as a soup ingredient. In Portugal, it is usually very thinly shredded as the principal ingredient of the traditional soup called "caldo verde".

This leafy vegetable has long petioles and large midrib bed leaves (Fonseca et al. 2002, Fig. 9.1a). It has excellent adaptation to climatic conditions and it is commonly integrated in the traditional small farming cropping system. The shredded Galega kale is commonly prepared at the retail level prior to display for sale at ambient store temperature, and its shelf life is very short—typically one day or less (Fonseca 2001). Shredded leaves may also be commercially available in polymeric bags with perforated or non-perforated or in trays covered by a flexible film (Beaulieu et al. 1997a).

If package selection is not adequate, fast spoilage of the product may happen due to the occurrence of favorable conditions to anaerobiosis. Some commercial packs of shredded Galega kale available in the market do not maintain their quality and, even worse, are not safe for health.

[1] CBQF – Centre of Biotechnology and Fine Chemistry, State Associated Laboratory, Faculty of Biotechnology, Catholic University of Portugal, Porto, Portugal.
[2] QOPNA – Organic Chemistry, Natural and Agrofood Products research unit, Department of Chemistry, University of Aveiro, Aveiro, Portugal.
* Corresponding author: tbrandao@porto.ucp.pt

Figure 9.1 Galega kale a) raw leaf b) shredded leaves.

Condensation is visible and critical levels of O_2, which would lead to anaerobic respiration, have been detected (Beaulieu et al. 1997b).

As mentioned, Galega kale is presented to consumers as raw or shredded (Fig. 9.1b) and usually only domestic preservation techniques are applied. This includes decontamination washings and refrigerated or frozen storage prior to cooking. The effect of several processes such as blanching, cooking and low temperature preservation on some bioactive and nutritive compounds of Galega kale has been reported in some works. Some interesting research has also arisen for dried shredded kale, showing promising results. Nevertheless, there is still a lack of scientific effort devoted to optimization of the processing methods, adequate treatments applied before storage and impacts on overall quality.

This chapter provides an overview of the nutritional profile of Galega kale (a rich source of healthy bio compounds), traditional and innovative treatments that can be applied to preservation, and potential impacts of processing and storage on the quality of fresh cut kale.

9.2 Composition

Many studies reported the nutritive value, high antioxidant and healthy potential of *Brassica* vegetables. Galega kale, in particular, has a high concentrations of vitamins, mineral compounds and phytochemicals, such as chlorophylls, carotenoids, flavonoids and glucosinolates that are known to have antioxidant and anticarcinogentic properties. The general chemical composition of Galega kale raw and cooked (boiled and drained without salt) is presented in Table 9.1. The nutrient composition of this vegetable compares favourably with other consumed brassicas. It has a high protein level, characterized by the great share of different amino acids. The basic part of carbohydrates is fibre, which are mainly insoluble hemicelluloses and soluble pectin. The fat content is mainly due to the presence of α-linolenic (which corresponds to nearly 50% of the total fatty acid content), linoleic acid and palmitic acid. These

Table 9.1 Major composition of Galega kale. Nutritional values are presented per 100 g of raw and cooked product (USDA Nutrient Database for Standard Reference).

Nutrient	Raw (value per 100 g)	Cooked (value per 100 g)
Energy, kcal/kJ	49/207	28/117
Water, g	84.04	91.20
Protein, g	4.28	1.90
Total lipid (fat), g	0.93	0.40
Ash, g	2.01	0.87
Carbohydrate, g	8.75	5.63
Dietary fibre, g	3.6	2.0
Total sugars, g	2.26	1.25
Minerals		
Ca, mg	150	72
Fe, mg	1.47	0.90
Mg, mg	47	18
P, mg	92	28
K, mg	491	228
Na, mg	38	23
Zn, mg	0.56	0.24
Cu, mg	1.50	0.16
Mn, mg	0.66	0.42
Se, mg	0.9	0.9
Vitamins		
Vitamin C (total ascorbic acid), mg	120.0	41.0
Thiamin, mg	0.110	0.053
Riboflavin, mg	0.130	0.070
Niacin, mg	1.000	0.500
Pantothenic acid, mg	0.091	0.049
Vitamin B-6, mg	0.271	0.138
Total folate, µg	141	13
Total choline, mg	0.8	0.4
Vitamin A, µg	500	681
β-carotene, µg	5927	8173
α-carotene, µg	54	0
β-cryptoxanthin, µg	81	0
Vitamin E (α-tocopherol), mg	1.54	0.85
Vitamin K (phylloquinone), µg	704.8	817.0

Table 9.1 cont....

Table 9.1 cont....

Nutrient	Raw (value per 100 g)	Cooked (value per 100 g)
Lipids		
Fatty acids (Total saturated), g	0.091	0.052
Fatty acids (Total monounsaturated), g	0.052	0.030
Fatty acids (Total polyunsaturated), g	0.338	0.193

first two polyunsaturated fatty acids, that are essential to human diet but which cannot be synthesized by humans, can satisfy some of the omega-3 fatty acid requirement (Ayaz et al. 2006). Among the mineral components, considerable amounts of calcium, potassium, phosphorus and magnesium are present. Moreover, some micronutrients such as copper, iron, manganese and zinc are present in higher amounts. Most of these elements are essential activators for enzyme catalyzing reactions. Galega kale also has a high content of vitamin C, β-carotene and phenolic compounds, which confers its characteristic high antioxidant activity.

9.3 Processing

As a vegetable, Galega kale is a highly perishable product that continues its metabolic processes after harvest. Lowering the temperature immediately after harvest is a key procedure that reduces respiration rates of the product, delays decay and prevents microbial growth. Maintaining refrigerated temperature levels from the field to consumer throughout the distribution chain should be a concern.

As referred in the introduction section, Galega kale is popular in Portugal and it is often sold in fresh cut shredded form. The fresh cut process inevitably alters the composition of the raw integral kale and an adequate package selection and temperature control are important issues to attain a safe product with high retention of quality parameters.

Galega kale is consumed cooked and domestic low temperature storage such as refrigeration and freezing are commonly procedures that inevitably have negative impacts on some nutritional characteristics and overall quality. Industrially, other more rapid freezing processes and novel technologies have been applied to vegetables. Yet, as far as our knowledge is concerned, Galega kale is not frozen at industrial scale.

The processes routinely applied to kale are herein described as well as their main impacts at quality characteristics. Reports on this nutritious vegetable are emerging and some innovative research about dried Galega kale has been recently addressed, aiming at obtaining a convenient, stable and valuable food ingredient.

9.3.1 Fresh cut

When thinly shredded, Galega kale is a fresh cut vegetable. Fresh cut vegetables are the ones that have been trimmed or cut into an entirely usable product, which is offered to consumers in convenient packages in ready-to-eat or ready-to-use forms (Rico et al. 2007). The added value of these products lies in their convenience and freshness, being sources of healthy promoting compounds.

Unavoidably the fresh cut processing disrupts cells, which results in the lysis of intracellular matter. Released enzymes accelerate products spoilage through tissues softening, browning, off-flavour production, decay of endogenous phenolic compounds and/or synthesis of others (Pradas-Baena et al. 2015). Those products usually have higher respiration rates than the corresponding intact raw products, with a more active metabolism and faster deterioration rates, which results in shorter shelf life. Fresh cut vegetables may also present a faster loss of acids, sugars and other components that regulate flavour and nutritive value (Gil and Allende 2012).

Fresh-cut processing involves washing steps for product decontaminations. Convenient modified atmosphere package and low temperature of storage may increase products' shelf life and maintain high values of nutrients and bioactive compounds.

9.3.2 Low temperature preservation

Vegetable deterioration can be retarded in low temperature environments. Refrigerated and frozen storage are examples of preservation methods for vegetables. Refrigerated temperatures maintain acceptable characteristics of the products only for a few days, and on the other hand freezing is a good long term preservation process but it causes significant sensory and nutritional quality degradation.

9.3.2.1 Refrigerated storage

A food that is stored at temperatures varying between −1 and 8°C, it is designated as a chilled food. Such temperature conditions slows down chemical and biological reactions, microbiological growth, water evaporation, food respiration rate and any further processes that affect food quality and reduces product shelf life (Campañone et al. 2002).

To extend the shelf life of vegetables some thermal processes such as blanching or pasteurization are often applied before chilling storage. However, and due the negative impact of heat on vegetable quality attributes, chilling can also be associated to non thermal processes with benefits (Alexandre et al. 2012). Chilling can also be associated to controlled and modified storage atmospheres with proven improvements (Fellows 2000).

The major physicochemical, physiological and enzymatic responses associated to chilling have been studied for a number of vegetables. In general, firmness is altered and weight loss is observed, ethylene and carbon dioxide production rates vary throughout storage, enzymes activity is reduced but occurs, and nutritive compounds are degraded.

Fundamental studies of refrigerated storage of Galega kale have not been attained yet. The process is not optimized and temperature abuses along the distribution chain of the product are certainly sources of quality decay that could be avoided with proper temperature control. Works focusing on the effect of chilling storage on microorganisms' loads and main important quality characteristics have not been found.

9.3.2.2 Frozen storage

Freezing is unquestionably the most efficient method for long-term preservation of foods. The process consists in decreasing the temperature to its freezing point, which decreases the deterioration of food products by turning water to ice, making it unavailable for bacterial survival and growth.

Freezing process involves four main stages: (i) pre-freezing stage—sensible heat is removed; (ii) super cooling—temperature falls below the freezing point; (iii) freezing—all water in the product turns into ice; (iv) sub freezing—the food temperature is lowered to the storage temperature.

The freezing times are directly dependent on the shape, size and composition of the product and on cooling medium conditions (Silva et al. 2008). According to freezing rates, freezing methods are usually classified into ultra rapid or cryogenic freezing (the highest freezing rates; higher than 10 cm/hour), rapid freezing (freezing rates between 1 and 10 cm/hour; fluidized bed freezers and plate freezers can be used), the normal freezing (freezing rates between 0.3 and 1 cm/hour; in the most air blast freezers), the slow freezing (freezing rate between 0.1 and 0.3 cm/hour) and the very slow freezing (the lowest freezing rates; below 0.1 cm/hour).

A rapid freezing process is desirable for maximum quality retention of the products because smaller ice crystals are formed. When the freezing process is too slow, the ice crystal growth is slower and nucleation sites are fewer. Consequently, larger ice crystals are formed, which usually causes mechanical damage to cell walls.

The formation of ice crystals leads to an increased concentration solution of salts, minerals, and other compounds, which alters the product's pH. Freezing also causes loss of water binding capacity that results in drip loss, protein changes that lead to toughening or dryness and loss of turgor in vegetables. Ice crystals disrupt cells membranes and cause softening of tissues. However, for most vegetables and in situations when freezing follows blanching, the freezing rate is of minor importance when compared to marked structural changes that result from the thermal process.

Although freezing broadly affects the texture and the colour of the products, different impacts on nutrients may be observed in different vegetables. Gonçalves et al. (2009) observed that chlorophylls and vitamin C levels in watercress were not affected by freezing and frozen storage while alterations in colour occurred and peroxidase residual activity decreased. For pumpkin, the freezing process and frozen storage degraded texture, colour and vitamin C (Gonçalves et al. 2011).

As previously mentioned, Galega kale can be stored under frozen conditions by consumers at home. Domestic freezers operate at temperatures around 0°F (–18°C). These conditions are unfavourable for microbes' survival and reactions of spoilage are retarded at such temperatures. The optimum storage temperature is usually slightly higher than the freezing point and it is important to avoid temperature fluctuations to minimize food deterioration (Tashtoush 2000).

Frozen vegetables are hardly associated with food-borne diseases, since pathogens do not proliferate at freezing temperatures and frozen products are commonly cooked. Nevertheless, for maximum hazard control and consumer defence disinfectant washings, thermal (blanching) treatments or alternative non thermal sanitation methods should be applied before freezing.

The effect of freezing in microorganisms depends on the type of the microorganism and its physiological state, food matrix and freezing rates. In general gram negative bacteria are more sensitive to freezing than gram positives (Lund 2000) and vegetative cells of micrococci, staphylococci and streptococci, in particular *Enterococcus faecalis*, are very resistant to freezing and frozen storage (Geiges 1996). Microorganisms can be injured by repeated freeze/thaw cycles but those conditions on most pathogens are not well documented.

In relation to Galega kale, some studies have been carried out regarding frozen storage and influence on some quality features. Korus (2013) observed that, after 12 months storage, frozen kale retained an attractive green colour; the level of chlorophylls fell by only 4 to 9% during the whole period of storage (–20 and –30°C, respectively). In another study, Korus and Lisiewska (2011) verified that after one year of storage, the total content of polyphenols in frozen kale was on average 12% lower than the one found in frozen products immediately after freezing. However, comparing with the raw product, the frozen Galega kale retained 22 to 45% polyphenols, 35 to 58% vitamin C and 43 to 56% of the initial antioxidant activity, depending on the type of preliminary processing (blanching and cooking) and storage temperature (–20 and –30°C). The level of amino acids were also analyzed after one year of frozen storage and the decrease in its content was around 6 to 15%, depending on the type of preliminary processing (Korus 2011).

9.3.3 MAP

Modified Atmosphere Packaging (MAP) is an atmosphere modification that relies on the interaction between the natural process of produce respiration and gas exchange through the package, leading to a buildup of CO_2 and depletion of O_2 (Fonseca et al. 2002). MAP retards respiration, ageing and oxidative reactions, and may suppress microbial growth (Gorris and Tauscher 1999).

It is known that respiration rate is greatly influenced by temperature and that the effect of gas composition increase with temperature. Therefore, refrigeration and MAP are important postharvest techniques in the extension of Galega kale shelf life (Fonseca et al. 2003). These authors evaluated the physiological response of shredded Galega kale under different low oxygen and high carbon dioxide concentrations. They concluded that an atmosphere of 1 to 2% (v/v) O_2 plus 15 to 20% (v/v) CO_2 will extend the shelf life of shredded Galega kale at 20°C to four days, compared to two days in air storage. Fonseca et al. (2005) established that the beneficial effects of low O_2 and high CO_2 on quality of shredded Galega kale was manifested by extension of the sensory characteristics, reduction in respiration rate, greater chlorophyll retention, and retardation of acid ascorbic degradation. Nevertheless, the same authors suggested that further research is necessary to assess microbial aspects before the best conditions for storage of shredded Galega kale.

9.3.4 Drying

Drying has been used for centuries and it is recognized as an efficient long term preservation process for foods. Drying is unavoidably accompanied by physical, biological and chemical alterations that lead to decay of overall quality of dried products. Physical characteristics such as texture and shape are significantly altered; shrinkage of the products is one important issue. The main significant chemical changes associated to drying are related to naturally heat-sensitive nature of most endogenous phytochemicals such as vitamins, antioxidants, minerals, pigments and other bioactive compounds (Devahastin and Niamnuy 2010). Maillard reactions, vitamins degradation and lipids oxidation affects perceived characteristics such as colour and flavour alterations. Nutrient losses associated to leaching that result from water removal from vegetables during the drying process are also a concern. An adequate design of drying conditions (exposure times and temperatures and air conditions) for maximum quality retention and minimum losses of pro healthy compounds is crucial. Additionally some treatments may be applied previously to the drying process. Blanching with steam, water or chemical solutions is the most commonly used, but power ultrasound, ohmic blanching, osmotic and edible coatings pre treatments have also been reported (Oliveira et al. 2015a).

Vegetables are commonly dehydrated by conventional convective drying. This process presents low heating rates and increases the temperature to attain a more rapid dehydration. Ohmic and microwave heating, fluidized bed, and vacuum or freeze drying techniques have also been used to remove water from vegetables.

Impacts of the drying process on nutritional and bioactive characteristics of vegetables are extensively exploited by food scientists (Oliveira et al. 2015a). Vitamins, sugars, proteins, phenolic compounds, carotenoids, pigments (chlorophylls and anthocyanins), antioxidant activity and volatiles and flavour have been reported to be affected by the drying processes.

Regarding Galega kale, some works have been found in literature. Korus (2011, 2013, 2014) studied the effect of air drying on amino acid content, protein, chlorophylls, carotenoids and antioxidants. Alibas (2009) compared microwave, vacuum, and convective drying of collard leaves. When using convective air drying, the colour was affected at the highest temperatures of the range 50 to 175°C and ascorbic acid retentions varied from 40 to 75% when temperatures of 50 and 175°C were used, respectively. Lefsrud et al. (2008) assessed the dry matter content and stability of pigments in kale submitted to freeze drying, vacuum drying and oven drying at 50 and 75°C. Lutein, β-carotene, and chlorophyll levels decreased over 70% as the drying temperature increased from –25 to 75°C.

Oliveira et al. (2015b) studied the temperature effect on some quality and bioactive characteristics of Galega kale dried by convective air drying in the temperature range from 35 to 85°C. Results showed that an increase of temperature caused a reduction of the drying time and an increase of the drying rate, with negative impacts on colour. Important nutritive losses were observed after drying, being the total phenolic compounds the most affected ones. Vitamin C and the antioxidant activity also suffered decay.

These results demonstrate that the convective drying of Galega kale without any pretreatment leads to quality deterioration and may limit consumer acceptance of the dried product, especially when high temperatures are applied.

9.3.5 Cooking

In the cooking process the temperature of the product is raised by applying a heat source for a certain period of time. In the process, the product undergoes physical, chemical, microbiological and organoleptic changes. The main achievement is a product with altered characteristics compared to crude: foods are more tender and digestible, they can acquire new flavours and textures, and if the temperature and process time are appropriate, they are safe from microbiological contamination.

The impact of the cooking process depends on characteristics of the cooked food product *per se* and the severity of the heat treatment applied, being the binomial process time/temperature at which the process runs, determining the quality and safety of the final product.

However, different types of cooking leads to more/less healthy food (from the perspective of nutritional content) and higher/lower retention of quality attributes (colour, aroma, flavour and texture are examples).

Boiling is the method of cooking that commonly uses water as a means of direct heating. The food is placed over a period of time in a given volume of water that reaches the boiling temperature (100°C at atmospheric pressure). A critical point of this process is the transfer of nutrients from food to water (i.e., leaching). An alternative is to use steam as the heating medium. Since the boiling water does not come into contact with the food, the reduction of its nutritional content by loss to the cooking water is minimized. The great application advantages of this process are: (i) retention of flavour, texture and colour of food, and (ii) minimization of the loss of nutrients, mainly vitamins and minerals by leaching.

A faster cooking can be achieved with water under pressure (commonly called the pressure cookers). Accordingly, the temperature required to boil water is greater than the one that occurs in atmospheric conditions (about 120°C). As a result, the foods are cooked at a higher temperature and the process time is considerably reduced compared with conventional cooking in water.

The cooking process affects greatly the composition of vegetables. In Table 9.1 are included the nutritional values of Galega kale after cooking and the impact is obvious. In general all components suffer a decline. However, the bioavailability of some vitamins such as vitamins A and K and carotenoids increased with cooking. This may be explained by being freed from the food microstructure (Hotz et al. 2007).

When proteins are heated they become denatured and structure is affected. In vegetables, the material becomes softer or more friable. The micronutrients, minerals, and vitamins may be destroyed or eluted by cooking. Vitamin C is especially prone to oxidation during cooking and may be completely destroyed by protracted cooking.

9.3.6 Processing impact on quality

Some studies were conducted to assess the qualitative and quantitative changes in Galega kale after blanching, cooking and also after preservation along time. In general, processed vegetables have lower nutritional value than the raw product. Sikora and Bodziarczyk (2012) evaluate the effect of the cooking process of Galega kale on different nutrient components. They concluded that the cooking process lower the antioxidant activity of the vegetable, by lowing especially vitamin C (89%), polyphenols (56%) and in less extent β-carotene (5%).

Korus and Lisiewska (2011) also investigate the changes in Galega kale after blanching and cooking and also after preservation by freezing and canning. In kale leaves, blanching brought about a 34% decrease in the content of vitamin C, a 51% decrease in polyphenols and a 33% decrease in antioxidant activity compared with raw material. When comparing the raw material with cooked samples, the levels of vitamin C, polyphenols and antioxidant activity were reduced by 57, 73 and 45%, respectively. This was expected, since the boiling process at temperatures above 95°C has been shown to decompose the antioxidant components of vegetables. They also concluded that freezing is a good method of preserving Galega kale since the loss of nutritive constituents is lower than in the case of canning.

Korus (2011) determined the effect of blanching, cooking and freezing versus canning on the content of amino acids in kale leaves. The author concluded that raw and processed (blanched or cooked) samples were a good source of amino acids. Moreover, freezing the blanched or cooked vegetable followed by one year storage resulted in decreases of 6 to 15% in total amino acids content, respectively.

The content of chlorophylls and carotenoids in fresh Galega kale leaves and its retention after blanching and cooking was evaluated by Korus (2013). It was observed that blanching did not significantly affect the content of these pigments. This may be due to the inactivation of enzymatic activity during blanching process. Cooking Galega kale did not significantly affect the level of total carotenoids and β-carotene but the longer heat processing time involved in cooking brought about a significant 19% reduction in the level of total chlorophylls compared with the raw vegetable. In sum, it was concluded that carotenoids proved to be more stable to food processing.

The glucosinolates content was also assessed after blanching and cooking by Korus et al. (2014). Significant glucosinolate losses in Galega kale were achieved after the thermal processes. In average, 30 to 42% reduction in total glucosinolates was obtained after blanching and cooking, respectively. The most commonly reported mechanisms for glucosinolates losses in thermal processes is through leaching into the water after cell lysis and thermal degradation.

9.4 Final remarks

Galega kale is undoubtedly a high nutritive vegetable with potential sources of bioactive compounds. Despite the popularity of this vegetable in Portugal, the potential market of shredded Galega kale is still underexploited by the food industry.

The preparation and preservation of shredded Galega kale is a challenging example in the fresh cut produce's technology and implementation of scientific research outcomes at industrial level seems valuable. The use of innovative non thermal technologies in kale decontamination may improve the quality of the product and preserve healthy promoting compounds that are affected by temperature. Even though the product is consumed cooked, the higher the quality of the raw vegetable the better the final processed product attained. Some promising results are arising for dried and shredded Galega kale. Dried kale can be a convenient ready-to-use ingredient with an extended shelf life.

Acknowledgements

Teresa R.S. Brandão, Fátima A. Miller, Elisabete M.C. Alexandre and Sara M. Oliveira gratefully acknowledge Fundação para a Ciência e a Tecnologia (FCT) and Fundo Social Europeu (FSE), the financial support through the Post-Doctoral grants SFRH/BPD/101179/2014, SFRH/BPD/65041/2009, SFRH/BPD/95795/2013 and SFRH/BPD/74815/2010, respectively.

This work was supported by National Funds from FCT - Fundação para a Ciência e a Tecnologia through project UID/Multi/50016/2013.

References

Alexandre, E.M.C., T.R.S. Brandão and C.L.M. Silva. 2012. Efficacy of non thermal technologies and sanitizer solutions on microbial load reduction and quality retention of strawberries. J. Food Eng. 108: 417–426.

Alibas, I. 2009. Microwave, vacuum, and air drying characteristics of collard leaves. Dry Technol. 27: 1266–1273.

Ayaz, F.A., R.H. Glew, M. Millson, H.S. Huang, L.T. Chuang, C. Sanz and S. Hayrloglu-Ayaz. 2006. Nutrient contents of kale (*Brassica oleraceae* L. var. *acephala* DC). Food Chem. 96: 572–579.

Beaulieu, J.C., F.A.R. Oliveira, T. Fernandes-Delgado, S.C. Fonseca and J.K. Brecht. 1997a. Fresh cut kale: quality assessment of Portuguese store supplied product for development of a MAP system. pp. 145. In: J.R. Gorny (ed.). Proceedings of the 7th International Controlled Atmosphere Research Conference. vol. 5, Davis, California, USA.

Beaulieu, J.C., F.A.R. Oliveira, S.C. Fonseca, T. Fernandes-Delgado and M.F. Poças. 1997b. A tecnologia pós-colheita como factor essencial no aproveitamento do potencial dos produtos hortofrutícolas nacionais, 17. In Livro de actas do 3°Encontro de Química de Alimentos, Faro, Portugal. In Portuguese.

Campañone, L.A., S.A. Giner and R.H. Mascheroni. 2002. Generalized model for the simulation of food refrigeration. Development and validation of the predictive numerical method. Int. J. Refrig. 25(7): 975–984.

Casteel, M.J., C.E. Schmidt and M.D. Sobsey. 2008. Chlorine disinfection of produce to inactivate hepatitis A virus and coliphage MS2. Int. J. Food Microbiol. 125(3): 267–273.

Devahastin, S. and C. Niamnuy. 2010. Modelling quality changes of fruits and vegetables during drying: a review. Int. J. Food Sci. Technol. 45(9): 1755–1767.

Fellows, P.J. 2000. Food Processing Technology: Principles and Practice. Woodhead Publishing Limited and CRC Press LLC, Cambridge, UK.

Fonseca, S.C. 2001. Development of perforation-mediated modified atmosphere packaging for extending the shelf life of shredded Galega kale. Ph.D. Thesis. Escola Superior de Biotecnologia, Universidade Católica Portuguesa, Portugal.

Fonseca, S.C., F.A.R. Oliveira, J.M. Frias, J.K. Brecht and K.V. Chau. 2002. Modelling respiration rate of shredded Galega kale for development of modified atmosphere packaging. J. Food Eng. 54(4): 299–307.

Fonseca, S.C., F.A.R. Oliveira, J.K. Brecht and K.V. Chau. 2003. Evaluation of the physiological response of shredded Galega kale under low oxygen and high carbon dioxide concentrations. pp. 389–391. *In*: Proceedings of the 8th International Controlled atmosphere research conference, Vols. I and II, Book Series: Ata Horticulturae, Issue 600.

Fonseca, S.C., F.A.R. Oliveira, J.K. Brecht and K.V. Chau. 2005. Influence of low oxygen and high carbon dioxide on shredded Galega kale quality for development of modified Atmosphere Packages. Postharvest Biol. Tec. 35(3): 279–292.

Gorris, L. and B. Tauscher. 1999. Quality and safety aspects of novel minimal processing technology. *In*: F.A.R. Oliveira and J.C. Oliveira (eds.). Processing of Foods: Quality Optimisation and Process Assessment. CRC Press, Boca Raton, USA.

Gil, M.I. and A. Allende. 2012. Minimal processing. *In*: V.M. Gómez-López (ed.). Decontamination of Fresh and Minimally Processed Produce. Wiley-Blackwell, Iowa, USA.

Gómez, P.L., S.M. Alzamora. M.A. Castro and D.M. Salvatori. 2010. Effect of ultraviolet-C light dose on quality of cut apple: Microorganism, color and compression behavior. J. Food Eng. 98(1): 60–70.

Gonçalves, E.M., R.M.S. Cruz, M. Abreu, T.R.S. Brandão and C.L.M. Silva. 2009. Biochemical and colour changes of watercress (*Nasturtium officinale* R. Br.) during freezing and frozen storage. J. Food Eng. 93(1): 32–39.

Geiges, O. 1996. Microbial processes in frozen food. Adv. Space Res. 18(12): 109–118.

Gonçalves, E.M., J. Pinheiro, M. Abreu, T.R.S. Brandão and C.L.M. Silva. 2011. Kinetics of quality changes of pumpkin (*Curcurbita maxima* L.) stored under isothermal and non-isothermal frozen conditions. J. Food Eng. 106(1): 40–47.

Gorny, J. 2006. Microbial contamination of fresh fruits and vegetables. *In*: G.M. Sapers, J.R. Gorny and A.E. Yousef (eds.). Microbiology of Fruits and Vegetables. CRC Press, Taylor & Francis group, Boca Raton, USA.

Hotz, C. and R.S. Gibson. 2007. Traditional food processing and preparation practices to enhance the bioavailability of micronutrients in plant-based diets. J. Nutr. 137(4): 1097–1100.

Knezevic, Z. and M. Serdar. 2009. Screening of fresh fruit and vegetables for pesticide residues on Croatian market. Food Control. 20(4): 419–422.

Korus, A. 2011. Effect of preliminary processing, method of drying and storage temperature on the level of antioxidants in kale (*Brassica oleracea* L. var. *acephala*) leaves. LWT-Food Sci. Technol. 44(8): 1711–1716.

Korus, A. 2012. Effect of technological processing and preservation method on amino acid content and protein quality in kale (*Brassica oleracea* L. var. *acephala*) leaves. J. Sci. Food Agr. 92(3): 618–625.

Korus, A. 2013. Effect of preliminary and technological treatments on the content of chlorophylls and carotenoids in kale (*Brassica oleracea* L. var. *acephala*). J. Food Process Pres. 37(4): 335–344.

Korus, A. 2014. Amino acid retention and protein quality in dried kale (*Brassica oleracea* L. var. *Acephala*). J. Food Process Pres. 38: 676–683.

Korus, A. and Z. Lisiewska. 2011. Effect of preliminary processing and method of preservation on the content of selected antioxidative compounds in kale (*Brassica oleracea* L. var. *acephala*) leaves. Food Chem. 129(1): 149–154.

Korus, A., J. Słupski, P. Gębczyński and A. Banaś. 2014. Effect of preliminary processing and method of preservation on the content of glucosinolates in kale (*Brassica oleracea* L. var. *acephala*) leaves. LWT-Food Sci. Technol. 59(2): 1003–1008.

Lefsrud, M., D. Kopsell, C. Sams, J. Wills and A.J. Both. 2008. Dry matter content and stability of carotenoids in kale and spinach during drying. Hortscience. 43: 1731–1736.

Lund, B.M. 2000. Freezing. *In*: B.M. Lund, T.C. Baird Parker and G.W. Gould (eds.). The Microbiological Safety and Quality of Food. Aspen Publishers, Gaithersburg, MD, USA.

Martínez, S., I. Olmos, J. Carballo and I. Franco. 2010. Quality parameters of *Brassica* spp. grown in northwest Spain. Int. J. Food Sci. Technol. 45: 776–783.

Oliveira, S.M., T.R.S. Brandão and C.L.M. Silva. 2015a. Influence of drying processes and pretreatments on nutritional and bioactive characteristics of dried vegetables: a review. Food Eng. Rev. (in press) doi: 10.1007/s12393-015-9124-0.

Oliveira, S.M., I.N. Ramos, T.R.S. Brandão and C.L.M. Silva. 2015b. Effect of air-drying temperature on the quality and bioactive characteristics of dried Galega Kale (*Brassica Oleracea* L. var. *Acephala*). J. Food Process Pres (in press) doi: 10.1111/jfpp.12498.

Pradas-Baena, I., J.M. Moreno-Rojas and M.D. Luque de Castro. 2015. Section 1. Effect of processing on active compounds in fresh cut vegetables. *In*: V.R. Preedy (ed.). Processing and Impact on Active Components in Food. Academic Press, New York, USA.

Rico, D., A.B. Martín-Diana, J.M. Baratand and C. Barry-Ryan. 2007. Extending and measuring the quality of fresh cut fruit and vegetables: a review. Trends Food SciTech. 18(7): 373–386.

Rodriguez-Romo, L.A. and A.E. Yousef. 2006. Microbial stress adaptation and safety of produce. *In*: G.M. Sapers, J.R. Gorny and A.E. Yousef (eds.). Microbiology of Fruits and Vegetables. CRC Press, Taylor & Francis Group, Boca Raton, USA.

Sikora, E. and I. Bodziarczyk. 2012. Composition and antioxidant activity of kale (*Brassica oleracea* L. var. *acephala*) raw and cooked. Acta Sci. Pol. Technol. Aliment. 11(3): 239–248.

Silva, C.L.M., E.M. Gonçalves and T.R.S. Brandão. 2008. Freezing of fruits and vegetables. *In*: J.A. Evans (ed.). Frozen Food Science and Technology. Blackwell Publishing Ltd., Oxford, UK.

Tashtoush, B. 2000. Natural losses from vegetable and fruit products in cold storage. Food Control. 11(6): 465–470.

Torres, C.M., Y. Picó and J. Mañes. 1996. Determination of pesticide residues in fruit and vegetables. J. Chromatogr. A. 754(1-2): 301–331.

USDA Nutrient Database for Standard References. [online], www.nal.usda.gov.

Waite, J.G. and A.E. Yousef. 2010. Overview of food safety. *In*: E. Ortega-Rivas (ed.). Processing Effects on Safety and Quality of Foods. CRC Press, Taylor & Francis Group, Boca Raton, USA.

Warriner, K. 2005. Pathogens in vegetables. *In*: W. Jongen (ed.). Improving the Safety of Fresh Fruit and Vegetables. Woodhead Publishing Limited, Cambridge, UK.

CHAPTER 10

Turkish Meat Products

Ismet Ozturk,[1,*] *Salih Karasu,*[2] *Osman Sagdic*[2] *and Hasan Yetim*[1]

10.1 Introduction

The popularity and consumption of meat products has increased over the years because of their important roles in human nutrition as nutritive sources of dietary protein with appealing sensory qualities or flavour components (Kayisoglu et al. 2003). Traditionally, most of the meat products are part of the cultural heritage that connects the past of the country to the modern day in culinary relevance (Kaban 2013). Since Turkey is situated between south west Asia and the Balkan region of south east Europe, it is a bridge between two continents. In respect of this strategic location, which embraces cultural diversity from the Oguz Turks, Ottoman, Western, and Islamic cultures (Kilic 2009), several kinds of traditional meat products such as Sucuk, Pastirma, Doner Kebab, Kavurma and raw meatballs (Cig Kofte) are produced and consumed in the country. In Turkey, these meat products can be dried, fermented, cured, cooked, emulsified or processed in different ways for human consumption. Although several types of traditional meat products are produced in Turkey, there has been limited research related to quality factors and production methods of these products (Kaban 2013, Kilic 2009). Therefore, this chapter presents considerable data about microbial, sensory, and physicochemical quality characteristics and the production methods of Sucuk, Pastirma, Kavurma and different types of meatballs.

[1] Department of Food Engineering, Faculty of Engineering, University of Erciyes, Kayseri, Turkey.
[2] Department of Food Engineering, Faculty of Chemical and Metallurgical Engineering, Yildiz Technical University, Istanbul, Turkey.
* Corresponding author: ismet@erciyes.edu.tr

10.2 Sucuk

Sucuk is one of the most widely consumed traditional Turkish meat products, which is dry, uncooked, cured, fermented sausage, produced from beef or water buffalo meat (Kaban and Kaya 2008). Non meat ingredients such as salt, garlic, nitrate, nitrite and some spices including cumin, pimento, black and red pepper are important factors contributing to the specific aroma and taste of sucuk (Kilic 2009). Sucuk is one of the most popular meat products in Turkey due to its unique taste and texture. It is commonly eaten as fried with egg in pan for breakfast or grilled before consumption. It is also used as an ingredient for snacks such as fast food toasts, pizza and sandwiches as well as in the consumption of breakfast (Kilic 2009). Sucuk and similar fermented meat products are produced using starter culture, salt and different spices and fat (beef tallow and/or sheep tail fat) (Gençcelep et al. 2008, Siriken et al. 2009). The Sucuk mixture is stuffed into artificial collagen or natural (beef small intestines) casings and ripened and dried for several weeks at different temperatures (18 to 25°C) and relative humidity (80 to 95%) conditions in a fermentation chamber under controlled conditions. In the production of Sucuk, starter cultures can be used to achieve the desired quality characteristic such as taste, aroma and texture. Also, it can be produced from natural microbiota without starter culture. Starter cultures used in dry fermented meat products are generally lactic acid bacteria (*Lactobacillus sakei, Lb. plantarum* and *Lb. curvatus, Pediococcus acidilactici, P. pentosaceus*), catalase positive cocci (*Staphylococcus xylosus, S. carnosus* and *Kocuria varians*), yeast (*Debaryomyces hansenii*) and mold (*Penicillium nalgiovense*). If starter culture is not used in Sucuk production, standard sensory and microbiological quality characteristics of sucuk may not develop during the fermentation processes. Although Sucuk production with starter culture is preferred in the meat industry, some small scale facilities still perform the natural fermentation by chance contamination using the indigenous microbiota (Kaban and Kaya 2008).

10.2.1 Production

Beef and water buffalo meats are commonly used as muscle material in Sucuk production and must be high in quality and freshness since quality characteristics of Sucuk are directly affected by the meat quality. The age of animals from which the meat is to be used in sausage production is important, and generally old and/or young animals are not preferred (the meat from old animals have more connective tissue and meats of young animals have high water composition). For this purpose, meats from three to five-year-old animals is preferred in the production of sucuk (Bozkurt and Belibağlı 2012). Besides, dark firm and dry (DFD) and pale soft exudative

(PSE) meats are not also preferred in the production of sucuk because these types of meats may cause a problem at the drying stage and may induce bacterial growth. The pH of meat used in Sucuk production should be lower than 5.9. Higher pH values in meats results in high water holding capacity which delays the drying process of the sucuk. In sucuk production, beef tallow and/or sheep tail fat is used as a fat source. These fats are generally used as chilled. If unchilled, the fats may stick to the grinder and the Sucuk dough is not homogenous in structure. Red pepper (hot and/or normal), black pepper, cumin and pimento are the most preferred spices in Turkish Sucuk formulation. Also cinnamon and clove may be used as a spice (Bozkurt and Erkmen 2002). Red pepper, black pepper, cumin and pimento are used at the proportion of 0.7%, 0.5%, 0.9% and 0.25%, respectively (Gençcelep et al. 2008, Gokalp et al. 2004, Ozturk 2014). Spice content of sucuk dough is usually lower than 3%. However, the amount of spice is optional and may vary from producer to producer in different regions of Turkey. Nitrate, nitrite, sugar (dextrose, saccharose, etc.), salt and garlic are other ingredients used in the production of Sucuk. Nitrate and nitrite are used at 300 ppm and 150 ppm to provide colour and other characteristics of Sucuk, respectively. While salt and garlic are used to give specific flavour and protection, sugar plays an important role in the fermentation of sucuk. Starter cultures have also important functions in fermented meat products such as acid production, colour and aroma formation, increasing colour stability, giving the desired flavour and delaying oxidation (Table 10.1). Types and combinations of starter culture may also vary according to fermented meat products produced in other countries.

Table 10.1 Functions of microorganisms used as starter culture in fermented sausages (Bozkurt and Belibağlı 2012, Dogbatsey 2010).

Microorganism	Species	Benefits to sucuk fermentation
Lactic Acid Bacteria	*Lactobacillus plantarum*	Inhibition of pathogenic and spoilage bacteria (bacteriocin and acid production), acceleration of colour formation and drying
	Lb. pentosus	
	Lb. sakei	
	Lb. curvatus	
	Pediococcus acidilactici	
	P. pentosaceus	
Catalase positive cocci (CPC)	*Kocuria varians (Micrococcus varians)*	Colour formation and stabilization
	Staphylococcus xylosus	Reduction of nitrate
	S. carnosus	Delay of rancidity Aroma formation
Yeasts	*D. hansenii*	Delay of rancidity Aroma formation
	C. famata	
Mold	*P. nalgiovense*	Colour stability
	P. chrysogenum	Aroma formation Delay of rancidity

Some starter culture combinations may provide fast fermentation of sausage, and some cultures provide desired aroma formation during the ripening. Effects of starter culture combinations in fermented sausages are shown in Table 10.2. Some LAB species such as *P. acidilactici* and *P. pentosaceus* provide fast fermentation and ripening in fermented sausage. Starter culture combinations containing *Pediococcus* sp. are generally used to produce fast fermented sausages (Dogbatsey 2010). However, some combinations of starters including *S. xylosus; P. pentosaceus* + *Lb. plantarum; S. carnosus; S. carnosus* + *Lb. pentosus* + *S. xylosus; P. pentosaceus* + *P. acililactici* + *S. carnosus; P. pentosaceus; S. xylosus* + *Lb. alimentarus; Lb. plantarum* + *S. xylosus* improved the quality of Sucuk (Bozkurt and Belibağlı 2012, Kaban and Kaya 2009). On the other hand, some commercial starter cultures may adversely affect the sensory properties of Sucuk (Bozkurt and Erkmen 2002). While ready-to-use starter cultures are commonly used in meat factories, small scale producers generally do not use starter culture in the sucuk production. Lactic acid bacteria (LAB) and catalase positive cocci (CPC) are also important microorganisms in the Sucuk, as is the

Table 10.2 Starter culture combinations used in fermented sausage production (Anonymous 2014).

Starter cultures and/or combinations/functions	
Lactobacillus sakei and *Staphylococcus carnosus* *Pediococcus pentosaceus* and *S. carnosus* *P. pentosaceus* and *S. xylosus* *Lb. plantarum* and *S. carnosus* *Lb. sakei* and *S. carnosus* *Lb. pentosus* and *S. carnosus*	Aroma formation and acidification
Lb. sakei and *S. carnosus* *Lb. sakei* and *S. carnosus* and *S. xylosus* *Lb. sakei* and *S. carnosus* *P. pentosaceus* and *S. xylosus* *P. pentosaceus* *Lb. curvatus* *Lb. curvatus* and Micrococcaceae sp. *Lb. curvatus* *Lb. farciminis* and *S. carnosus* and *S. xylosus* *P. acililactici* and *P. pentosaceus* *P. acililactici* and Micrococcaceae sp. *P. acililactici* *P. acililactici*	Fast fermentation and ripening
S. xylosus, P. acidilactici, Lb. curvatus	Acidification and protection with bacteriocin production
S. carnosus *S. xylosus* Micrococcaceae sp.	Flavour and colour enhancing cultures
Penicillium nalgiovense	Uniform drying and flavour enhancing

case with other fermented sausages. These bacteria contribute to several quality characteristics of Sucuk such as texture, flavour and colour. *Lb. plantarum, Lb. sakei* and *Lb. curvatus* are the LABs isolated and identified in sucuks produced in Turkey. However, *Lb. brevis, Lb. buchneri* and *Lb. paracasei* are *Lactobacillus* species which are less common and isolated in Turkish Sucuk (Başyiğit et al. 2007). It needs to be emphasized that *P. pentosaceus, S. carnosus* and *L. sakei* have an important role in improving the hygienic quality, colour and flavour properties of the Sucuk (Bingol et al. 2014a).

As shown in the flow chart below, the production of Sucuk takes about 10 to 20 days (Fig. 10.1). Briefly, meat, fat, spices and other ingredients are first mixed and conditioned for 12 hours at 0 to 4°C. Then the mixture is kneaded by machine or hand and filled to natural or artificial casings.

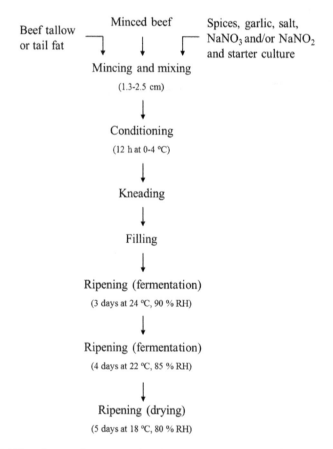

Figure 10.1 Flow chart on the process of traditional Sucuk production (Bozkurt and Erkmen 2004, Kaban and Kaya 2009, Ozturk 2014).

Sucuks are fermented, ripened and dried in accordance with the following procedures: the first three days at 24 ± 1°C and 90 ± 2% relative humidity (RH), following four days at 22 ± 1°C and 85 ± 2% RH and finally for five days at 18 ± 1°C and 80 ± 2% RH in fermentation chambers (Kaban and Kaya 2009, Ozturk 2014).

Sucuk (Fig. 10.2a,b) is commercially produced in Turkey, and a typical flow chart of the production process is presented in Fig. 10.3.

Figure 10.2 (a) Ripening process, (b) Sucuk.

Commercial Sucuk is produced by applying heat treatment to prevent spoilage of the product and thus extending its shelf life.

Sucuk is fermented for 18 to 24 hours at 25 to 29°C and 98% relative humidity after the filling process. Starter cultures used in the Sucuk production to regulate the acidity and pH values of sucuk to below 5.4. Then, the Sucuks are washed using pressurized water (18 to 20°C) and dried for one to three days at 10 to 15°C to meet the desired moisture level (<40%). As mentioned before, the starter culture is used in both traditional and commercial Sucuk production, and fermentation time in commercial Sucuk production is very short. Unlike traditional production, starter culture combinations used in the commercial production can somewhat provide aroma. Starter cultures decrease the pH of Sucuk with acid formation and this leads to the drying of Sucuk. The removal of water from the Sucuk depends on pH and other fermentation parameters, like RH, temperatures, air flow etc. (Bozkurt and Belibağlı 2012).

10.2.2 Quality characteristics

Quality characteristics of traditional Sucuk have been determined in Turkish Standard (TS 1070) and Microbiological Criteria Regulations of the Turkish Food Codex as shown in Table 10.3. According to both Turkish Standard and Turkish Food Codex, the moisture content of Sucuk should be <40% and its pH range from 4.7 to 5.4 or <5.4, respectively. Both moisture and pH levels are critical quality parameters in Sucuk, otherwise, extreme

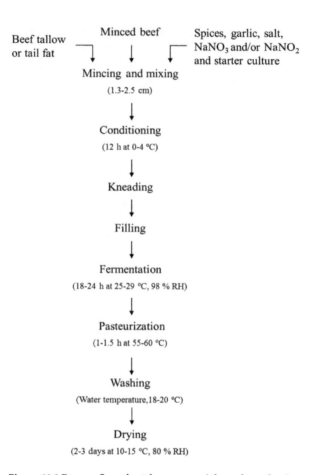

Figure 10.3 Process flow chart for commercial sucuk production.

pH and moisture contents may lead to spoilage of the meat product, e.g., sucuk (Bozkurt and Belibağlı 2012). Existing regulatory standards approve the incorporation of nitrate in Sucuk production, and its level should be lower than 50 mg/kg and 150 mg/kg in Turkish Food Codex (2012), Turkish Standard (2002), respectively. Nitrite is used to contribute colour and taste and to prevent *Clostridium botulinum* and hygienic quality of Sucuk. But the primary purpose of nitrite is to contribute to the formation of sucuk colour. Sucuks should not contain pathogenic bacteria such as *Salmonella* sp., *Listeria monocytogenes* and *Escherichia coli* O157:H7. However, *Staphylococcus aureus* and yeast mold counts should be 10^2–10^3 cfu/g or less in 2 of 5 samples. Although *E. coli*, *Clostridium perfringens* and total coliform limits are presented in Turkish Standard, there is no standard for them in Turkish Food Codex (Turkish Food Codex 2009).

Table 10.3 Regulatory standards on some quality traits of traditional Sucuk.

	Turkish Standard[1]	Turkish Food Codex[2,3]
Physicochemical properties	Limits	Limits
Moisture	<40%	<40%
Salt	<5%	-
pH	4.7–5.4	<5.4
Protein	>22%	>16%
Fat	<35%	-
Nitrite	<50 mg/kg	<150 mg/kg
Hydroxyproline	<220 mg/100 g	-
Microbiological properties		
Escherichia coli (cfu/g)	5×10^1–1×10^2 in 1 out of 5 samples	-
Escherichia coli O157:H7 (cfu/g)	None	None in 25 g sample
Staphylococcus aureus (cfu/g)	5×10^2–5×10^3 in 1 out of 5 samples	10^2–10^3 in 2 out of 5 samples
Clostridium perfringens (cfu/g)	10–100 in 2 out of 5 samples	-
Listeria monocytogenes (cfu/g)	None	None in 25 g sample
Salmonella sp. (cfu/g)	None in 25 g sample	None in 25 g sample
Yeast and Mold (cfu/g)	10–100 in 1 out of 5 samples	10^2–10^3 in 2 out of 5 samples
Coliform (cfu/g)	<10	-

Source: [1](Turkish Standard 2002), [2](Turkish Food Codex 2009) and [3](Turkish Food Codex 2012).

10.2.3 Microbiological properties

Some microorganisms which are the most significant elements during the fermentation of sucuk are *Lactobacillus* sp., *Micrococcus* sp., and *Staphylococcus* sp. and these influence the Sucuk's unique colour, taste, aroma and structure. Since they prevent the growth of most of the pathogenic and spoilage microorganisms, starters have an effective role in Sucuk safety, quality and shelf life of the product (Gönülalan et al. 2004, Sameshima et al. 1998). It is also essential to understand that the microbial populations are responsible for the ripening processes to protect fermented foods during the fermentation (Rantsiou and Cocolin 2006). In several studies, microorganisms of the sucuk from different regions of Turkey were determined by using phenotypic or genotypic techniques. In a study, the traditional fermented Sucuks produced with same formulations same fermentation conditions and without starter culture, obtained from eight local factories in Kayseri and Afyon in Turkey were analyzed. It was observed that 10 different species of LAB which are *Lb. sakei, Lb. curvatus, Lb. brevis, Lb. alimentarius, Lc. piscium, W. viridescens, W. halotolerans, S. succinus, S. vitulinus*; among them *Lb. sakei, W. viridescens* and *Lb. curvatus* were frequently identified in the samples (Kesmen et al. 2012). In a similar study, 140 *Lactobacillus* species were isolated from Turkish fermented sausage (Sucuk), and their percentages were *L. sakei* (92.1%), *L. curvatus* (7.1%) and *L. cornis* (0.7%) (Özdemir et al. 1995). However, lactic acid bacteria microbiota of traditional Sucuk were *Lb. plantarum, P. pentosaceus, Lactococcus lactis* ssp. *lactis, Lb. curvatus* ssp. *curvatus, Lb. brevis, Lb. fermentum, Weisella viridescens, Lb. delbrueckii* ssp. *delbrueckii, W. confusa, Lb. collinoides* and *Leuconostoc* mesenteroides ssp., *mesenteroides/dextranicum* which were determined by genotypic method and *Lb. plantarum* were the dominant LAB species (Adiguzel and Atasever 2009). In a similar study, *Lb. plantarum* (45.7%) was the dominant LAB species and was followed by *Lb. curvatus* (10.9%), *Lb. fermentum* (9.3%). In addition to LAB, *S. xylosus* (41.5%) and *S. saprophyticus* (28.5%) were catalase positive cocci species isolated from mostly in traditional sucuk (Kaban and Kaya 2008). In another study, *Lactobacillus* sp. including *Lactobacillus sake, Lb. curvatus, Lb. alimentarius, Lb. plantarum,* and *Lb. brevis* were identified as the most predominant LAB species in Turkish dry sausage, sucuk (Gürakan et al. 1995). In the study of Yüceer and Özden Tuncer (2015), *P. acidilactici* (47.7%), *Enterococcus faecium* (36.9%), *Lb. sakei* sp. *carnosus* (4.6%), *Lb. sakei* sp. *sakei* (4.6%), *P. pentosaceus* (3.1%), *E. faecalis* (1.5%) and *W. viridescens* (1.5%) were determined as LAB microbiota of traditional Sucuk samples. Güleren (2008) determined *Lb. brevis* and *Lb. plantarum* as the most predominant LAB species in the sucuk samples. These studies showed that LAB microbiota of traditional Sucuk is considerably varying. Yeast mycobiota of traditional Turkish fermented sucuk and their functional and technological properties were determined by Ozturk and Sagdic (2014), and they analyzed 35 different sucuk samples from different regions of Turkey and obtained 255 yeast isolates— eventually, *Candida zeylanoides* and *Debaryomyces hansenii* were identified as the dominant strains in the Sucuk samples. Sucuk is accepted as a safe fermented meat product, pathogenic and saprophytic bacteria can not usually grow or survive due to substances such as organic acids and antimicrobial peptide (bacteriocin) produced by LAB in Sucuk. However, some pathogenic and saprophytic bacteria were also determined in the studies conducted in Turkish Sucuks. In a study, pathogenic bacteria including *E. coli, S. aureus, Salmonella* sp. and *L. monocytogenes* were determined in the fermented Sucuk (Kök et al. 2007). In a similar study, *L. monocytogenes* was determined in 35 of 300 Sucuk samples (Colak et al. 2007). Yalçın and Can (2013) determined *L. monocytogenes, S. aureus* (Coagulase positive) and coliform bacteria in the sucuk. These results show that the hygienic quality of Sucuks may not be good in some cases.

Although Turkish Sucuks are consumed by applying heat treatment, so it may have some health risks if not properly cooked before consumption.

10.3 Pastirma

Pastirma (Fig. 10.4a,b) is a very popular dry-cured meat product produced from beef and water buffalo muscles, and it can be considered as an delicatessen intermediate moisture meat product (Aksu et al. 2005b, Gök et al. 2008). Although Pastirma has been produced in Turkey as a traditional meat product since the 12th century, it is also produced in many parts of the world like Middle East, Middle Asia and some Mediterranean and European countries. The consumption of this product has rapidly increased due to its high nutritional value, special taste and aroma from cemen paste (Aksu et al. 2005a). In a Pastirma production, all the exterior fat and connective tissue is removed from the meat and then the muscles are shaped, cured, dried, pressed, re-dried, re-pressed and coated with a cemen paste that is prepared with garlic, red pepper, paprika, ground fenugreek seeds (*Trigonella foenum-graecum*) and water, and dried again to meet a maximum 40% moisture level (Yetim et al. 2006b, Aksu and Erdemir 2014).

Figure 10.4 (a) Ripening process, (b) Pastirma.

10.3.1 Production

Raw material used in Pastirma production significantly affects the quality. Pastirma is produced from whole muscles, which are obtained from certain parts of beef and buffalo carcasses. In addition to beef and buffalo carcasses, camel, sheep, turkey, and occasionally chicken can also be used as a raw material in Pastirma production. It was reported that frozen meat can also be used in Pastirma production, and this product has similar properties with the Pastirma produced from fresh meat (Aksu et al. 2005a). Due to excessive moisture loss during the drying process, meat from young animals is not preferred due to their high moisture content. In the same vein, meat from very old animals is also not preferred because of their coarse and tough texture (Obuz et al. 2012). However, fresh meat from cow or water buffalo within three to six-year-olds is considered ideal for Pastirma production. The first step of production is fabrication of meat obtained from carcass; an important step affecting final product quality. Approximately 20 types of Pastirma can be produced from a carcass, and each has different shape and quality traits (Ceylan and Aksu 2011, Obuz et al. 2012). For example, Sirt (loin) Pastirma has a rectangular shape 60 to 70 cm long, 15 to 20 cm wide and 3 to 4 kg in weight (Ceylan and Aksu 2011).

The next stage is dry curing in pastirma production. Salt is the major ingredient in the curing mixture with levels of 8 to 10% (Kaban 2013). The role of the salt in the curing mixture is in the inhibition of the growth of undesirable microorganisms, contribution to salty taste and increase of myofibrillar protein solubility. Medium size grain salt is used in Pastirma production. Coarse salt is not preferred because it cannot be adequately absorbed by the meat and may negatively affect the Pastirma quality (Obuz et al. 2012). In addition to salt, several curing agents can be used in the curing mixture such as nitrite, nitrate, ascorbic acid, sucrose and glucose (Kaban 2013). Nitrite is responsible for a stable red colour, has a strong antioxidant activity, prevents or retards microbial growth and gives a pleasant flavour (Honikel 2008). Salt is applied to deeply meat cut surfaces by hand, and the temperature for curing varied from 4 to 10°C. After the salting, meat is piled up about 1 metre high and the pieces are turned upside down for two days (Obuz et al. 2012).

After the curing process, meat strips are washed to remove excess salt adhering to the surface and dried by hanging (Kaban 2009, Obuz et al. 2012). If the Pastirma is produced via traditional method, drying parameters such as temperature and relative humidity depend on environmental factors. The months of September to November are therefore preferred for traditional production of Pastirma. A period of three to five days is suitable for drying Pastirma in summer. On the other hand, drying times are lower in air conditions where the drying parameters such as temperature, air speed and relative humidity are controlled (Gokalp et al. 2004, Kaban 2009, Obuz et al. 2012). The drying temperature in a controlled condition is about 15°C (Kaban 2013). After the drying period, meat pieces are pressed with heavy weight (about 250 kg). The pressing period ranges from six to 15 hours (Obuz et al. 2012). After pressing, meat pieces are dried again for three days at 15 to 20°C (Kaban 2013). After the second drying process, the second pressing is made. This step is also called shot pressing. Fats on the meat surface diffuse to the interior part and usually very tender meat is obtained after this second pressing step (Obuz et al. 2012). After the drying and pressing process, the meat surface is covered by cemen paste which gives characteristic taste, colour, aroma and flavour to Pastirma. Cemen paste is a mixture of ground fenugreek seeds, crushed fresh garlic, ground paprika and water. The thickness of cemen paste is about 3 to 4 mm on the surface. Cemen paste not only gives a specific aroma to Pastirma but also has a protective effect against the some pathogenic bacteria and molds (Yetim et al. 2006b). Pastirma can be stored for about nine months without refrigeration. The stability of the pastirma is related to low water activity and the cemen paste which has garlic and some other spices. Processing steps of a classical Pastirma production are summarized in Fig. 10.5.

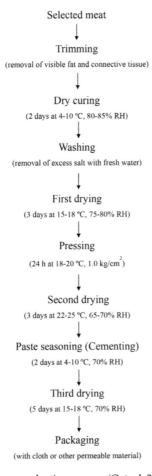

Selected meat

Trimming
(removal of visible fat and connective tissue)

Dry curing
(2 days at 4-10 °C, 80-85% RH)

Washing
(removal of excess salt with fresh water)

First drying
(3 days at 15-18 °C, 75-80% RH)

Pressing
(24 h at 18-20 °C, 1.0 kg/cm^2)

Second drying
(3 days at 22-25 °C, 65-70% RH)

Paste seasoning (Cementing)
(2 days at 4-10 °C, 70% RH)

Third drying
(5 days at 15-18 °C, 70% RH)

Packaging
(with cloth or other permeable material)

Figure 10.5 Pastirma production process (Ozturk 2015, Gök et al. 2008).

10.3.2 Quality characteristics

The quality of the Pastirma can be evaluated based on the microbiological, physicochemical and sensorial properties. According to Turkish standards, it should have maximum fat of 40%, salt of 8.5%, pH 4.5 to 5.8, maximum moisture content of 50%, maximum cemen level of 10%, no rancidity and zero tolerance to pathogens (TS 2002b). It was also reported that Pastirma should contain an average of 45% moisture, 30% protein, 15% fat, 5% carbohydrate (from paste) and 5% ash (Yetim et al. 2006b). As previously indicated, Pastirma is an intermediate moisture food, and its stability might be due to low water activity (a_w). a_w value of the Pastirma is reported lower than 0.90 and can be stored at room temperature. Colour is also an important quality parameter for Pastirma—redness (a^*) is considered as an indicator for colour quality. This value is affected by several factors such as storage period, curing agents and packaging technology. Pastirma quality is also assessed based on the muscles coming from different regions of carcasses, which are classified as first, second and third class. Quality characteristics and the classifying system of the pastirma is shown in Table 10.4.

Table 10.4 Quality characteristic and classifying system of Pastirma (Gokalp et al. 2004).

Quality characteristic	First class	Second class	Third class
Marbling structure	High	Low	No marbling structure
Intermuscular fat	Low	Intermediate	High
Colour	Bright red colour in cross section surface and has no visible crust	Dark red colour on cross section surface and has visible crust	Dark red brownish colour on cross section surface and significant crust formation
Texture	High tenderness	Intermediate tenderness	No tenderness, very dry and coarse structure
Cemen thickness	<4 mm	>4 mm and no homogenity	Excessive amount

10.3.3 Microbiological properties

Lactic acid bacteria, *Micrococcus* and *Staphylococcus* (non-pathogenic) are main microorganisms of Pastirma flora. These microorganisms have an important role in production and quality. In addition to low proteolytic activities, lactic acid bacteria affects sensory and textural properties of Pastirma by producing lactic acid (Kaban 2009). Catalase positive cocci (*Micrococcus* and non-pathogenic *Staphylococcus)* are responsible for colour stability, aroma formation and delaying of oxidation (Kaban 2013). Therefore, some authors suggested that these microorganisms could be used as a starter culture in Pastirma production. Different researchers reported lactic acid bacteria count as 3.92 to 6.01 log cfu/g (Kaban and

Kaya 2006), 5.18 to 6.81 log cfu/g (Çakıcı et al. 2014) and 2.78 to 7.89 log cfu/g (Aksu and Kaya 2001). *Micrococcus* and *Staphylococcus* counts were reported to be 5.63 to 7.18 log cfu/g (Cakıcı et al. 2014) and found to be 4.62 to 7.05 log cfu/g (Kilic 2009). Çakıcı et al. 2014 reported that both lactic acid bacteria and catalase positive cocci were dominant microbiota in Pastirma but the lactic acid bacteria count was not as high as catalase positive cocci counts.

Total aerobic mesophilic bacteria (TAMB) count was determined to be 6.20 to 7.59 log cfu/g (Cakici et al. 2014), and in another research, it varied from 5 to 8.39 log cfu/g (Aksu and Kaya 2001). Since Pastirma is an intermediate moisture food with relatively low a_w and pH, microbiological problems associated with pathogen microorganisms is rare. It was reported that cemen paste and all its ingredients showed inhibitory effects on *E. coli*, *S. aureus* and *Yersinia enterocolitica* (Yetim et al. 2006a). In this study, *Y. enterocolitica* was the most resistant bacterium to the cemen paste. Similar results also reported that cemen paste had positive effects on the microbiological quality of Pastirma (Dogruer et al. 1998). It was reported from some other studies that *Enterobacteriaceae* counts of Pastirma samples were low <2 log cfu/g (Aksu et al. 2005a, Gök et al. 2008, Çakıcı et al. 2014). According to results of a study (Aksu and Kaya 2002), mold and yeast count increased in initial stage of production while the counts decreased after final drying and cemen paste covering. Mold and yeast counts were also reported to be at a low level in pastirma (Kaban 2013). It has been suggested that the applying of MAP (Modified Atmosphere Packaging) might also increase the microbiological quality of Pastirma (Aksu et al. 2005b). Ozturk (2015) conducted a study to investigate change in the yeast dynamic during Pastirma processing and the technological properties of yeast strain. It was reported that the initial population of yeast decreased from 4.42 log cfu/g to 3.61 log cfu/g during the curing process, and total 100 isolates were identified according to their genotype. *Candida zeylanoides* (58%), *Candida deformans* (12%) and *Candida galli* (11%) were the dominant yeast species in Pastirma. It can be concluded that Pastirma has long shelf life as a meat product due to its lower a_w and cemen paste coverings and it can have longer shelf-life if the appropriate conditions are provided.

10.4 Kavurma

Kavurma (Fig. 10.6) is one of the most popular traditional Turkish cooked and fried meat products. Kavurma is also produced in many regions of the world such as the Middle East, Middle Asia, some Mediterranean and some European countries (Aksu 2009); it can be produced from beef, water buffalo, or mutton. Kavurma has higher amount of fat content and a long

Figure 10.6 Kavurma.

shelf life among the ready-to-eat meat products. In the past, Kavurma was used only to conserve meat while recently, it has been produced in modern meat processing plants and is commercially available in vacuum-packaged forms (Yetim et al. 2006a). In production, meat is cut into small pieces and mixed with salt, beef fat, tallow or tailed fat and is collectively cooked and then packaged (Kayaardi et al. 2005). Since Kavurma is generally consumed without further processing or cooking, it is considered as a ready-to-eat meat product.

10.4.1 Production

Figure 10.7 presents the flow diagram of Kavurma production, according to the procedure described in Turkish Standard for Kavurma (TS 2002a) and related publications (Aksu and Kaya 2005, Kayaardi et al. 2005). In the first stage of production, beef is trimmed for excessive fat and connective tissue. After that, the meat is diced, cut into small pieces of (4×5×6 cm) and mixed with table salt in the amount of 2 to 5% of meat weight (Kayaardi et al. 2005, Kilic 2009). Then, approximately 30% of tallow is added to salty meat pieces, and the cooking process is started slowly at 50 to 60°C in an open cover boiler for about 30 minutes. After the cooking process, the tallow required for Kavurma production is added and the temperature is raised to 105°C at 4 to 5 atm pressure. Cooking takes about three to four hours. Meat pieces are mixed during this cooking stage for the uniform cooking. Cooking is finished when the meat colour turns from red to dark gray, elasticity of meat is lost and the moisture content reaches approximately 40%. After cooking, Kavurma is cooled to about 55 to 60°C and finally, the Kavurma is vacuum packaged and stored at 4°C (Aksu and Kaya 2005, Kayaardi et al. 2005).

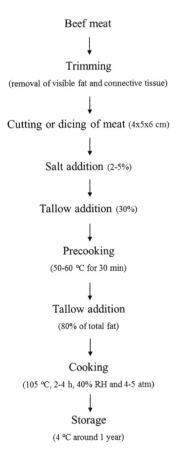

Figure 10.7 Flow chart on the Kavurma production process (Aksu and Kaya 2005, Kayaardi et al. 2005).

10.4.2 Quality characteristics

According to Turkish standards, Kavurma must have maximum 35% fat, 40% moisture, 7% salt and pH of 4.5 to 5.5 (TS 2002a). Additionally, Kavurma must have no rancidity and show no pathogenic bacteria. Since Kavurma has a high amount of animal fat compared to other meat products, lipid oxidation and colour stability are serious concerns during the storage at market (Kilic 2009). In their report, Aksu and Kaya (2005) found that the use of butylated hydroxyanisole (BHA) and α-tocopherol as antioxidative agents significantly reduced, lipid oxidation and improved colour stability of the vacuum packaged Kavurma sample stored at 4°C. In another study the addition of rosemary or *Hibiscus sabdariffa* enhanced the colour properties of Kavurma during the cooking (Bozkurt and Belibagli 2009). Kayaardi et al. (2005) suggested that using some natural condiments like sage, thyme and ginger improved oxidative stability and the sensory properties of Kavurma. Yetim et al. (2006a) reported that use of nitrite at the level of 100 ppm might limit lipid oxidation and increase antimicrobial activity. But addition of nitrite is not recommend due to unusual color in Kavurma. Appearance of the Kavurma must be uniform and in marbled structure. The colour of the outer surface must be dark brown; grey colour of the outer surface is not preferred. The colour of interior parts of the Kavurma must be brown. Red colour in outer surface is not desired because it is indicator of insufficient heat treatment. Textural quality of the Kavurma can be evaluated by hand and fingers; it should show a resistance to finger force, and elastic structure is not preferred as it indicates the insufficient cooking too (Gokalp et al. 2004).

10.4.3 Microbiological properties

Kavurma is a meat product treated with heat, and its shelf life is longer than that of the other ready-to-eat meat products. In spite of the application of heat treatment in Kavurma production, microbiological deterioration may occur due to the prevalence of small microbial load. Some spore forming bacteria such as *C. botulinum* are considered to be a potential health risk due to their heat-stable and deadly toxins (Yetim et al. 2006a).

The microbiological quality of the Kavurma sold in the Turkish market was studied and total aerobic mesophilic bacteria, yeasts and molds and coliform count were between 2.00 to 7.40 log cfu/g, 1.00 to 5.48 log cfu/g, 1 to 3.75 log cfu/g respectively while *E. coli*, *Salmonella* and *S. aureus* are not detected (Cetin et al. 2005). Yetim et al. (2006a) conducted a research to determine the influence of nitrite and the traditional cooking process on the survival and proliferation of *C. botulinum* and the autoxidation properties of the Kavurma. In the study, *Clostridium sporogenes* was used due to similar characteristics with *C. botulinum*. They observed that *C. sporogenes* could survive during the traditional cooking process and its count decreased from 3.21 log cfu/g to 2.73 log cfu/g during the storage period. Kivanc et al. (1992) conducted a study to determine *Aspergillus parasiticus* or *Aspergillus flavus* in some meat products including Kavurma sold in the market, and reported that Kavurma did not contain these microorganisms and their toxins. However, they stated that Kavurma may be a favourable substrate for aflatoxin production by *A. parasiticus* at 25°C.

10.5 Kofte

Meatballs are another widely consumed meat product in Turkey that is also referred as Kofte. There are different types of meatballs; their peculiar characteristics may be depending on meat type, ingredients, additives and processing methods. Meatballs in Turkey take their name from the place where they are produced (Sivas, Tekirdag, Inegol, Akcaabat and others), production technique (Kadınbudu, Icli Kofte) and material used (Cig Kofte) (Saricaoglu and Turhan 2013). Meatballs are produced from minced meat and some other ingredients such as onions, bread crumbs, bulgur and spices. The differences between meatball types produced in Turkey and other countries resulted from the use of different meats and, different additives like onions, bread crumbs, spices, garlic and their production techniques (Sarıcaoglu and Turhan 2013).

10.5.1 Production

Meatball production may vary according to the types of meatballs. Generally, beef is used as a main ingredient. Before production, all the subcutaneous fat and intermuscular fats nervous and visible connective tissues of meat are removed and used as the fat source for the recipe. After grinding of the lean and fat, different spices (red pepper, black pepper, cumin, etc.) and some other ingredients such as onions, bread crumbs, garlic and salt are added to form the meatball mixture. And then the mix is homogenized by kneading for about 15 minutes to obtain meatball dough. Then the dough is shaped based on the meatball type and stored at 4°C until the use (Serdaroglu 2006, Yilmaz 2005). Akcaabat, Sultanahmet, Sivas, Inegol, Cine, Tekirdag, Tire, Kadınbudu Kofte, Islama Kofte, Cig Kofte and Icli Kofte are considered to be the most popular meatball types produced in Turkey.

As mentioned before, several types of meatballs are produced, and they vary due to ingredients and production technology. For example, in the production of Akcaabat meatballs, bread crumbs, salt and garlic are used in the ground beef. Only garlic is used as a spice in the production of this method (Sarıcaoglu and Turhan 2013). However, in the production of Tekirdag meatballs, manufacturers generally prefer veal as a meat source. Veal is ground and several kinds of spices (0.1% ground black pepper, 2% red pepper and 0.4% cumin) and some other additives are mixed in. Then, the mix is kneaded for about 30 minutes and reground. After one day storage period in a cold room, the meatball dough is shaped into 2 cm diameters with a weight of 18 to 20 g each (Yilmaz et al. 2002).

Sivas meatballs have a similar production method to the Tekirdag meatballs (Tornuk et al. 2010). Here, only salt is used during the preparation of the mixture (Can et al. 2013). Production flow chart of Inegol meatballs, which is one of the most popular meatballs in Turkey, is shown in Fig. 10.8 according to the procedure described in the literature (Temelli et al. 2011a). As can be seen, only salt is used in addition to the ground lamb and veal. Again, in a production of the Cine meatballs, which is another traditional meatball, ground veal or rib is used as meat source. In the production of

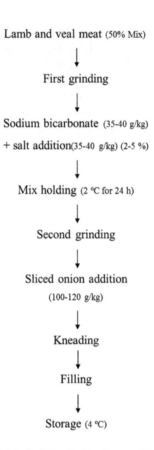

Lamb and veal meat (50% Mix)

↓

First grinding

↓

Sodium bicarbonate (35-40 g/kg)

+ salt addition(35-40 g/kg) (2-5 %)

↓

Mix holding (2 °C for 24 h)

↓

Second grinding

↓

Sliced onion addition

(100-120 g/kg)

↓

Kneading

↓

Filling

↓

Storage (4 °C)

Figure 10.8 Inegol Meatball Production Process (Temelli et al. 2011a).

Cine Kofte, ground meat is mixed with salt, sliced onions and cumin. No other ingredients are used in this recipe.

Cig Kofte (raw meatballs) are popular as appetizers all over Turkey, especially in the south east and east Anatolia regions (Yıldırım et al. 2005). These are produced from raw ground meat, bulgur, tomato paste, fresh onions, garlic, and several other spices especially cumin, hot pepper (local name: isot) and black pepper. All ingredients according to the recipe is weighed, and they are kneaded by hand or machine in a big bowl and then shaped by hand. There is no heating or cooking process in Cig Kofte production (Cetinkaya et al. 2012). It is generally prepared and freshly consumed, its storage life is very limited.

Kadınbudu is another traditional meatball produced in Turkey, which is also produced commercially. Here, ground meat is mixed with some other ingredients such as salt and several types of spices. After mixing, the dough is shaped and the meatballs are covered with batter prepared using bread crumbs and egg. Then, the coated and shaped meatball is fast deep-fried and baked in a humid steam convection oven (Temelli et al. 2011b).

Icli Kofte (stuffed meatballs) are also popularly consumed as a hot appetizer before other kebabs. These meatballs have two parts—internal and external. Fillings (internal part) contain ground veal or lamb, walnut, parsley, onions, pine nuts, some spices and salt. The outer part includes bulgur and spices such as red pepper, cumin and salt. All ingredients of the filling material are precooked until the minced meat loses all its water. Bulgur and other ingredients are mixed and kneaded to form the outer part of the meatballs. After stuffing, the meatballs are fried and served hot. Icli Kofte is usually consumed in southern Turkey.

10.5.2 Quality characteristics

The quality of meatballs is affected by several factors such as production method, meat type, fat, salt content and spices used. According to Turkish standards, moisture, fat and salt content must not exceed 65%, 25% and 2% respectively in the uncooked form (TS 2010). Sarıcaoglu and Turhan (2013) conducted a survey to investigate physicochemical and textural properties of Akcaabat meatballs produced in the Black Sea region of Turkey. In their study, the average value of the moisture, protein, and salt contents ranged 48.38 to 54.53%, 14.15 to 15.14%, 19.09 to 22%, 1.23 to 1.51% respectively. As can be seen, meatballs have a considerable amount of fat and are susceptible to oxidation. It is well known that oxidation causes quality losses in stored meat products, especially minced or ground meat products, by reducing nutritional value of the lipid, changing colour and undesirable flavour and aroma as a results of formation of secondary volatiles (Akarpat et al. 2008). It was suggested that vacuum or MAP and some antioxidants could inhibit lipid oxidation during the storage period (Karpinska and Tymoszczyk 2013).

Meatball quality is highly affected from the meat and fat source used in meatball formulation. Generally, ground beef is used as a meat source in the formulation. On the other hand, type of meat used in meatball formulation is changed according to the different kinds of meatballs. For example, veal is used in Tekirdag meatballs while lamb and veal mix are preferred in Inegol meatball formulation. Sensory and textural properties of ground meat products are highly related to fat content and moisture binding capacity, which is the critical factor for cooking quality. Reduction of the meat in formulation may cause undesirable effects on water and fat binding properties, texture and sensory quality and increase cooking loss (Serdaroglu 2006). For example, tenderness, juiciness and flavour ratings decreased and shear force increased with decreasing fat content in ground meat products. Serdaroglu and Degirmencioglu (2004) conducted a study to investigate the effects of fat level (5%, 10% and 20%) and corn flour (CF 0%, 2% and 4%) on chemical composition, cooking characteristics and sensory properties of Turkish meatballs. They reported that texture and overall palatability scores decreased with decreasing fat content. However, CF had no detrimental effect on sensory properties, except for on appearance.

Some authors have also conducted studies on low fat ground meat products due to the adverse effects of fat on health (Serdaroglu 2006, Serdaroğlu et al. 2005, Tekin et al. 2010). In one study, meatballs are formulated with blackeye bean flour, chickpea flour, lentil flour and rusk at a level of 10% (Serdaroğlu et al. 2005). It was found that all meatball samples showed high sensory scores while blackeye bean flour and chickpea flour had higher water holding capacity. Tekin et al. (2010) reported that formulation of wheat bran in meat patty samples increased fat retention, water holding capacity and cooking yield. Serdaroglu (2006) reported that addition of whey powder in meatball formulation increased cooking yield regardless of the fat level, fat and moisture retention. However, whey powder addition had no effect on the juiciness of the meatballs (Yılmaz 2005), it was reported that formulation of wheat bran instead of fat increased firmness properties. In another study, some gums were added to meatball formulation and locust bean gum exhibited the highest sensory properties in some Turkish meatballs (Demirci et al. 2014).

Salt content is another important ingredient affecting significantly the overall quality of meatball samples. Addition of the salt in ground meat increased protein solubility and water holding capacity. It was reported that salt content increased water holding capacity at moderate levels. However, higher salt content decreased the water holding capacity due to denaturation and unfolding of protein (Tekin et al. 2010). It can be concluded that formulation of the fat, salt and other non meat ingredients should be optimized to produce better textural, sensory and cooking quality of meatballs, regardless of the meatball types.

10.5.3 Microbiological properties

Meat and meat products are generally sold in small butcher shops either as in diced or in ground form, which is generally preferred in Turkish markets. Formulation of the meatballs is mainly composed of raw ground meat, which is the suitable environment for microbial growth. Therefore, shelf life of meatballs is highly dependant on the ground meat and microbiological quality of other ingredients. Since meatballs are consumed after cooking process, the processing technique is also an important factor for the microbiological quality of final products. According to Turkish standards, total mesophilic bacteria, psychrophilic bacteria and yeast and mold counts are not allowed higher than 10^6 cfu/g, 10^5 cfu/g and 10^2

cfu/g, respectively. Additionally, *E. coli*, *S. aureus*, *Salmonella* and sulphide-reducing anaerobes should not be found in meatballs (TS 2010). Recently, some studies were conducted to determine microbiological quality of ground meat in terms of total bacteria and coliform bacteria counts. It was revealed that the microbiological quality of the ground meat was very poor (Yilmaz et al. 2005, Can et al. 2013). For example, in the study of Can et al. (2013), ready-to-eat Sivas meatball samples were analyzed to determine total mesophilic aerobic bacteria, Enterobacteriaceae, *E. coli*, coagulase positive *S. aureus*, *Salmonella* sp. and psychrophilic bacteria. Total mesophilic aerobic bacteria, Enterobacteriaceae, coagulase positive *S. aureus*, psychrophilic bacteria counts of the Sivas meatball were 2.7–4.9 log cfu/g, <10–2.1 log cfu/g, <10–1.9 log cfu/g, 1.6–3.8 log cfu/g, respectively, and *E. coli* and *Salmonella* sp. were not found in any of the meatball samples.

In another study, Yilmaz et al. (2005) reported that the average counts of total mesophilic aerobic bacteria, *S. aureus*, yeast and mold and *E. coli* were 2.4×10^6 cfu/g, 4.8×10^3 cfu/g, 2.2×10^4 cfu/g, and 2×10^2 cfu/g, respectively. They also investigated the changes in the microbiological quality of Tekirdağ meatballs at 4°C after different heating systems. It was stated that *E. coli* O157:H7 was destroyed by all cooking techniques and 2–3 log cycles reduction was observed after grill cooking or the microwave cooking process. A similar study was conducted (Yilmaz et al. 2002), and the effects of different cooking processes (grilling, oven and microwave cooking) on the microbial flora of the raw Tekirdag meatballs were investigated. The number of total bacteria, psychrophilic bacteria, yeast and mold, coliforms, *E. coli*, total Staphylococcaceae, *S. aureus* count of raw Tekirdag meatballs were 6.02×10^6 cfu/g, 1.36×10^5 cfu/g, 2.46×10^5 cfu/g, 1.16×10^5 cfu/g, 1.06×10^2 cfu/g, 3.36×10^2 cfu/g, 85 cfu/g respectively. In this study, *Salmonella* was detected only in one sample while *C. perfringens* was not found in any samples. The microbial flora of the samples decreased two to three log cycles after grilling (71.8°C) and oven cooking (79.8°C), but three to four log cycles decreased in the microwave oven (97.8°C) process.

In another research, the effect of nisin and bovine lactoferrin on the microbiological quality of Tekirdag meatballs was studied (Colak et al. 2008). It was reported that total aerobic bacteria, *coliform*, *E. coli*, *total psychrophilic* bacteria, *Pseudomona* ssp., yeast and mold counts were significantly reduced when the lactoferrin and its combination with nisin were applied to the meatball samples. The largest reduction in these microbial counts was observed from the meatballs treated with a mixture of lactoferrin (200 mg/g) and nisin (100 mg/g). This combination increased the refrigerated shelf life of the naturally contaminated Turkish meatballs up to 10 days compared to three days for control samples.

In another study, thermal resistance of *L. monocytogenes* in Inegol meatballs was investigated (Soyutemiz and Cetinkaya 2005). In the study, three different heat treatments (30 minutes at 63°C, 4 minutes at 80°C and 4 minutes at 85°C) were applied. In the results, the heat treatment for four minutes at 85°C adequately eliminated *L. monocytogenes* while the heat treatment for four minutes at 80°C and 30 minutes at 63°C did not eliminate the high number of *L. monocytogenes* in meatballs.

Since microbial growth in raw meat products can be rapid during storage, microbial spoilage is one of the major problems of the meat industry. Some alternative storage methods are necessary to overcome this problem. MAP is one of the effective methods to increase the shelf life of meat products. Yilmaz and Demirci (2010) conducted a study to determine effects of MAP on physiochemical and microbiological quality of meatballs. In that study, meatball samples was enclosed by polyethylene film (PS packs) in polystyrene trays, vacuumed and modified atmosphere packaged (MAP) (65% N_2, 35% CO_2), and then stored under refrigerated conditions (4°C) for 8, 16 and 16 days for PS packs, vacuum and MAP, respectively. They suggested that shelf life of the meatballs was suitable with MAP and vacuum packaging. Again, Bingol et al. (2012) investigated the effect of MAP on the shelf life of Tekirdag meatballs. They reported that microbial growth could be retarded by increasing CO_2 level, and shelf life of the meatball samples could be increased upto eight days.

Cig Kofte is very popular raw meatballs consumed as appetizers in Turkey. Since no heating or cooking process is applied during any stage of manufacturing, the possibility of microbiological growth during storage is high. It is a well known fact that as a major ingredient, ground meat has potential microbial health risks being a suitable environment for a variety of food-borne pathogenic bacteria and zoonotic parasites (Can et al. 2013).

Several studies showed that microbiological quality of the cig kofte was very poor (Cetin et al. 2010, Cetinkaya et al. 2012). Ardic and Durmaz (2008) carried out a study to determine effects of different storage temperatures on the microbiological characteristics and pH values of Cig Kofte during storage. They reported that their microbial load at refrigerated storage was lower than that of the samples stored at room temperature. They did not detect *Listeria* and *Salmonella* sp. for both refrigerator temperature and room temperature storages. They also suggested that for the best microbial quality for minced meat and ingredients, personnel hygiene was necessary to minimize microbial growth. For this purpose, some studies were conducted to determine effects of sodium lactate on shelf life of the cig kofte samples (Bingol et al. 2013, Bingol et al. 2014b). They concluded that sodium lactate could be used to improve shelf life. Dikici et al. (2013) studied the effects of some essential oil compounds such as saline, cineole, limonene, carvone, linalool or eugenol for the inactivation of *L. monocytogenes* and *E. coli* O157:H7 in cig kofte. It was suggested that eugenol, carvone and linalool could be used for the inactivation of *E. coli* O157:H7 and *L. monocytogenes* in cig kofte.

References

Adiguzel, G.C. and M. Atasever. 2009. Phenotypic and genotypic characterization of lactic acid bacteria isolated from Turkish dry fermented sausage. Rom. Biotech. Lett. 14: 4130–4138.

Akarpat, A., S. Turhan and N.S. Ustun. 2008. Effects of hot-water extracts from myrtle, rosemary, nettle and lemon balm leaves on lipid oxidation and colour of beef patties during frozen storage. J. Food Process. Pres. 32: 117–132.

Aksu, M.I. 2009. Fatty acid composition of beef intermuscular, sheep tail, beef kidney fats and its effects on shelf life and quality properties of kavurma. J. Food Sci. 74: 65–S72.

Aksu, M.I. and E. Erdemir. 2014. A survey of selected minerals in ready-to-eat pastirma types from different regions of Turkey using ICP/OES. Turk. J. Vet. Anim. Sci. 38: 564–571.

Aksu, M.I. and M. Kaya. 2001. Some microbiological, chemical and physical characteristics of Pastirma marketed in Erzurum. Turk. J. - Vet. Anim. Sci. 25: 319–326.

Aksu, M.I. and M. Kaya. 2002. Potasyum nitrat ve starter kültür kullanılarak üretilen pastirmaların bazı mikrobiyolojik ve kimyasal özellikleri. Turk. J. Vet. Anim. Sci. 26: 125–132.

Aksu, M.İ. and M. Kaya. 2005. The effect of α-tocopherol and butylated hydroxyanisole on the colour properties and lipid oxidation of Kavurma, a cooked meat product. Meat Sci. 71: 277–283.

Aksu, M.İ., M. Kaya and H.W. Ockerman. 2005a. Effect of modified atmosphere packaging, storage period, and storage temperature on the residual nitrate of sliced-pastirma, dry meat product, produced from fresh meat and frozen/thawed meat. Food Chem. 93: 237–242.

Aksu, M.I., M. Kaya and H.W. Ockerman. 2005b. Effect of modified atmosphere packaging and temperature on the shelf life of sliced pastirma produced from frozen/thawed meat. J. Muscle Foods. 16: 192–206.

Anonymous. 2014. Starter Cultures for Making Fermented Sausages. http://www.meatsandsausages.com/sausage-types/fermented-sausage/cultures.

Ardic, M. and H. Durmaz. 2008. Determination of changes occurred in the microflora of cig Kofte (raw meatballs) at different storage temperatures. Int. J. Food Sci. Technol. 43: 805–809.

Başyiğit, G., A.G. Karahan and B. Kılıç. 2007. Functional starter cultures and probiotics in fermented meat products. Turkish Bulletin of Hygiene. 64: 60–69.

Bingol, E.B., O. Cetin, H.C. Uzum and H. Hampikyan. 2013. Effects of sodium lactate on the presence of *Staphylococcus aureus* and enterotoxins in Cig Kofte (raw meatball). Turk. J. Vet. Anim. Sci. 37: 719–726.

Bingol, E.B., G. Ciftcioglu, F.Y. Eker, H. Yardibi, O. Yesil, G.M. Bayrakal and G. Demirel. 2014a. Effect of starter cultures combinations on lipolytic activity and ripening of dry fermented sausages. Ital. J. Anim. Sci. 13: 776–781.

Bingol, E.B., H. Colak, O. Cetin and H. Hampikyan. 2014b. Effects of sodium lactate on the shelf life and sensory characteristics of Cig Kofte—a Turkish traditional raw meatball. J. Food Process Pres. 38: 1024–1036.

Bingol, E.B., H. Colak, O. Cetin, T. Kahraman, H. Hampikyan and O. Ergun. 2012. Effects of high-oxygen modified atmosphere packaging on the microbiological quality and shelf life of Tekirdag Kofte: a Turkish type meatball. J. Anim. Vet. Adv. 11: 3148–3155.

Bozkurt, H. and K.B. Belibagli. 2009. Use of rosemary and *Hibiscus sabdariffa* in production of Kavurma, a cooked meat product. J. Sci. Food Agric. 89: 1168–1173.

Bozkurt, H. and K.B. Belibağlı. 2012. Sucuk: Turkish Dry-Fermented Sausage. pp. 663–683. *In*: Y.H. Hui and E. Özgül Evranuz (eds.). Handbook of Animal-based Fermented Food and Beverage Technology. CRC press, Boca Raton, FL, USA.

Bozkurt, H. and O. Erkmen. 2002. Effects of starter cultures and additives on the quality of Turkish style sausage (Sucuk). Meat Sci. 61: 149–156.

Bozkurt, H. and O. Erkmen. 2004. Effect of nitrate/nitrite on the quality of sausage (Sucuk) during ripening and storage. J. Sci. Food Agric. 84: 279–286.

Çakıcı, N., M.I. Aksu and E. Erdemir. 2014. A survey of the physico-chemical and microbiological quality of different Pastirma types: a dry-cured meat product. CyTA - J. Food (in Press).

Can, O.P., S. Şahin, M. Erşan and F. Harun. 2013. Sivas Kofte and examination of microbiological quality. Biotech. Anim. Husbandry 29: 133–143.

Cetin, B., S. Sert and H. Yetim. 2005. Microbiological quality of the Kavurma samples marketed in Erzurum, Turkey. Ann. Microbiol. 55: 27–31.

Cetin, O., E.B. Bingol, H. Colak, O. Ergun and C. Demir. 2010. The microbiological, serological and chemical qualities of mincemeat marketed in Istanbul. Turk. J. Vet. Anim. Sci. 34: 407–412.

Cetinkaya, F., T.E. Mus, R. Cibik, B. Levent and R. Gulesen. 2012. Assessment of microbiological quality of Cig Kofte (raw consumed spiced meatballs): Prevalence and antimicrobial susceptibility of salmonella. Food Control. 26: 15–18.

Ceylan, S. and M.İ. Aksu. 2011. Free amino acids profile and quantities of 'sırt', 'bohca' and 'sekerpare' pastirma, dry cured meat products. J. Sci. Food Agric. 91: 956–962.

Colak, H., H. Hampikyan, E.B. Bingol and H. Aksu. 2008. The effect of nisin and bovine lactoferrin on the microbiological quality of Turkish-style meatballs (Tekirdag Kofte). J. Food Safety 28: 355–375.

Colak, H., H. Hampikyan, B. Ulusoy and E.B. Bingol. 2007. Presence of *Listeria monocytogenes* in Turkish style fermented sausage (Sucuk). Food Control. 18: 30–32.

Demirci, Z.O., I. Yilmaz and A.S. Demirci. 2014. Effects of xanthan, guar, carrageenan and locust bean gum addition on physical, chemical and sensory properties of meatballs. J. Food Sci. Technol. Mysore 51: 936–942.

Dikici, A., O.I. Ilhak and M. Calicioglu. 2013. Effects of essential oil compounds on survival of *Listeria monocytogenes* and *Escherichia coli* O157:H7 in Cig Kofte. Turk. J. Vet. Anim. Sci. 37: 177–182.

Dogbatsey, F.K. 2010. The combined effects of *Pediococcus acidilactici* and *Lactobacillus curvatus* on *Listeria monocytogenes* ATCC 43251 in dry fermented sausages. *In*: Food & Nutritional Sciences The Graduate School, University of Wisconsin-Stout, USA.

Dogruer, Y., M. Nizamlioglu, U. Gurbuz and S. Kayaardi. 1998. The effects of various cemen mixtures on the quality of pastrami II: Microbiological quality. Turk. J. Vet. Anim. Sci. 22: 221–229.

Gençcelep, H., G. Kaban, M.I. Aksu, F. Öz and M. Kaya. 2008. Determination of biogenic amines in Sucuk. Food Control. 19: 868–872.

Gokalp, H.Y., M. Kaya and O. Zorba. 2004. Meat products processing (Et Ürünleri İşleme Mühendisliği). Atatürk Univ. Publ. No: 786, Faculty of Agric. No.: 320, Erzurum, Turkey.

Gök, V., E. Obuz and L. Akkaya. 2008. Effects of packaging method and storage time on the chemical, microbiological, and sensory properties of Turkish Pastirma—a dry-cured beef product. Meat Sci. 80: 335–344.

Gönülalan, Z., A. Arslan and A. Köse. 2004. Effects of different starter culture combinations on fermented sausages. Turkish Journal of Veterinary and Animal Sciences 28: 7–16.

Güleren, A. 2008. Molecular and biochemical idenfication of *Lactobacillus sake* and *Lactobacillus curvatus* strains and determinations of some technological properties. *In*: Department of Nutrition Hygiene and Technology, Uludag University, Bursa, Turkey.

Gürakan, G.C., T.F. Bozoglu and N. Weiss. 1995. Identification of Lactobacillus strains from Turkish-style dry fermented sausages. LWT - Food Sci. Technol. 28: 139–144.

Honikel, K.O. 2008. The use and control of nitrate and nitrite for the processing of meat products. Meat Sci. 78: 68–76.

Kaban, G. 2009. Changes in the composition of volatile compounds and in microbiological and physicochemical parameters during pastirma processing. Meat Sci. 82: 17–23.

Kaban, G. 2013. Sucuk and pastirma: Microbiological changes and formation of volatile compounds. Meat Sci. 95: 912–918.

Kaban, G. and M. Kaya. 2006. Pastirmadan katalaz pozitif kokların izolasyonu ve identifikasyonu. In: Türkiye 9. Gıda Kongresi, Bolu, Turkiye.

Kaban, G. and M. Kaya. 2008. Identification of lactic acid bacteria and gram-positive catalase-positive cocci isolated from naturally fermented sausage (Sucuk). J. Food Sci. 73: 385–388.

Kaban, G. and M. Kaya. 2009. Effects of *Lactobacillus plantarum* and *Staphylococcus xylosus* on the quality characteristics of dry fermented sausage "sucuk". J. Food Sci. 74: 58–63.

Karpinska-Tymoszczyk, M. 2013. The effect of oil-soluble rosemary extract, sodium erythorbate, their mixture, and packaging method on the quality of Turkish meatballs. J. Food Sci. Technol. Mysore 50: 443–454.

Kayaardi, S., F. Durak, A. Kayacier and M. Kayaardi. 2005. Chemical characteristics of Kavurma with selected condiments. Int. J. Food Prop. 8: 513–520.

Kayisoglu, S., I. Yilmaz, M. Demirci and H. Yetim. 2003. Chemical composition and microbiological quality of the doner kebabs sold in Tekirdag market. Food Control. 14: 469–474.

Kesmen, Z., A.E. Yetiman, A. Gulluce, N. Kacmaz, O. Sagdic, B. Cetin, A. Adiguzel, F. Sahin and H. Yetim. 2012. Combination of culture-dependent and culture-independent molecular methods for the determination of lactic microbiota in Sucuk. Int. J. Food Microbiol. 153: 428–435.

Kilic, B. 2009. Current trends in traditional Turkish meat products and cuisine. LWT - Food Sci. Technol. 42: 1581–1589.

Kivanc, M., S. Sert and I. Hasenekoglu. 1992. Production of aflatoxins in sausage, salami, Sucuk and Kavurma. Nahrung-Food 36: 293–298.

Kök, F., G. Özbey and A. Muz. 2007. Aydın İlinde Satışa Sunulan Fermente Sucukların Mikrobiyolojik Kalitelerinin İncelenmesi. Fırat Üniv. Sağlık Bilimleri Vet. Derg. 21: 249–252.

Obuz, E., L. Akkaya and V. Gök. 2012. Turkish Pastirma: a dry cured beef product. pp. 637–646. *In*: Y.H. Hui (ed.). Hand Book of Meat and Meat Processing. 2nd ed. Boca Raton, Fl.: CRC Press.

Ozturk, I. 2015. Antifungal activity of propolis, thyme essential oil and hydrosol on natural mycobiota of Sucuk, a turkish fermented sausage: monitoring of their effects on microbiological, colour and aroma properties. J. Food Process. Pres. 39: 1148–1158.

Ozturk, I. 2015. Presence, changes and technological properties of yeast species during processing of Pastirma, a Turkish dry cured meat product. Food Control. 50: 76–84.

Ozturk, I. and O. Sagdic. 2014. Biodiversity of yeast mycobiota in Sucuk, a traditional Turkish fermented dry sausage: Phenotypic and genotypic identification, functional and technological properties. J. Food Sci. 79: 2315–2322.

Ozdemir, H., H.T. Çelik, I. Erol and A. Yurtyeri. 1995. Some physiological and biochemical characterization of lactobacilli isolated from Turkish fermented sausage Sucuk. Vet. J. Ankara Univ. 42: 317–321.

Rantsiou, K. and L. Cocolin. 2006. New developments in the study of the microbiota of naturally fermented sausages as determined by molecular methods: a review. Int. J. Food Microbiol. 108: 255–267.

Sameshima, T., C. Magome, K. Takeshita, K. Arihara, M. Itoh and Y. Kondo. 1998. Effect of intestinal *Lactobacillus* starter cultures on the behaviour of *Staphylococcus aureus* in fermented sausage. Int. J. Food Microbiol. 41: 1–7.

Sarıcaoglu, F.T. and S. Turhan. 2013. Chemical composition, colour and textural properties of Akçaabat meatball: a traditional Turkish meat product. Gida. 34: 191–198.

Serdaroglu, M. 2006. Improving low fat meatball characteristics by adding whey powder. Meat Sci. 72: 155–163.

Serdaroglu, M. and O. Degirmencioglu. 2004. Effects of fat level (5%,10%,20%) and corn flour (0%,2%,4%) on some properties of Turkish type meatballs (Kofte). Meat Sci. 68: 291–296.

Serdaroğlu, M., G. Yıldız-Turp and K. Abrodímov. 2005. Quality of low-fat meatballs containing legume flours as extenders. Meat Sci. 70: 99–105.

Siriken, B., O. Cadirci, G. Inat, C. Yenisey, M. Serter and M. Ozdemir. 2009. Some microbiological and physico-chemical quality of Turkish Sucuk (sausage). J. Anim. Vet. Adv. 8: 2027–2032.

Soyutemiz, G.E. and F. Cetinkaya. 2005. Thermal resistance of *Listeria monocytogenes* in Inegol meatballs. Turk. J. Vet. Anim. Sci. 29: 319–323.

Tekin, H., C. Saricoban and M.T. Yilmaz. 2010. Fat, wheat bran and salt effects on cooking properties of meat patties studied by response surface methodology. Int. J. Food Sci. Technol. 45: 1980–1992.

Temelli, S., M.K.C. Sen and S. Anar. 2011a. Assessment of microbiological changes in fresh uncooked Inegol meatballs stored under two different modified atmosphere packaging conditions. Ank. Univ. Vet. Fak. Derg. 58: 273–278.

Temelli, S., M.K.C. Sen and S. Anar. 2011b. Microbiological evaluation of chicken Kadinbudu meatball production stages in a poultry meat processing plant. Ank. Univ. Vet. Fak. Derg. 58: 189–194.

Törnük, F., Z. Kesmen and H. Yetim. 2010. "Sivas Köftesi (Tombul Köfte) ve Geleneksel Gıda Kültürümüzdeki Yeri", 380-381. First International Symposium on Traditional Foods from Adriatic to Caucasus, Tekirdag, TÜRKIYE.

TS. 2002a. Kavurma standardı (Kavurma standard). Ankara, Turkey.

TS. 2002b. Pastirma standardı (Turkish Pastirma standard). Ankara, Turkey.

TS. 2010. Köfte ve Hamburger standardı (meatball and hamburger standard). Ankara, Turkey.

Turkish Food Codex. 2009. Türk Gıda Kodeksi Mikrobiyolojik Kriterler Tebliği Tebliğ No.: 2009/68, Agriculture and Rural Affairs Minister, Ankara, Turkey.

Turkish Food Codex. 2012. Türk Gıda Kodeksi Et Ürünleri Tebliği. In: Tebliğ No.: 2012/74, Agriculture and Rural Affairs Minister, Ankara, Turkey.

Turkish Standard. 2002. Turkish Sucuk (No.: TS 1070). Turkish Standards Institute, Ankara, Turkey.

Yalçın, H. and O.P. Can. 2013. An Investigation on the Microbiological Quality of Sucuk Produced by Traditional Methods. Kafkas Univ. Vet. Fak. Derg. 19: 705–708.

Yetim, H., A. Kayacier, Z. Kesmen and O. Sagdic. 2006a. The effects of nitrite on the survival of *Clostridiumsporogenes* and the autoxidation properties of the Kavurma. Meat Sci. 72: 206–210.

Yetim, H., O. Sagdic, M. Dogan and H.W. Ockerman. 2006b. Sensitivity of three pathogenic bacteria to Turkish cemen paste and its ingredients. Meat Sci. 74: 354–358.

Yıldırım, I., S. Uzunlu and A. Topuz. 2005. Effect of gamma irradiation on some principle microbiological and chemical quality parameters of raw Turkish meatballs. Food Control. 16: 363–367.

Yilmaz, I. 2005. Physicochemical and sensory characteristics of low fat meatballs with added wheat bran. J. Food Eng. 69: 369–373.

Yilmaz, I., M. Arici and T. Gumus. 2005. Changes of microbiological quality in meatballs after heat treatment. Eur. Food Res. Technol. 221: 281–283.

Yilmaz, I. and M. Demirci. 2010. Effect of Different Packaging Methods and Storage Temperature on Microbiological and Physicochemical Quality Characteristics of Meatballs. Food Sci. Technol. Int. 16: 259–265.

Yilmaz, I., H. Yetim and H.W. Ockerman. 2002. The effect of different cooking procedures on microbiological and chemical quality characteristics of Tekirdag meatballs. Nahrung-Food. 46: 276–278.

Yüceer, Ö. and B. Özden Tuncer. 2015. Determination of Antibiotic Resistance and Biogenic Amine Production of Lactic Acid Bacteria Isolated from Fermented Turkish Sausage (sucuk). J. Food Safety 35: 276–285.

CHAPTER 11

Greek Dairy Products

Ekaterini Moschopoulou and *Golfo Moatsou**

11.1 Introduction

In Greece, the manufacture of dairy products, especially of cheeses, is a
tradition of centuries. Greek dairy products are produced from sheep, goat
or cow milk or mixtures of them, compiling the traditional manufacturing
steps and the practices of the modern food industry. Greece is among
the leading countries worldwide in terms of annual production of small
ruminants' milk per habitant, i.e., about 116 kg per person per year. In
2013, according to the Hellenic Organization of Milk and Meat, about
608.000, 520.000 and 124.000 tonnes of cow, sheep and goat milk were
processed respectively (www.elog.gr). Cow milk is used mainly for the
production of drinking milk, i.e., pasteurized, ESL, UHT and evaporated,
and of fermented milk products, i.e., yoghurt, yoghurt desserts and
buttermilk-type products. Only 10 to 15% of cow milk is exploited by the
cheese industry. In contrast, over 90% of sheep and goat milk is utilized
exclusively for cheese and yoghurt production. Apart from yoghurt and
cheese, goat milk is also used for the production of drinking milk. Sheep
and goat milk is mostly produced from animals of indigenous breeds
and have a remarkably superior quality with regard to quantitative and
qualitative characteristics of their fat and casein fractions (Alichanidis and
Polychroniadou 2008, Moatsou et al. 2008). Consequently, products like
traditional yoghurt and well-known Protected Denomination of Origin
(PDO) cheeses such as Feta, Graviera, Kasseri, all made from ewe or/and
goat milk, exhibit exceptional sensory properties due to the composition of
the milk that is used. Among the Greek dairy products, cheese and yoghurt
are the most popular and widely consumed by the Greek population.

Laboratory of Dairy Research, Department of Food Science and Human Nutrition,
Agricultural University of Athens, Athens, Greece.
* Corresponding author: mg@aua.gr

Moreover, considerable quantities of these products are exported to all parts of the world. The aim of this review is to describe the manufacturing technology of various Greek dairy products and to provide information on their compositional and particular nutritional characteristics.

11.2 Cheese varieties

The production of cheese in Greece has been a diachronic way to exploit and preserve the valuable constituents of raw milk. A great variety of added-value cheeses are produced nowadays mostly from small ruminants' milk according to traditional manufacturing steps. More than 25 kg cheese per capita are consumed annually in Greece mainly as table cheeses; about two thirds of them are Feta cheese. Twenty-one Greek cheeses are Protected Designation of Origin (PDO) products (European Union 1996, 2002, 2011) manufactured in particular regions of Greece under specified standards.

Cheese varieties can be grouped according to composition, mainly moisture, or according to processing, i.e., cheesemaking technology, or according to texture. In fact, processing configures the composition; therefore various cheese varieties can be produced from the same cheesemilk by changing the manufacturing conditions. In turn, cheese composition is related to shelf life, organoleptic characteristics and nutritional value. The latter is based on the nutritive and biological value of cheese substances. It is well established that various and numerous biological activities in relation to cheese protein and fat emerge which are enhanced during cheese ripening (e.g., López-Exposito et al. 2012, Tsorotioti et al. 2014).

In the present text, Greek cheese varieties are grouped according to technological features. Emphasis is given to manufacturing steps and composition. Furthermore, particular nutritional properties of their lipid and inorganic fractions are also reviewed due to their significance for human nutrition. Microbiology of Greek cheeses is not part of the present text; relevant reviews have been published recently (Litopoulou-Tzanetaki and Tzanetakis 2011, Angelidis and Govaris 2012).

11.2.1 Brined cheeses

The production of brined cheeses, i.e., cheeses ripened and stored in brine for a considerable amount of time, has been a traditional practice in the eastern Mediterranean area. Brine was a mean of conservation of cheese in areas with relatively high ambient temperatures and in eras without refrigeration facilities and transportation. According to Alichanidis and Polychroniadou (2008), brined cheeses can be divided into "white brined cheeses", the curds of which are not subjected to any heat treatment and "miscellaneous brined cheeses", the curds of which are subjected to various heat treatments. This classification is also used in the present chapter. The compositional characteristics of Greek brined cheeses are summarized in Table 11.1 and substances with nutritional significance in Table 11.2.

11.2.1.1 White brined cheeses

The most famous white brined cheese worldwide is Greek Feta, a PDO cheese, which is the major way for the exploitation of small ruminants' milk in Greece (Moatsou and Govaris 2011). According to PDO Standards, Feta is made from sheep milk or from its mixtures with goat milk up to 30%, produced in continental Greece and Lesvos Island and ripened for at least two months. Its moisture must be ≤56% and fat-on-dry matter ≥43%. According to the Greek Code of Foods (2012), the milk used for PDO cheeses should be raw milk from local animals, while addition of milk powder, concentrated milk, milk proteins, sodium caseinates, colours, preservatives and antimicrobial substances is forbidden. Among food additives only $CaCl_2$ at a ratio 20 g/100 kg of milk is allowed.

Numerous studies have been published about PDO Feta cheese (e.g., Alichanidis et al. 1984, Vafopoulou et al. 1989, Litopoulou-Tzanetaki et al. 1993, Mallatou et al. 1994, Michaelidou et al. 1998, 2003, 2005, Bintsis et al. 2000, Valsamaki et al. 2000, Kandarakis et al. 2001, Moatsou et al. 2002, 2004a, Georgala et al. 2005, Kondyli et al. 2013, Zoidou et al. 2015). Several review papers compiling research findings have been published during the last decade (Anifantakis and Moatsou 2006, Alichanidis 2007, Alichanidis and Polychroniadou 2008, Moatsou and Govaris 2011). Compositional and nutritional parameters of Feta are presented in Tables 11.1 and 11.2.

The main points of the current production procedure of Feta are presented in Fig. 11.1. Feta cheesemaking is based on the traditional one in regard to (i) milk kind, (ii) cutting of curd in rather big pieces, (iii) lack of scalding, (iv) draining by gravity, (v) dry salting, and (vi) ripening in "weak" brine, in two distinct phases for at least two months. As a result, (a) Feta's colour is pure white due to milk kind, (b) its taste is slightly acid because the high moisture content and the lack of scalding of the curd facilitates the decrease of pH during draining, and (c) it has no rind due to ripening in brine. Furthermore, Feta mass is smooth, has no gas holes, except for small uneven mechanical openings that result from the way by which curd cubes are stacked into the mould during draining (Fig. 11.2).

Traditionally, fresh yoghurt was used as starter, a practice that sometimes is also followed nowadays. In industrial manufacture, combinations of commercial mesophilic or thermophilic cultures are utilized, e.g., *Lactococcus lactis* subsp. *lactis* and *Lc. lactis* subsp. *cremoris*, *Lactococcus lactis* subsp. *lactis* and *Lb. bulgaricus*. Moulding and draining

Table 11.1 Average composition of brined Greek cheeses, expressed as percentages on cheese mass. FDM, fat-on-dry matter; ripening index, percentage of soluble nitrogen on total nitrogen; S/M, salt-in-moisture, i.e., salt×100/(salt+moisture).

Cheese	Days	pH	Moisture	Fat	FDM	Protein	Ripening index	NaCl	S/M	Ash	References
Feta[a]	120	4.53	54.99		51.3	16.35	20.0		5.15		Anifantakis and Moatsou 2006
Feta[b]		4.68	53.21	25.8	55.1	17.55	13.8	2.34	4.20	3.20	Nega and Moatsou 2012
Kalathaki Limnou[b]		4.48	56.07	23.8	54.1	15.23	15.8	3.30	5.55	4.08	Nega and Moatsou 2012
Telemes[c]	120	4.17	55.78	23.8	53.9	15.17	12.5	2.77			Pappa et al. 2006a,b
Telemes[d]	120	4.33	56.55	20.5	47.1	15.17	9.4	2.84			Pappa et al. 2006a,b
Telemes[e]	120	4.28	57.70	21.2	50.4	15.05	9.7	2.91			Pappa et al. 2006a,b
Touloumotyri[b]		5.17	54.1					3.86			Anifantakis 1991
Sfela[b]		5.07	42.2	28.4	49.1	21.1	13.9	5.75	12.02	7.58	Nega and Moatsou 2012
Batzos[f]	60	5.76	57.1	20.3					8.82		Nikolaou et al. 2002
Batzos[g]	~70	5.63	39.6						11.3		Psoni et al. 2003
Batzos[b]		4.8	43.4	19.6		23.2		5.4			Anifantakis 1991

[a] experimental cheeses; [b] market cheeses; [c] traditionally manufactured in spring from raw sheep's milk; [d] traditionally manufactured in spring from raw goat's milk; [e] experimental cheeses from sheep's milk; [f] experimental cheeses from goat's milk; [g] experimental cheeses from cow's milk.

Table 11.2 Particular nutritional characteristics of brined Greek cheeses. Fat expressed as percentage of cheese mass; SFA, saturated; MUFA, monounsaturated; PUFA, polyunsaturated fatty acids expressed as percentages of fatty acids; CLA, conjugated linoleic acid mostly C9, t11-C18:2 expressed as percentage of fatty acids; CHOL, cholesterol expressed in mg per 100 g cheese; Ca, P, Mg, K, Na expressed in mg per 100 g cheese; Se expressed in µg per 100 g cheese.

Lipid fraction

Cheese	Fat	SFA	MUFA	PUFA	CLA	CHOL	References
Feta						72.5	Fletouris et al. 1998
Feta	18.8	70.2	20.0	4.7	0.92		Zlatanos et al. 2002
Feta	25.5	67	18	3.1		68.1	Andrikopoulos et al. 2003
Feta	21.1	71.8	20.6	3.8	0.72		Zlatanos and Laskaridis 2009
Feta		75.8	27.9	3.5	0.44		Laskaridis et al. 2013
Telemes						61.9	Fletouris et al. 1998
Telemes	23.9	66	20.5	2.8		58.9	Andrikopoulos et al. 2003

Inorganic fraction

Cheese	Ca	P	Mg	K	Na	Se	References
Feta	430	302	25		1036		Kandarakis et al. 2002
Feta	318	248	22	70	940		Nega et al. 2012
Feta						5.18	Pappa et al. 2006c
Telemes						7.62	Pappa et al. 2006c
Kalathaki Limnou	321	270	25	64	1371		Nega et al. 2012
Sfela	621	459	43	66	2296		Nega et al. 2012

of curd pieces are carried out in cylindrical or rectangular perforated molds in relation to the shape of the container used in temporary and final packaging. Dry salting starts when cheese pieces are molded and can continue during the first three to five days when Feta pieces are temporary packaged (Fig. 11.2). Salt promotes further draining and soon cheese pieces are immersed in salty whey. During this period, pH decreases further, i.e., from pH ~4.9 to 5.0 at day 2 to pH <4.6, moisture decreases and salt-in-moisture content of cheeses reaches its final level that is ≥5%. Cheeses pieces are not expected to take additional salt from the storage brine during the second stage of ripening, although exchange phenomena involving mainly proteins and minerals occur (Bintsis et al. 2000, Zoidou et al. 2015). The greatest part of ripening changes that result in the particular sensory characteristics of Feta take place during the first stage of ripening at higher temperatures.

Traditionally, Feta pieces were packaged in wooden barrels with capacity of about 40 to 50 kg of cheese. Sphenoid Feta pieces were tightly

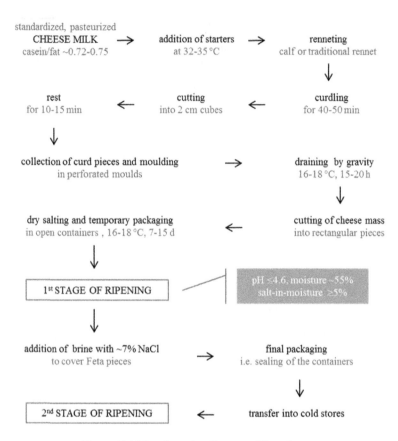

standardized, pasteurized
CHEESE MILK → addition of starters → renneting
casein/fat ~0.72-0.75   at 32-35 °C   calf or traditional rennet

rest ← cutting ← curdling
for 10-15 min   into 2 cm cubes   for 40-50 min

collection of curd pieces and moulding → draining by gravity
in perforated moulds   16-18 °C, 15-20 h

dry salting and temporary packaging ← cutting of cheese mass
in open containers , 16-18 °C, 7-15 d   into rectangular pieces

1st STAGE OF RIPENING   pH ≤4.6, moisture ~55%
salt-in-moisture ≥5%

addition of brine with ~7% NaCl → final packaging
to cover Feta pieces   i.e. sealing of the containers

2nd STAGE OF RIPENING ← transfer into cold stores

Figure 11.1 Manufacturing diagram of Feta cheese.

Figure 11.2 Feta mold.

layered one on top of another with dry salt between them and no brine was added. A very thin layer of salted defatted whey cheese (Myzithra) may be placed on the bottom of the barrel, between pieces and on the upper surface of the stack; a two cm void space is necessary on the top since cheeses may expand during ripening (Moatsou et al. 2002). Packaging in barrels is also used today, despite the difficulties to handle them; however, the most common packaging for Feta is in rectangular tinned or lacquered metal containers holding 17 to 18 kg each (Alichanidis and Polychroniadou 2008). Ripening in barrels is considered to result in more rich and piquant flavour of the final product. It is reported that accumulation of low molecular proteolysis compounds and lipolysis products was more intense in Feta ripened in wooden barrels compared to industrial Feta (Moatsou et al. 2002, Georgala et al. 2005). Also according to Kondyi et al. (2012), Feta ripened in wooden barrels had significantly higher levels of particular volatiles than Feta in tin containers. Intense and piquant flavour has been also related to the use of traditional rennet from kids or lambs abomasa (Moschopoulou et al. 2007). In fact, it is reported that cheese made with traditional rennet contained significantly more C4:0 (butyric acid) and C10:0 (capric acid) free fatty acids compared to Feta made with commercial calf rennet (Moatsou et al. 2004a). During the last years, Feta-type cheeses with low fat or medium fat content are produced, following all the essential steps of the traditional manufacture (e.g., Katsiari and Voutsinas 1994a). The feasibility of reducing the Na content of traditional ewe milk Feta cheese by using mixtures of NaCl and KCl (3:1 or 1:1 w/w) has been also investigated (Katsiari et al. 2000a,b).

A cheese very similar to Feta in terms of cheesemaking technology and sensory characteristics is the PDO Kalathaki Limnou. Kalathaki has a particular embossed surface because draining and acidification of the curd takes place in special cylindrical basket-like molds. It is produced in the Island of Limnos and its compositional and biochemical characteristics are similar to that of Feta cheese (Table 11.2).

Another white brined cheese variety produced in Greece is Telemes, a soft rindless white cheese matured and kept in brine. Telemes originated from Romania and introduced during the first years of the 20th century by the Greek refugees of East Romylia (Anifantakis 1991, Anifantakis and Moatsou 2006). Telemes and Feta share many manufacturing steps but differ in their sensory features. In fact, the flavour of Telemes is more acidic and sometimes more rancid and its texture is rather crumbly and harder than that of Feta. Moreover, the colour of Feta is always 'clear' white, whilst Telemes can be slightly yellowish depending on the ratio of cow milk mixed with sheep milk (Zerfiridis 2001a).

These differences are due to differences in manufacturing steps as summarized by Moatsou and Govaris (2011).

- Telemes is often manufactured from cow milk or from mixtures of cow, sheep and goat milk.
- Telemes curd drains under pressure by applying a weight equal to the curd weight after the transfer of Telemes curd into the molds.
- Telemes pieces, immediately after the pressing stage are salted in a brine bath (~18%) at 15 to 18°C for ~20 hours. Brine salting results in immediate salt absorption into Telemes mass in contrast to the slow salt penetration in Feta cheese, which apparently affects the evolution of microbiota and of biochemical changes.

Recent research studies focus on the effect of milk kind and combination of starters on the characteristics of Teleme (Mallatou et al. 2003, 2004, Mallatou and Pappa 2005, Massouras et al. 2006, Pappa et al. 2006a,b, 2007). According to the findings, goat milk Telemes has lower moisture and harder texture compared to its sheep milk counterpart and primary and secondary proteolysis proceed more slowly in goat milk cheese. Composition of various Telemes cheeses is presented in Table 11.1. Acidification proceeds faster when mesophilic starters are used in comparison to thermophilic starters. In this case also higher hardness, fracturability, shortness and higher sensory scores have been observed regardless of the cheese milk kind.

Touloumotyri or Touloumissio meaning cheese in skin bag, is a white brined cheese that today is produced sporadically locally in many regions of Greece, in small quantities. It is produced from raw small ruminants' milk according to the manufacturing steps of Feta. After salting, the cheese pieces are tightly placed in skin bags and brine with 10% salt or previously boiled salted milk is added. Cheese ripens in the skin bag up to two months. Special attention is given to the preparation of skin bags. Skins salted with a lot of grainy salt, folded and pressed are kept for 10 to 15 days. The dehydrated skins are shaved and washed with brine and reversed in order the hairy part to be inside the bag. The preparation of skin bags and the treatments of cheeses during ripening are laborious and time-consuming. The skin bags filled with cheese pieces must be overturned daily during the first days of ripening and brine or milk addition is needed from time to time (Anifantakis 1991).

11.2.1.2 Miscellaneous brined cheeses

Two other brined PDO cheeses with particular characteristics produced in Greece locally in small quantities are Sfela and Batzos, the curds of which are heat treated after cutting.

Sfela, means strip which is the shape of cheese pieces. It is a semi-hard salty rindless cheese with many small holes in its mass and very intense flavour, called also 'Feta of Fire' (Table 11.1; Fig. 11.3). According

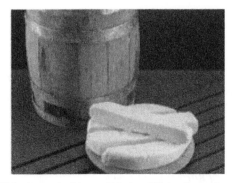

Figure 11.3 Sfela cheese (archives of National Dairy Committee of Greece).

to PDO standard, it is produced in the prefectures of Messinia and Laconia (Southern Peloponnese) from small ruminants' milk, it ripens in brine, either in wooden barrels or in tins, for at least three months and its moisture must be ≤45% and fat-on-dry matter ≥40%. Raw or pasteurized non standardized milk is coagulated at 30 to 32°C using traditional artisanal rennet (Anifantakis 1991, Moschopoulou et al. 2007). After 40 minutes, the curd is cut into rice-size pieces, stirred continuously, and scalded slowly to 38 to 40°C under continuous stirring. After a resting period of 15 minutes, the curd pieces in the bottom of the cheese vat are pressed by hand. The uniform cheese mass is divided into large pieces which are collected with cheese cloths and remain hung for about 60 minutes to drain. Then the drained cheese mass is pressured on a cheese table for another hour. According to traditional practice, the pressed cheese mass is cut into strips, e.g., 4 to 7 cm wide, dry salted for 24 hours at room temperature before placement to barrels or tins with brine (18 to 20% salt concentration). Cheese strips are carefully and tightly placed side by side in rows in the container. Each layer of cheese pieces is packed crosswise to the previous layer to avoid void spaces between the cheese strips. After one to two weeks, the containers are sealed and after a short stay at room temperature, they are transferred into the cold stores at 4 to 6°C, where they can be preserved up to two years (Anifantakis 1991, Litopoulou-Tzanetaki and Tzanetakis 2011).

Batzos is a semi-hard to hard low fat brined cheese (Table 11.1), which is slightly acidic, piquant, very salty in taste, and has many holes. It is traditionally manufactured from small ruminants' milk in Western and Central Macedonia and Thessaly. The unique feature of its manufacture is the cutting of the curd into very small pieces as soon as the coagulation occurs (~10 min after addition of the rennet), using a specially designed wooden tool. This procedure is repeated until the completion of curdling as presented by Nikolaou et al. (2002) and Psoni et al. (2003, 2006). This vigorous treatment of the curd aims to the production of whey with high fat content (>2.5%) suitable for Manouri cheese production. Curdling is completed within 50 minutes. The curd is cut into small pieces and scalded up to 42–45°C. Curd granules are collected with cheese cloths and drained hung for 24 hours. The drained curd is left at room temperature for 24 hours until the appearance of gas holes ("blowing"). Then, it is cut into pieces and dry salted. The next day, the cheese pieces are packed in metal containers filled with brine (10 to 12%), and kept in the cold stores for at least three months.

11.2.2 Hard and very hard type cheeses

Many hard type cheese varieties are currently produced in many regions of Greece. Compositional and nutritional characteristic of these categories are presented in Tables 11.3 and 11.4. The manufacture of the most part of them is based on the traditional cheese making procedure of the very popular Kefalotyri cheese. According to Greek Code of Foods (2012) the moisture of hard cheeses is ≤38% and that of very hard cheeses is ≤32%. The fat-on-dry matter of both categories is ≥32% and their ripening lasts at least three months. Veinoglou and Anifantakis (1981), Anifantakis (1991), Zerfiridis (2001a) and recently Litopoulou-Tzanetaki and Tzanetakis (2011) provide information about origin, technology and characteristics of various cheese varieties of this category than are often consumed as grated cheeses. As reported, written documents about Kefalotyri appeared in 1899; the cheese name consists of two words, i.e., "kefali" meaning head or "kefalos" meaning a particular type of Greek hat and "tyri" meaning cheese. Traditional Kefalotyri is a very hard, salty cheese with uneven openings and strong flavour and its shape is usually similar to a big wheel with diameter 32 to 34 cm and height 12 to 14 cm. Apparently, low moisture and high salt content favoured the long shelf life of the product under non refrigerated conditions. Its characteristics are consistent to a grating cheese, although in Greece it has been also consumed as a table cheese. Traditionally, it was manufactured from sheep milk or from mixtures of sheep and goat milk but during the last decades cow milk is also used. This cheese type is traded under the name of the region where it is manufactured, e.g., Kefalotyri of Crete, Kefallonia, Naxos, Epiros, etc. Nowadays, its quality has been improved mainly by reducing the salt content. An indicative manufacturing diagram of Kefalotyri is shown in Fig. 11.4. Partially skimmed cheese milk, i.e., by the removal of up to 25% of the initial fat content, is used, with casein-to-fat ratio 0.82–0.85. In this respect, sheep and cow milk standardization to 5.6% and 3% fat respectively is applied. Mixtures of thermophilic or mesophilic and thermophilic strains, are added to pasteurized cheesemilk. The quantity of rennet added at 35 to 36°C is in accordance to curd cutting after 35 to 40 minutes. Traditional

Table 11.3 Average composition of hard and very hard type Greek cheeses, expressed as percentages on cheese mass. FDM, fat-on-dry matter; ripening index, percentage of soluble nitrogen on total nitrogen; S/M, salt-in-moisture, i.e., salt×100/(salt+moisture).

Cheese		pH	Moisture	Fat	FDM	Protein	Ripening index	NaCl	S/M	Ash	References
Kefalotyri[a]	at 90 d	5.50	36.34	28.76	45.26	26.65	24.50	3.93			Anifantakis 1991
Kefalotyri[b]			35.91	31.54	49.11	24.57		4.3		6.57	Kandarakis et al. 2002
Orinotyri[a]	at 90 d	5.79	36.40					2.59	6.64		Prodromou et al. 2001
Kefalotyri Kefalonias[a]	in brine		34.29	30.59	46.67	26.67		8.19	19.28	10.77	Massouras, personal communication
	at 18°C		32.97	33.59	49.73	26.65		4.76	18.23	7.35	
	at 4°C		33.65	32.50	49.24	26.65		5.11	18.52	7.65	
Arseniko Naxou[a]		5.30	35.60	33.77	52.46	26.29	16.33	1.84		4.09	Kandarakis et al. 2005a
Ladotyri Mytilinis[b]		5.52	32.58	35.57	53.22	26.60	14.40	2.17	6.25	4.86	Nega and Moatsou 2012
Ladotyri Zakynthou[a]			37.02	30.7	48.8	26.7	22.2	2.84		5.23	Massouras, personal communication
San Michali[b]		5.69	34.18	26.23	38.65	32.5	17.0	3.04	8.19	6.17	Nega and Moatsou 2012
Xinotyri Naxou[a]	at 90 d	4.29	16.43		59	30.7		2.57			Bontinis et al. 2008
Manura Sifnou[a]	at 100 d	4.79	30.0						9.2		Gerasi et al. 2003
Melichloro Limnou[a]		4.49	31.40						5.8		Litopoulou-Tzanetaki and Tzanetakis 2011
Kefalograviera[b]			37.62	30.68	50.44	23.89		3.81		6.22	Kandarakis et al. 2002
Kefalograviera[b]		5.46	37.77	31.25	50.17	24.46	18.5	3.51	8.49	5.96	Nega and Moatsou 2012

[a] experimental cheeses; [b] market cheeses.

Table 11.4 Particular nutritional characteristics of hard and very hard Greek cheeses. Fat expressed as percentage of cheese mass; SFA, saturated; MUFA monounsaturated; PUFA, polyunsaturated fatty acids expressed as percentages of fatty acids; CLA, conjugated linoleic acid mostly C9, t11-C18:2 expressed as mg per 100 g cheese; CHOL, cholesterol expressed as mg per g cheese; Ca, P, Mg, K, Na expressed in mg per 100 g cheese; Se expressed as µg per 100 g cheese.

Lipid fraction

Cheese	Fat	SFA	MUFA	PUFA	CLA	CHOL	References
Kefalotyri						85.7	Fletouris et al. 1998
Kefalotyri	32.8	63	19.3	3.8		95.8	Andrikopoulos et al. 2003
Kefalotyri	27.8				0.68		Zlatanos et al. 2002
Ladotyri	35.6	61	17.2	4.7		89.3	Andrikopoulos et al. 2003
Ladotyri	28.6				0.97		Zlatanos et al. 2002
Kefalotyri Kefalonias	32.5	75.8	21.1	3.1		96	Massouras, personal communication
Ladotyri Zakynthou	30.7	69	27.5	3.5		101	Massouras, personal communication
Arseniko Naxou	36.6	76.8	20.2	3.0		91.6	Kandarakis et al. 2005a
Kefalograviera	34.5	65	18.1	3.6		84.2	Andrikopoulos et al. 2003

Inorganic fraction

Cheese	Ca	P	Mg	K	Na	Se	References
Kefalotyri						6.7	Pappa et al. 2006c
Kefalotyri	791	528	45		1692		Kandarakis et al. 2002
Ladotyri Mytilinis	895	576	59	78	803	7.62	Nega et al. 2012
Kefalotyri Kefalonias	699		52	64	2314		Massouras, personal communication
Ladotyri Zakynthou	747	551	42	68	957		Massouras, personal communication
Arseniko Naxou	785		43	103	656		Kandarakis et al. 2005a
Kefalograviera	756	510	42		1499		Kandarakis et al. 2002
Kefalograviera	851	574	56	81	1510	5.18	Nega et al. 2012

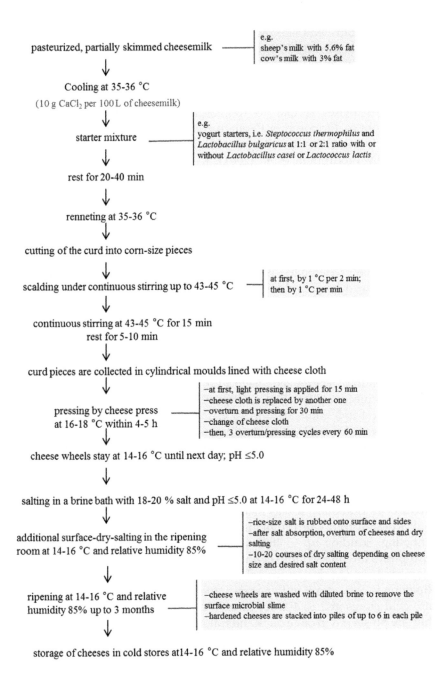

pasteurized, partially skimmed cheesemilk ———| e.g.
sheep's milk with 5.6% fat
cow's milk with 3% fat

↓

Cooling at 35-36 °C

(10 g CaCl₂ per 100 L of cheesemilk)

↓

starter mixture ———| e.g.
yogurt starters, i.e. *Steptococcus thermophilus* and
Lactobacillus bulgaricus at 1:1 or 2:1 ratio with or
without *Lactobacillus casei* or *Lactococcus lactis*

↓

rest for 20-40 min

↓

renneting at 35-36 °C

↓

cutting of the curd into corn-size pieces

↓

scalding under continuous stirring up to 43-45 °C ———| at first, by 1 °C per 2 min;
then by 1 °C per min

↓

continuous stirring at 43-45 °C for 15 min
rest for 5-10 min

↓

curd pieces are collected in cylindrical moulds lined with cheese cloth

↓

pressing by cheese press
at 16-18 °C within 4-5 h ———| –at first, light pressing is applied for 15 min
–cheese cloth is replaced by another one
–overturn and pressing for 30 min
–change of cheese cloth
–then, 3 overturn/pressing cycles every 60 min

↓

cheese wheels stay at 14-16 °C until next day; pH ≤5.0

↓

salting in a brine bath with 18-20 % salt and pH ≤5.0 at 14-16 °C for 24-48 h

↓

additional surface-dry-salting in the ripening
room at 14-16 °C and relative humidity 85% ———| –rice-size salt is rubbed onto surface and sides
–after salt absorption, overturn of cheeses and dry
salting
–10-20 courses of dry salting depending on cheese
size and desired salt content

↓

ripening at 14-16 °C and relative
humidity 85% up to 3 months ———| –cheese wheels are washed with diluted brine to remove the
surface microbial slime
–hardened cheeses are stacked into piles of up to 6 in each pile

↓

storage of cheeses in cold stores at14-16 °C and relative humidity 85%

Figure 11.4 Manufacturing diagram of Kefalotyri cheese.

rennet from lambs' and kids' abomasa (Moschopoulou et al. 2007) may be used alone or in mixture with calf rennet for a stronger and more piquant flavour of the final product. Kefalotyri is salted by combining brine and dry surface salting. Cutting, scalding, salting and ripening conditions are presented in Fig. 11.4.

Many varieties similar to Kefalotyri are produced regionally in small quantities. Orinotyri or Oreinotyri meaning mountainous cheese is produced in mountainous villages of Epirus and its features have been reported by Prodromou et al. (2001). It is made from raw sheep milk without starter and the curd is cut into corn-size pieces 45 minutes after renneting at 30°C. Scalding is under continuous stirring up to 47°C and then pressure is applied in two kg cheese wheels for about two hours. Cheese is firstly salted in brine for 24 hours at about 15°C and then two or three dry saltings take place within 10 days. It is stored at 4°C for three months.

Kefalotyri Kefalonias is a Kefalotyri variant produced in the homonymous island of Ionian Sea. Originally it was very hard and extremely salty, because traditionally it is salted and ripened in dense brine with 18 to 20% NaCl for more than two months. In this respect, it could be considered as a brined cheese. However, alternative schemes of ripening are applied today to avoid excessive saltiness; cheeses can ripen wooden shelves at room or lower temperatures (Table 11.3, T. Massouras, personal communication). Thermized sheep milk is used and intense scalding of small curd pieces at 49 to 51°C within 40 minutes takes place. Pressure is applied for 4 to 5 hours at 16 to 18°C and the next day, cheese is placed into brine with 18 to 20% NaCl to ripen up to three months. Alternatively, after salting cheese wheels ripen on wooden shelves at temperature lower than 18°C.

Another Kefalotyri-type cheese is Arseniko Naxou. Arseniko means masculine and it is produced in Naxos Island from sheep and goat milk mixture. Traditionally, it was made from raw milk but nowadays cheese milk is pasteurized. Curd is cut into corn-size pieces about 30 minutes after renneting. Scalding up to 48°C within 40 minutes is applied. Collected curd pieces are pressed into cylindrical molds by hand. The day after the cheesemaking, the cheeses are salted in brine with 20% NaCl for 48 hours. Ripening is carried out in two stages, firstly at 16°C for 30 days and then into cold stores at 2 to 4°C for at least another two months. Sometimes they are preserved in olive oil (Kandarakis et al. 2005a, Aktypis et al. 2011).

Other varieties in this group are cheeses called Ladotyri meaning (olive) oil cheese because traditionally they were ripened and stored in olive oil (Anifantakis 1991, Nega and Moatsou 2012). The most well known is Ladotyri Mytilinis, a PDO cheese manufactured in Lesvos Island in Northern Aegean Sea from sheep milk or from its mixtures with goat milk; the latter up to 30% in the mixture. It is a hard cheese with small uneven openings and moderately salty taste marketed in small cylinders of 1.2 to 1.5 kg with 10 cm diameter and 10 to 12 cm height (Fig. 11.5). According to PDO standards (Greek Code of Foods 2012), its moisture must be ≤38% and fat-on-dry matter ≥40%. Manufacturing steps are described by Veinoglou and Anifantakis (1981), Anifantakis (1991) and in the Greek Code of Foods (2012). Cheese curd is cut at two stages; at first into cubes of 1.5 cm edge and after a short resting period the curd pieces are cut into rice-size pieces. Scalding at 42 to 45°C is applied for five to 10 minutes and after partial removal of the whey the curd pieces are pressed by hand. The resulted cheese mass is divided in small sub-parts that can be placed into specific weavy molds called "tyrovolia". After additional intense pressing of the cheeses into the molds by hand, they are wetted by whey overturned and left for two to three hours. Cheeses are drawn off the molds, dry salted with fine salt and placed into the molds again till next day. Then, the small cheese cylinders are transferred to a ripening room at 12 to 18°C and 85% relative humidity. Salting is completed either by remaining for 24 hours in a brine bath with 20% NaCl or by three additional dry saltings. The slimy layer appearing on the cheese surface is washed using warm water and cheeses ripen up to three months. Thereafter, cheeses after washing may be preserved in olive oil or alternatively into cold stores.

Figure 11.5 Ladotyri Mytilinis cheese (archives of National Dairy Committee of Greece).

Ladotyri Zakynthou is produced in small quantities in the Zakynthos Island in Ionian Sea in the Western Greece. Cheesemilk is a mixture of sheep and goat milk at a ratio 80:20, which is thermized at 62 to 63°C. Usually, traditional rennet is used. The curd is hooped into 20×20×15 cm rectangular molds and is dry salted with coarse salt. On the next day, cheeses are overturned, salted again and placed on wooden shelves to drain for 10 to 12 hours. Finally, they are put in layers into tanks with olive oil paste to ripen for at least three months at room temperature (T. Massouras, personal communication, Aktypis et al. 2011).

San Michali is a very much appreciated PDO hard cheese produced in Syros Island of Cyclades from partially skimmed pasteurized cow milk. Curd is cut into corn-size pieces which are scalded to 48 to 50°C. Intense mechanical pressing is applied for many hours with several changes of cheesecloths and overturns. Then, cheeses are salted into brine with 18 to 20% NaCl for 10 to 12 days at 14 to 16°C. Ripening takes place at 14 to 16°C and 85% relative humidity for at least four months (Anifantakis 1991, Greek Code of Foods 2012).

Other very hard cheeses produced locally in small quantities from raw small ruminants' milk without starters in Aegean Islands are Xinotyri of Naxos, Manura of Sifnos and Melichloro of Limnos. They are low-pH cheeses, their cheesemaking technology does not have similarities with Kefalotyri and they ripen usually under uncontrolled conditions. Xinotyri is an extremely hard cheese produced in farms in Naxos Island from raw goat milk. As described by Bontinis et al. (2008), no starters are used and curdling lasts about 24 hours at room temperature. Then curd drains into cheesecloths for two or three hours. Dry salt is incorporated manually and uniformly into cheese mass in a ratio of about 15 g NaCl per kg. The salted cheese mass is transferred into truncated conical molds and remain for three to four days to harden with daily turnovers. After the removal of the molds, cheeses ripen on wooden shelves for about 30 to 45 days at uncontrolled room temperature. They are turned upside down regularly and surface molds are removed by washing with brine. If not consumed, Xinotyri cheeses can be stored in cold stores.

Manura Sifnou is a farmhouse cheese produced in farms in Sifnos Island of Cyclades from raw sheep or mixture of sheep and goat milk without starters (Gerasi et al. 2003, Litopoulou-Tzanetaki and Tzanetakis 2011). It is a very hard cheese covered by reddish wine sediment marketed as small wheels of 600 to 700 g, with about 12 cm diameter and 8 cm height. No scalding is applied and the draining of the nut-size curd pieces is enhanced by thorough stirring. Then, the curd is put into baskets called tyrovolia to drain further. Dry salting is carried out within two or three days using coarse salt. Cheeses are placed on straw lattices to dry and ripen up to three or four months. Afterwards, they are placed into barrels with red wine for approximately 10 days and finally they are placed in other barrels and covered with wine sediment for 24 hours. Finally, they are stored in clean barrels.

Melichloro cheese is a traditional cheese of Limnos Island made from raw ewe milk produced at the end of lactating season, i.e., at the end of spring. According to Litopoulou-Tzanetaki and Tzanetakis (2011) and Papanikolaou et al. (2012), the raw milk is renneted warm within one hour after milking by traditional rennet derived from small ruminants' abomasa. Curd is cut into very small pieces and after a short resting period it is transferred into baskets on wooden shelves to drain. After several hours of draining, which is facilitated with regular overturns, cheeses are dry salted several times in every side. Then, fresh cheeses in the baskets are put into wooden lattices hung on trees to dry outdoors for up to five days. After the withdrawing of baskets, the small cheese wheels of approximately 500 g each are stored up to two or three months depending on their hardness. Finally, cheeses are washed with sea water and dried with a cloth, before consumption.

Kefalograviera is a very popular PDO cheese made from sheep milk or from its mixtures with goat milk at a ratio of 90:10. It "stands" between Kefalotyri and Graviera in regard of cheesemaking and composition (Table 11.3). It is produced in semi-industrial scale in the Western Greek mainland, i.e., Western Macedonia, Epirus, Etoloakarnania and Evritania. According to PDO standards (Greek Code of Foods 2012), it has moisture ≤40%, fat-on-dry matter ≥40% and ripens for at least three months. The curd is cut into small corn-size pieces and scalded at about 48°C for 30 minutes. After intense mechanical pressing, the rather big cheese wheels are salted in brine and by additional dry surface salting (Fig. 11.6). The production steps of Kefalograviera in detail (Anifantakis 1991, Katsiari and Voutsinas 1994b) are presented below.

- Pasteurization of cheese milk.
- Addition of starters (usually yoghurt culture) at 35 to 38°C.
- Addition of $CaCl_2$ (~25 g/100 g milk).
- Renneting, using calf rennet at 32 to 34°C.
- Curd cutting ~35 minutes after renneting, into 2–3 cm cubes.
- Rest for 2–3 minutes.
- Further cutting of curd pieces to the size of corn grains by gentle stirring for 15 minutes.
- Scalding up to 47 to 48°C within 20 to 25 minutes under continuous stirring.
- After the end of cooking, stirring proceeds for 5–10 minutes.
- Curd is left to settle on the bottom of cheese-vat.
- The curd is wrapped with cheese-cloths and it is placed into cylindrical molds, e.g., ~12 cm height, ~30 cm diameter.
- Pressing (~0.22 kg/cm^2) for 2–4 hours with intermediary changes of cheese cloths.
- Removal of cheese cloths.
- Cheeses left into the molds without pressing at 14 to 16°C and 90% RH.
- Salting begins the next day by submerging cheese wheels in 18 to 20% salt brine, for two days at 14 to 16°C; brine volume/cheese weight ~2.0.
- Dry-salting follows brine-salting of cheeses: 5–10 saltings, duration ~15 to 20 days at 14 to 16°C and 90% RH.

- About one month after their manufacture, cheeses are washed with brine with 12 to14% NaCl and then dried using a cheesecloth.
- Packaging in plastic bags under vacuum may be applied.
- Ripening up to at least 90 days after manufacture is continued at 6 to 10°C.

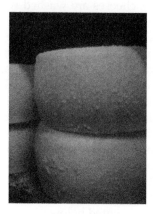

Figure 11.6 Dry surface salting of Kefalograviera cheese wheels.

11.2.3 Gruyere type cheeses

The name Gruyere or Graviera in Greek is used to describe hard type cheeses that are subjected more or less to propionic acid fermentation and have special properties. In general, they come from intensively heat treated curds and they have low salt contents contrast to Kefalotyri-type cheeses. Graviera cheeses were initially produced in the second decade of the 20th century and in different regions of Greece from sheep milk, but nowadays this type of cheese is made also either from cow or goat milk or from mixtures of them with sheep milk. At least 10 different Greek Graviera cheeses are traded and distinguished from the name of the region where they are manufactured, e.g., Graviera Paros, Graviera Mitilinis, while three of them, i.e., Graviera Kritis, Graviera Agrafon and Graviera Naxou belong to the PDO family.

Graviera Kritis or Gruyere of Crete (Fig. 11.7), the most famous PDO hard cheese in Greece, is produced in Crete Island from sheep milk or mixtures of it with goat milk up to 20%. According to the Greek Code of Foods, Graviera Kritis must contain moisture ≤38%, fat-on-dry matter ≥40% and salt ≤2%. Compositional and nutritional characteristics of Graviera Kritis are shown in Tables 11.5 and 11.6 respectively. Its exceptional characteristics are mainly attributed to the quality of the milk of the local breeds as well as to the experience of the local cheesemakers (Anifantakis 1991). Traditional Graviera Kritis has a distinct flavour, a surface with a

Figure 11.7 Graviera Kritis cheese.

smear rind and exhibits openings with variable size characterized rather slits or holes than eyes (Moatsou et al. 2004b). General manufacturing steps of Graviera Kritis are shown in Fig. 11.8.

Traditionally, Graviera Kritis is made from insufficiently pasteurized milk without starter cultures. Milk coagulates at 34 to 36°C and curd is cut usually in corn-size pieces after about 30 minutes. Another practice is the curd division in two phases. At the beginning, curd is cut into 2 cm cubes, rests for 15 minutes and then is further cut in corn-size pieces (Moatsou et al. 2015). After cutting, curd is scalded under continuous stirring up to 50 to 52°C. During scalding a common practice is the curd washing in order to improve textural characteristics. This happens by replacing part of whey, i.e., about 10%, with equal amount of water to limit high acidification of the curd as low pH values are consistent with textural defects. After scalding, curd is placed into cylindrical molds and is pressed for about 24 hours. Next day, the fresh cheese wheels are transferred into the ripening room (14 to 16°C, 85% RH) where they rest for one day and then, depending on their size, they are salted in brine solution with 18–10% NaCl be for two to five days. Additional salt is given to the cheese at the beginning of ripening by surface dry salting for about 10 times for large cheese wheels with turnovers. Graviera Kritis ages in ripening rooms at 14 to 18°C and 85 to 90% RH. Ripening at 18 to 20°C increases the level of proteolysis and decreases its quality (Moatsou et al. 1999, 2004b). In practice, various combinations of time/temperature for Graviera cheese ripening, e.g., 17°C for 45 days and then at 9°C for 1 month (Moatsou et al. 2015) or 15°C for 2 months and then at 4°C (Kandarakis et al. 1998) are used. During ripening progress, a special microbiota consisting mainly by yeasts and molds grows on the cheese surface, considering contributing to ripening process and cheese organoleptic characteristics. This type of cheese should be ripened for at least three months and then be stored in cold rooms.

There is little published information concerning composition and quality of Graviera Kritis. Although this cheese is traditionally

Table 11.5 Average composition of Gruyere-type Greek cheeses, expressed as percentages on cheese mass. FDM, fat-on-dry matter; ripening index, percentage of soluble nitrogen on total nitrogen; S/M, salt-in-moisture, i.e., salt×100/(salt+moisture).

Cheese	Days	pH	Moisture	Fat	FDM	Protein	Ripening index	NaCl	S/M	Ash	References
Greek Gruyere[a]	90	5.76	35.08		54.57			1.31			Zerfiridis et al. 1984
Greek Gruyere[a]	180	5.94	33.96		55.29			1.49			Zerfiridis et al. 1984
Graviera Kritis[b]	90	5.52	35.70		63.75				5.0	4.19	Kandarakis et al. 1998
Graviera Kritis[b]	210	5.76	33.59						3.64	4.34	Moatsou et al. 2004b
Graviera Kritis[b]	90	5.07	33.60						2.47		Litopoulou-Tzanetaki and Tzanetakis 2011
Graviera Kritis[c]		5.56	33.76	34.72	52.42	26.97	17.3	1.78	5.02	4.47	Nega and Moatsou 2012
Graviera Naxou[c]		5.45	35.82	32.32	49.80	27.29	18.7	1.46	3.93	3.95	Nega and Moatsou 2012
Graviera Agrafon[c]	90		36.90	31.7	45.00	27.8					Fletouris et al. 2014
Graviera Paros[c]			38.00	32.50	45.00	28.44		2.00		4.56	Kandarakis et al. 2005b

[a]experimental cheeses from cow milk; [b] experimental cheeses; [c] market cheeses.

Table 11.6 Particular nutritional characteristics of Gruyére-type Greek cheeses. Fat expressed as percentage of cheese mass; SFA, saturated; MUFA monounsaturated; PUFA, polyunsaturated fatty acids expressed as percentages of fatty acids; CLA, conjugated linoleic acid mostly C9, t11-C18:2 expressed as percentage of fatty acids; CHOL, cholesterol expressed in mg per 100 g cheese; Ca, P, Mg, K, Na expressed in mg per 100 g cheese; Se expressed in µg per 100 g cheese.

Lipid fraction

Cheese	Fat	SFA	MUFA	PUFA	CLA	CHOL	References
Graviera						97.9	Fletouris et al. 1998
Graviera	36.4	62.1	21.2	3.3		110.2	Andrikopoulos et al. 2003
Graviera Kritis	26.5				0.96		Zlatanos et al. 2002
Graviera	29.9	70	22.6	3.6	0.75		Zlatanos and Laskaridis 2009
Graviera Agrafon	31.7	68.1	23.8	4.3	1.03		Fletouris et al. 2014
Graviera Paros	32.5			24.3	3.1	107	Kandarakis et al. 2005b

Inorganic fraction

Cheese	Ca	P	Mg	K	Na	Se	References
Graviera	917	560	42		812		Kandarakis et al. 2002
Graviera						8.9	Pappa et al. 2006c
Graviera Kritis	963	624	58	95	642		Nega et al. 2012
Graviera Paros	976		40	118	705		Kandarakis et al. 2005b

produced without starter cultures, the use of mixed lactic acid bacteria and *P. freundreichii* at different ratios does not significantly affect its chemical composition, its organoleptic characteristics as well as its ratio of nitrogen fractions during ageing (Kandarakis et al. 1998, Moatsou et al. 1999). Moreover, the use of a mixture of lactic acid and propionic acid starter cultures in combination with a curd wash can improve the textural characteristics, i.e., more elastic curd with evenly distributed eyes in the cheese mass (Moatsou et al. 2004b). On the other hand, microbial quality of matured Graviera Kritis made with starter cultures is better as no coliforms and significantly lower counts of enterococci are found (Kandarakis et al. 1998).

Graviera Agrafon (PDO) is produced in Agrafa region, in West-central Greece. It is a hard cheese made from sheep milk or mixture of it with goat milk up to 30%. Cheesemilk for Graviera Agrafon is either thermized or pasteurized and coagulates at 34 to 36°C with traditional rennet. Curd is cut after 25 to 35 minutes, cooked at 48 to 52°C under continuous stirring and is then placed in molds and pressed for one day. After pressing, cheeses are removed from the molds and rest onto wooden shelves for

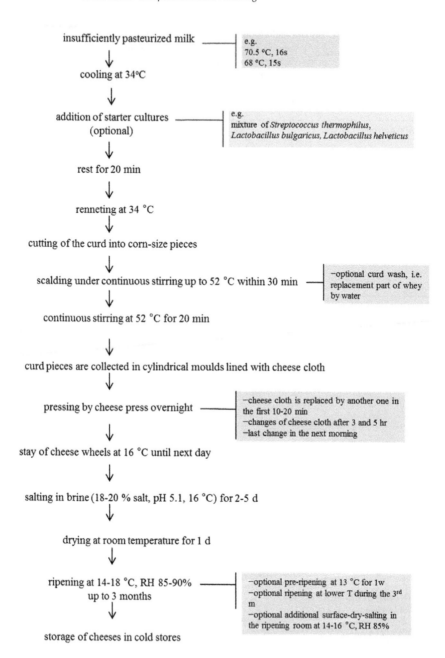

insufficiently pasteurized milk ——— e.g.
70.5 °C, 16s
68 °C, 15s

↓

cooling at 34°C

↓

addition of starter cultures ——— e.g.
(optional) mixture of *Streptococcus thermophilus*,
Lactobacillus bulgaricus, Lactobacillus helveticus

↓

rest for 20 min

↓

renneting at 34 °C

↓

cutting of the curd into corn-size pieces

↓

scalding under continuous stirring up to 52 °C within 30 min ——— −optional curd wash, i.e.
replacement part of whey
by water

↓

continuous stirring at 52 °C for 20 min

↓

curd pieces are collected in cylindrical moulds lined with cheese cloth

↓

pressing by cheese press overnight ——— −cheese cloth is replaced by another one in
the first 10-20 min
−changes of cheese cloth after 3 and 5 hr
−last change in the next morning

↓

stay of cheese wheels at 16 °C until next day

↓

salting in brine (18-20 % salt, pH 5.1, 16 °C) for 2-5 d

↓

drying at room temperature for 1 d

↓

ripening at 14-18 °C, RH 85-90% ——— −optional pre-ripening at 13 °C for 1w
up to 3 months −optional ripening at lower T during the 3rd
m
−optional additional surface-dry-salting in
the ripening room at 14-16 °C, RH 85%

↓

storage of cheeses in cold stores

Figure 11.8 Manufacturing diagram of Graviera Kritis cheese.

two days to dry and then they are salted by immersion in brine 19 to 20% for two to four days. Cheese ripening begins at 12 to 15°C, 85% RH and in parallel surface-dry salting occurs in order for cheeses to have 2% salt content. First stage of ripening ends when dry salting has finished, then it continues at 16 to 18°C for a month and completes at 12 to 15°C, 90 to 95% RH. Ripening lasts three months. During ripening, microbiota grows on the cheese surface, contributing thus to ripening process and cheese organoleptic characteristics. Graviera Agrafon has a slightly sweet taste and rich aroma and should contain moisture up to 38% and fat-on dry matter ≥40%. Compositional and nutritional characteristics of Graviera Agrafon are shown in Tables 11.5 and 11.6 respectively.

Graviera Naxou (PDO) is produced in Naxos Island of Cyclades exclusively from locally produced cow milk or a mixture of it with sheep and/or goat milk. According to the Greek Code of Foods (2012), Graviera Naxou must have moisture content up to 38% and fat-on-dry matter ≥40%. The latter should be up to 20%. Cheese milk, which can be raw or pasteurized, is coagulated at 36 to 37°C, after 30 to 40 minutes the curd is cut into the size of rice grains and is scalded at 50°C for 60 minutes under continuous stirring. The curd is then transferred into molds lined with cheesecloth and is pressed for 3 to 4 hours. During pressing period, a few changes of cheesecloths occur. After pressing, the cheese wheel is salted in brine 20Be at 14 to 15°C for two to five days depending on its size. After salting, cheeses are dried and placed in ripening rooms (15°C) for 70 to 80 days. Graviera Naxou has a pleasant taste and slight aroma and its composition is shown in Table 11.5.

Graviera Paros is produced in Paros Island of Cyclades and at present is not a PDO cheese. It is made exclusively from cow milk and exhibits very pleasant organoleptic characteristics and a piquant taste. Cow milk is heat treated at 67°C for 10 minutes and cooled at 36°C. Starter cultures and rennet are added, curd is cut and cooked at 49 to 51°C for 40 minutes. After scalding, curd is transferred to molds and is pressed for a day. Next day, cheeses are salted in brine (20 to 22%) for two days and then ripen at 16°C for 30 days and at 4°C for a rest period of two months. Compositional and nutritional characteristics of Graviera Paros are shown in Tables 11.5 and 11.6 respectively.

11.2.4 Pasta Filata and other semi hard cheese varieties

The most famous pasta filata cheese in Greece is Kasseri cheese. It is a PDO cheese produced in regions of Northern Greece, i.e., Macedonia, Thessaly and prefectures of Xanthi and Lesvos. According to PDO standards (Greek Code of Foods 2012), it is made from sheep milk or from its mixtures with goat milk up to 80:20. It has a close structure, rather strong flavour, moisture 45% at maximum, fat-on-dry matter 40% at minimum and it is mostly marketed as cheese wheels of four to five kg or 0.8 to 1 kg. The particular feature of Kasseri cheesemaking technology is the kneading and stretching at hot water (~75°C) of biologically acidified cheese mass (pH ~5.2) called "kasseromaza" meaning Kasseri mass or "baski" in very hot water that results in particular textural characteristics. In this respect, Kasseri belongs to the same group with the Italian Provolone and Mozzarella and Balcan Kashkaval. According to Anifantakis (1991) this cheese was introduced to Greece from other Balkan regions in late 19th century, to which it was introduced from Italy named as Cacciocavallo. The preparation of cheese mass in the mountainous premises of the shepherds was a very convenient way to preserve their milk, reducing also the volume that had to be transported to cheese plants, in which further processing took place.

Therefore, cheesemaking procedure of Kasseri is divided in two stages, i.e., the manufacture of "baski" and the manufacture of the final product, i.e., Kasseri; nowadays both of them are carried out in the same cheese plant. The aim of the first stage is the production of a semi-hard cheese mass facilitating the draining by mild scalding and pressing by hands. Traditionally, raw milk and no starters were used and the acidification was due to milk native microbiota. However, safety reasons and uncontrolled acidification in terms of duration and quality of fermentation have emerged the pasteurization of cheese milk and the use of starters, e.g., mixtures of *Lactoccus lactis* subsp. *lactis*, *Streptococcus thermophilus* and *Lactobacillus bulgaricus* or mixtures of *Streptococcus thermophilus*, *Lactobacillus lactis* and *Lactobacillus helveticus*. Milk fat standardization resulting in a casein to fat ratio of about 0.85 is suggested to avoid fat expulsion during ripening. The curd is cut into 0.5 to 0.7 cm cubes, 30 to 40 minutes after renneting. Ten minutes after, moderate scalding up to 40 to 42°C within 30 minutes under continuous stirring is applied. After a 10 minute rest, cheese curd is collected with cheese cloths. The curd is pressed by hand or by a weight equal to cheese mass to drain and to form a compact cheese mass. Usually, "baski" stays at room temperature until the next day to be acidified at pH ~5.2. Proper biological acidification is very critical for the successful kneading and stretching of the cheese mass. When specially designed starter mixtures are used 4 to 6 hours are enough for acidification. Then, "baski" is cut into long slices of 0.5 cm thickness; Cheddar mills may be used for this purpose. Before kneading the ability of the cheese mass to stretch and strain is assayed by a manual test. A small quantity equal to the size of an egg is put into hot water at 75 to 80°C and kneaded with the aid of a spoon. The cheese mass can be processed in hot water if a glace and continuum cheese string of about one metre long can be formed by straining of the hot cheese mass by hand. "Baski" pieces can be kneaded manually into a wooden basket immersed into hot water at 75 to 80°C called "kofinello" with the aim of a specific wooden spatula or scoop until a hot homogeneous elastic mass is formed. A portion of the curd sufficient to fill a cylindrical mould with capacity of four to five kg is formed. Fine-grade salt can be incorporated into the hot cheese mass. Hooping is a very delicate procedure because anadiplosis and blisters jeopardize the successful ripening of cheeses, being also sensory defects. Today, this procedure is carried out mechanically. Kneading is continuous by means of a warm screw, which rotates in a hot water bath of ~80°C. Finally, hot cheese pieces are derived directly into the hoops. Moulds are overturned after 15 minutes and then less frequently; five or six overturns take place within two days. Additional dry surface salting with medium grain salt may be applied, which facilitates the formation of a dry rind. Traditionally, 10 to 12 dry salting sessions were put into practice, but nowadays, if any, they are much reduced. Cheeses ripen uncovered at about 15°C for 10 to 20 days; then they are waxed or packaged in plastic film under vacuum up to approximately two months. Finally, they are transferred to cold stores to complete ripening up to three months as defined in the PDO standard. The second stage of Kasseri manufacture that is the processing of acidified cheese mass is shown in Fig. 11.9. Composition and nutritional characteristics of Kasseri are shown in Tables 11.7 and 11.8 respectively.

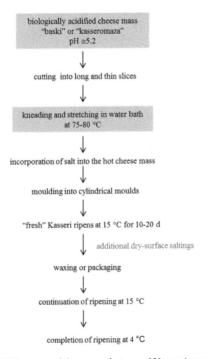

Figure 11.9 Diagram of the second stage of Kasseri manufacture.

Table 11.7 Average composition of pasta filata and semi hard Greek cheeses, expressed as percentages on cheese mass. FDM, fat-on-dry matter; ripening index, percentage of soluble nitrogen on total nitrogen; S/M, salt-in-moisture, i.e., salt×100/(salt+moisture).

Cheese	Days	pH	Moisture	Fat	FDM	Protein	Ripening index	NaCl	S/M	Ash	References
Kasseri[b]			43.22	25.30	44.47	25.78		2.29		4.39	Kandarakis et al. 2002
Kasseri[a]	90	5.69	43.90	23.98		25.50	10.3			4.73	Kaminarides et al. 1995
Kasseri[a]	120	5.50	44.0			28.78	19.0				Moatsou et al. 2001
Kasseri[b]		5.67	40.86	28.74	48.49	25.75	18.3	1.65	3.92	4.07	Nega and Moatsou 2012
Metsovone[b]		5.48	41.80	25.90		26.80		2.80			Anifantakis 1991
Metsovone[b]		5.50	40.31	29.05	48.65	26.24	17.0	2.26	5.32	3.93	Nega and Moatsou 2012
Formaella Parnassou			≤50	≥40							Greek Code of Foods 2012
Krassotyri or Possias[b]		4.50	46.5	28.7				2.2			Anifantakis 1991
Tyraki Tinou	2	4.38	40.0	33.38	57.15	21.86	4.57	2.23		2.56	Zoidou et al. 2011

[a] experimental cheeses; [b] market cheeses.

Table 11.8 Particular nutritional characteristics of pasta filata and semi hard Greek cheeses. Fat expressed as percentage of cheese mass; SFA, saturated; MUFA monounsaturated; PUFA, polyunsaturated fatty acids expressed as percentages of fatty acids; CLA, conjugated linoleic acid mostly C9, t11-C18:2 expressed as mg per 100 g cheese; CHOL, cholesterol expressed as mg per g cheese; Ca, P, Mg, K, Na expressed in mg per 100 g cheese; Se expressed as µg per 100 g cheese.

Lipid fraction							
Cheese	Fat	SFA	MUFA	PUFA	CLA	CHOL	References
Kasseri						86.2	Fletouris et al. 1998
Kasseri	27.9	65.6	19.9	3.7		90.8	Andrikopoulos et al. 2003
Kasseri	21.55				0.82		Zlatanos et al. 2002
Metsovone	33.3	65.2	19.5	3.3		79.0	Andrikopoulos et al. 2003
Formaella Parnassou	30.25				0.77		Zlatanos et al. 2002

Inorganic fraction							
Cheese	Ca	P	Mg	K	Na	Se	References
Kasseri						8.6	Pappa et al. 2006c
Kasseri	842	560	38		901		Kandarakis et al. 2002
Kasseri	858	561	48	74	599	7.62	Nega et al. 2012
Kasseri	680	540					Kaminarides et al. 1995
Tyraki Tinou		131	8.8	63	1050		Zoidou et al. 2011

Pasta filata cheeses are also produced in Greece from cow milk, like Victoria cheese with moisture <45% and fat-on-dry matter >45% produced in the prefecture of Thessaloniki. Each cheese has a long cylindrical shape of diameter 11 to 12 cm and height 26 to 28 cm and weight of about 3 kg. It is manufactured similarly to Kasseri cheese but salting in brine is applied (Zerfiridis 2001a). Other pasta filata cheeses from cow milk are produced in Greece mainly for cooking use.

The group of pasta filata cheeses includes some locally produced smoked cheeses. The most well known is Metsovone cheese (PDO), produced in the mountainous small town Metsovo in Epirus, at an altitude of 1100 m in north western Greece. It is a semi hard to hard cheese with slightly salty and piquant flavour with a thin, dry almost nut-brown rind. It is produced from cow milk or from its mixtures with small ruminants' milk, in which the former is more than 80%. Metsovone has moisture ≤38% and fat-on-dry matter ≥40% (Table 11.7). Before renneting, acidified whey of the previous day is added to the cheesemilk at a ratio of 3 to 4% (w/w). Curd cut into pieces of a hazelnut-size is scalded to 46 to 48°C for 15 to 20 minutes. The hot curd pieces are collected and drain by means of cheese cloths. When cheese mass is adequately acidified, i.e., pH 5.1 to 5.3 and the outcome of the manual kneading assay is successful, it is cut into slices and kneading and stretching in water at 75 to 80°C is carried out. Hot cheese mass is placed into long cylindrical moulds, which are placed into cold water. Next day, cheeses become steady and firm and they are salted in a brine bath with 18 to 20% NaCl. The duration of salting in days is equal to cheese weight in kg. Salted cheeses are tied all around with string, similarly to Provolone; so they can ripen hung without contacting any surface (Fig. 11.10). They stay at 15 to 17°C and 85% relative humidity for at least three months (Anifantakis 1991). Mature cheese weighing from about 1.5 to 4 kg are smoked by burning vine twings and aromatic leaves and plants, waxed and preserved at low temperature (Foundation of barone Michail Tositsa 1993).

Figure 11.10 Metsovone cheese (archives of National Dairy Committee of Greece).

Other semi hard, non pasta filata cheeses are produced locally in very low quantities; their characteristics are presented in Tables 11.7 and 11.8. Formaella Parnassou is a semi-hard PDO cheese produced in the mountainous town Arachova in the skiing area of Parnassos Mountain in Central Greece. It is made from goat or sheep milk or their mixtures and its moisture is <50% and fat-on-dry matter >40% (Greek Code of Foods 2012). It has a small cylindrical shape with height higher than diameter and mass of about 400 g (Fig. 11.11). Often, it is consumed fried or grilled. Traditionally, cheeses five days after cheesemaking were transferred to caves to ripen (Anifantakis 1991). Artisanal traditional rennet from small ruminants' abomasa is mostly used, which results in cheese with piquant flavour. Curdling lasts two hours at about 32°C and the curd is cut into corn-sized pieces, which are scalded up to 40°C for 10 minutes. Curd is left to settle in the cheese vat and it is pressed by hand to form a continuous mass, which is then cut into pieces of about 400 g. Each piece is placed into cylindrical mold called "tyrovoli" with wavy walls and with height longer than diameter. This is a Formaella cheese, which is then placed in whey bath at 60°C for about 60 minutes to drain. After removal of molds from the whey bath, Formaellas are overturned and placed again in whey bath at 75 to 80°C for another hour. Then cheeses with molds remain to dry for 24 hours at room temperature and surface dry salting in both sides is applied. The next day, Formaellas without the molds stay on shelves for 4 to 5 hours to dry and then they may be transferred to cold rooms to ripen for three months. Salting can be accelerated by piercing the cheese mass before salting.

Figure 11.11 Formaella Parnassou cheese (archives of National Dairy Committee of Greece).

A very particular semi hard cheese is Possias or Krassotyri; the latter means cheese in wine. It is made from sheep and goat milk mixtures in Kos Island of Dodecanese (Anifantakis 1991, Litopoulou-Tzanetaki and Tzanetakis 2011). Cheese milk is boiled sheep and goat milk coming from afternoon milking naturally defatted by staying in swallow and broad pots overnight. After partial removal of the fat, fresh raw milk from the morning milking is added. Curdling lasts 60 to 90 minutes. Small curd pieces are stirred for about 15 minutes and no scalding is applied. Cheese mass of about one kg is formed into a cylindrical mold with much greater height than diameter. After the draining and the removal of molds, dry salting with medium size salt is carried out. Cheeses are put into containers at low temperature and shortly after brined whey covers cheese surface. After 20 to 30 days, the cheeses are placed onto shelves to dry. Then, cheeses are put into containers, where they are covered with wine sediment or with a mixture of sediment and wine, called "possia" for at least one week before consuming. As a result, cheese surface acquires a reddish colour, whereas the cheese interior is white.

Tyraki Tinou is a semi hard cheese produced in Tinos Island of Cyclades (Zoidou et al. 2011). Tyraki is made from cow milk without starters and coagulation is achieved by combination of low quantities of rennet and acid produced by the metabolism of native milk microbiota. According to nowadays practice, cow milk is pasteurized and raw milk is added at 8 to 12%, which serves as a source of native milk microbiota. Renneting is carried out with low quantity of rennet at 25°C and curdling is carried out until next day at about 22°C. Then, the cut curd is collected in sacs where it drains within the next two days with intermediary changes of their placements. After draining, salt is incorporated at a ratio of 3% into the cheese mass by gentle kneading. Homogenized cheese mass is schematized in small cylindrical molds lined with cheese cloths and it is pressed until next day. Finally, the small cheese wheels are packaged in plastic bags under vacuum and stored in cold stores.

11.2.5 Soft-type cheeses with spreadable texture

Greek soft type cheeses without rind and spreadable, rather creamy texture are low-pH cheeses with moisture up to 75%. Their shelf life is limited and they are consumed fresh without ripening, since in most cases no ripening takes place. Compositional characteristics of cheeses of this group are presented in Table 11.9.

The most well known soft spreadable cheese in Greece is Galotyri cheese, which traditionally was manufactured for household needs from sheep milk at the end of the lactation period in summer. The traditional procedure is presented by Anifantakis (1991) and Litopoulou-Tzanetaki and Tzanetakis (2011). It is a PDO cheese produced in Epirus and Thessaly regions from sheep milk or its mixtures with goat milk with moisture ≤75% and fat-on-dry matter ≥40%. Boiled milk stays in containers for 24 hours at room temperature. After the addition of salt in a ratio of 3 to 4%, it is left for further 48 hours by stirring periodically to be acidified. The biologically acidified salted milk is transferred into fabric or skin bags or wooden barrels; rennet at low concentration may be added at this stage. This procedure is repeated in consecutive days until the fill of bag or barrel. Melted fat may cover cheese surface and the containers are transferred to cold, dry rooms at <8°C, where they can stay for two to three months. Traditional cheesemakers used to incorporate into the acidified milk small pieces of well-drained fresh Feta cheese, instead of a small quantity of rennet. There are several research studies about Galotyri cheese with emphasis to microbiota, safety and shelf-life, e.g., Rogga et al. (2005), Lekkas et al. (2006), Kondyli et al. (2008), Kondyli et al. (2013).

Similar to Galotyri is Katiki Domokou cheese, which is a low pH PDO variety, produced in Domokos area in a plateau of Central Greece from goat milk or from its mixtures with sheep milk with moisture ≤75% and fat-on-dry matter ≥40%. Mesophilic starter and low quantity of rennet may be added at 27 to 28°C. Cheese milk is left at room temperature overnight and the acidified curd is placed into fabric bags to drain. The drained curd

Table 11.9 Average composition of soft spreadable Greek cheeses, expressed as percentages on cheese mass. FDM, fat-on-dry matter; ripening index, percentage of soluble nitrogen on total nitrogen; S/M, salt-in-moisture, i.e., salt×100/(salt+moisture).

Cheese	Days	pH	Moisture	Fat	FDM	Protein	Ripening index	Lactose	NaCl	S/M	Ash	References
Galotyri[b], sheep milk		3.80	76.9	9.1		8.3			1.8			Rogga et al. 2005
Galotyri[b], sheep milk		4.00	75.8	9.1		7.7			1.8			Rogga et al. 2005
Galotyri[a], sheep milk	2	4.40	73.9	10.7	40.8	10.1	6.09	3.0	1.6	2.17	2.38	Kondyli et al. 2008
Galotyri[a], sheep milk	15	4.54	74.1	11.1	42.7	10.8			1.6	2.15	2.29	Katsiari et al. 2009
Anevato[a], goat milk		4.40	62.3		46.6					3.19		Xanthopoulos et al. 2000
Pichtogalo Chanion[b]	3–6	4.36	61.6		54.0	14.2			1.02			Papageorgiou et al. 1998
Xygalo Siteias			≤75		33–46	≥31.5			≤1.5			European Union 2010, 2011
Kopanisti[a]	16	4.58	52.5	22.5		18.7	16.53					Kaminarides et al. 1990

[a] experimental cheeses; [b] market cheeses.

is salted by incorporation of fine-grade salt in the cheese mass and then it is packaged in containers of various capacities and stored in cold stores. It is consumed fresh since its storage time is limited due to high moisture.

Anevato, is another spreadable PDO cheese with low pH and particular granular texture produced traditionally in Western Macedonia from raw sheep or goat milk or their mixtures. Its moisture is ≤60% and fat-on-dry matter ≥45%. Local shepherds used to inoculate the milk collected during the morning milking with a small quantity of rennet. In the evening, when they came back to their premises a curd had been formulated floating into the whey; that's why it was called Anevato, meaning something that rises. As described in the PDO Standards (Greek Code of Foods 2012), a small quantity of rennet is added at 18 to 20°C into previously biologically acidified milk to induce curdling within 12 hours. Currently, Anevato is manufactured also in cheese plants and starters may be added to control acidification. As described by Xanthopoulos et al. (2000) heated milk either pasteurized (72°C for 15 seconds) or thermized (63°C for 10 minutes) is inoculated with starters and after 30 minutes, rennet is added. When curd is formed it is cut crosswise (2×2 cm) and left to rest for four to five hours. The curd drains into cheesecloth until the next day. Salt is added to the drained curd by thorough mixing and cheese is packaged in plastic containers stored thereafter in cold stores.

Two PDO cheeses of this category are produced in the different regions of Crete Island from raw or pasteurized small ruminant's milk named Pichtogalo Chanion, produced in the prefecture of Chania and Xygalo Siteias produced exclusively in the region of Siteia in the prefecture of Lasithi. Their characteristics and manufacturing steps have been specified in Greek Code of Foods (2012) and European Union (2010, 2011). Both of them are spreadable cheeses, white with slightly acid taste and strong aroma resulted from the acid coagulation of milk. These cheese types differ in regard to moisture and fat due to differences in the cheese making procedure. Pichtogalo means dense or thick milk and Xygalo means sour milk. Traditionally both of them were produced by shepherds at home or in the "mitata", which were extemporaneous cheese plants in the milking premises onto the mountains. Pichtogalo has <65% moisture and >50% fat-on-dry matter. Therefore, its texture is more dense than that of Xygalo mainly due to differences in fat content. The latter is made mostly from goat milk that has lower fat content than sheep milk. When sheep milk participates in the cheese milk, a part of milk fat is removed. Furthermore, during the acidification of the curd, a portion of the surface fat layer is removed. Pichtogalo is made from milk that curdles at room temperature for two hours. The curd is acidified biologically by the native microbiota or by starters until next day and then it is placed without cutting in bags or cheesecloths to drain. Salt at about 1% ratio is incorporated into the cheese mass. Xygalo is made from salted milk at a ratio of 2% and curdling and biological acidification take place in containers at 15 to 20°C within seven to 10 days. Ripening is carried out in the same containers for about a month, without cutting or disturbing the curd. Finally, cheese is separated from the whey, gathered on the bottom of the container. It is then packaged in barrels or containers stored at 4°C. When raw milk is used, Xygalo is held for at least two months at 4°C before consumption.

Kopanisti is a soft spreadable cheese with very particular organoleptic characteristics, i.e., creamy texture, piquant and salty flavour with light yellow to tan colour due to intense yeast and mould activity (Fig. 11.12). It is a PDO cheese produced in the islands of Cyclades from cow, sheep, goat milk or their mixtures, which contains moisture ≤56% and fat-on-dry matter ≥43%. According to the standards the rennet is added to milk at 28 to 30°C in order curdling to take place within two hours. Curd is left in the cheese vat until next day to be acidified and then it is cut and placed into bags, where it is squeezed to drain. The drained cheese mass is called "petroma". It is salted, i.e., 4 to 5% salt is incorporated into the cheese mass by hand, and stays in an open vat in a rather cool room with high relative humidity. Alternatively, according to Kopanisti PDO amendment (European Union 2013) fresh butter up to 15% may be added at the stage of salting to improve texture and flavour. After approximately 10 days, when a yeast and mold layer is apparent on the surface, the cheese mass is kneaded with the aim to distribute uniformly the surface microbiota. This procedure is repeated three or four times until the end of Kopanisti ripening, i.e., 30 to 40 days after cheesemaking. According to a traditional practice a quantity of good quality mature Kopanisti called "mana" is mixed with acidified curd in a ratio of 10% to control and accelerate ripening (Anifantakis 1991, Tzanetakis et al. 1987). The strong flavour of this cheese type is mainly due to lipolysis, which is seven-folded from eight to 32 days (Kaminarides et al. 1990, Karali et al. 2013).

Figure 11.12 Kopanisti cheese (archives of National Dairy Committee of Greece).

11.2.6 Whey cheeses

Greek Whey cheeses are mostly soft cheeses with very limited shelf life resulted from the thermal coagulation of whey proteins induced by the heat treatment of cheese whey. The composition of whey is very important for the yield and the quality of whey cheeses and it is configured by the type of cheese from which it is derived. The composition of wheys derived from different cheesemaking procedures applied in Greece has been determined by Kandarakis (1981) and is presented in Table 11.10. Sheep cheesemilk and application of severe curd cutting and intense scalding and stirring during the manufacture of hard type and Gruyere type cheeses enriches the whey with total solids, especially with fat. Composition and particular nutritional features are presented in Tables 11.11 and 11.12 respectively.

Table 11.10 Composition of whey from the manufacture of Greek cheeses, expressed as percentage on whey mass (Kandarakis 1981). SM, sheep milk; CM, cow milk.

Whey constituents	Feta SM	Telemes CM	Graviera SM	Graviera CM	Kefalotyri SM	Kefalotyri CM
Total solids	7.87	6.69	8.74	6.90	7.48	6.55
Fat	0.39	0.40	1.26	0.60	1.26	0.40
Total proteins	1.61	0.85	1.52	0.90	1.52	0.80
Lactose	5.33	4.90	5.27	4.90	5.27	4.85
Ash	0.78	0.52	0.50	0.50	0.50	0.50

The most widely produced whey cheese in Greece is Myzithra, a high-moisture rindless cheese with grainy texture, not spreadable. In fact, Myzithra mass consists of clusters of denatured whey proteins. The manufacturing steps of Myzithra are shown in Fig. 11.13.

Whey is filtered before heating to remove any curd pieces and is heated under continuous stirring in two stages. Firstly, slow heating is applied up to 80 to 82°C; at this point small whey protein flakes become apparent. During the second stage heating is accelerated up to 88 to 92°C, stirring slows down and finally stops when a Myzithra coagulum appears on the surface (Fig. 11.14). The final temperature is attained within 40 to 45 minutes and remains almost steady for 15 to 30 minutes. Acidification of whey to pH 5.2 by the addition of citric acid or of whey of the previous day is necessary when it comes from cow milk cheeses. However, acidification is not applied to sheep and goat whey (Anifantakis 1991). Whey for Myzithra can be supplemented with 3 to 5% milk at 65 to 70°C. Myzithra can be unsalted but 1 to 1.5% salt can be added to the whey at 73 to 75°C. The Myzithra curd is gradually transferred into cheesecloths or into truncated cone-shaped perforated molds to drain within three to

Table 11.11 Average composition of Greek whey cheeses, expressed as percentages on cheese mass. FDM, fat-on-dry matter; ripening index, percentage of soluble nitrogen on total nitrogen; S/M, salt-in-moisture, i.e., salt×100/(salt+moisture).

Cheese	pH	Moisture	Fat	FDM	Protein	Lactose	NaCl	S/M	Ash	References
Myzithra[b]		74.62	3.0	11.82	14.63		2.95		4.14	Kandarakis et al. 2002
Dried Myzithra[b]		36.82	20.92	33.2	24.42		9.05		10.94	Kandarakis et al. 2002
Anthotyros[b]		67.54	13.88	42.75	12.42		0.57		1.75	Kandarakis et al. 2002
Manouri[b]		45.28	41.75	76.53	11.04		1.56		2.05	Kandarakis et al. 2002
Manouri[a]		52.62	29.93		12.93	3.2			1.65	Kandarakis 1981
Anthotyros[b]	6.17	65.24						1.46		Kalogridou-Vassiliadou et al. 1994
Myzithra[a]		68.04	15.7	49.03	11.25	4.0			0.77	Kaminarides et al. 2013
Myzithra[c]		65.29	17.0	49.50	11.37	3.9			1.10	Kaminarides et al. 2013

[a] experimental cheeses; [b] market cheeses; [c] 10% milk was added to whey.

Table 11.12 Particular nutritional characteristics of Greek whey cheeses. Fat expressed as percentage of cheese mass; SFA, saturated; MUFA monounsaturated; PUFA, polyunsaturated fatty acids expressed as percentages of fatty acids; CLA, conjugated linoleic acid mostly C9, t11-C18:2 expressed as mg per 100 g cheese; CHOL, cholesterol expressed as mg per g cheese; Ca, P, Mg, K, Na expressed in mg per 100 g cheese; Se expressed as μg per 100 g cheese.

Lipid fraction

Cheese	Fat	SFA	MUFA	PUFA	CLA	CHOL	References
Myzithra	15.2				0.83		Zlatanos et al. 2002
Anthotyros	18.3				0.41		Zlatanos et al. 2002
Manouri	32.5				0.87		Zlatanos et al. 2002
Myzithra	16.9	64.5	16.6	2.7		42.7	Andrikopoulos et al. 2003
Anthotyros	16.3	65.6	19.6	3.7		56.2	Andrikopoulos et al. 2003
Manouri	50.1	66.1	19.1	2.8		99.5	Andrikopoulos et al. 2003
Manouri	46.2	72.5	20.3	3.4	0.54		Zlatanos and Laskaridis 2009

Inorganic fraction

Cheese	Ca	P	Mg	K	Na	Se	References
Whey cheeses						42.4	Pappa et al. 2006c
Myzithra	314	190	26				Kandarakis et al. 2002
Dried Myzithra	397	294	37		3563		Kandarakis et al. 2002
Anthotyros	371	200	24		227		Kandarakis et al. 2002
Manouri	136	120	14		614		Kandarakis et al. 2002

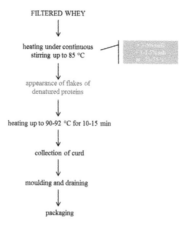

FILTERED WHEY

↓

heating under continuous stirring up to 85 °C

appearance of flakes of denatured proteins

↓

heating up to 90-92 °C for 10-15 min

↓

collection of curd

↓

moulding and draining

↓

packaging

Figure 11.13 Diagram of the manufacture of Myzithra from sheep and goat milk.

Figure 11.14 Collection of Myzithra curd.

five hours, after intense squeezing by hand. Myzithra is stored in cold stores and it is marketed soon after, as a fresh product. Myzithra can be preserved after intense drying in "dried Myzithra", which can be stored in cold rooms for months and is utilized as grated cheese. For this purpose Myzithra is salted and remains hung to well ventilated rooms until moisture decreases to <40%.

Anthotyros is a variant of Myzithra made exclusively from sheep and goat hard type cheese whey like Kefalotyri, and supplemented with sheep and goat milk at about 10%. Salt is also added. This type of whey along with added milk results in a product with very rich taste and smooth texture. Initially, it was produced in Crete but nowadays is produced in many regions of Greece (Kalogridou-Vassiliadou et al. 1994, Anifantakis 1991).

Xinomyzithra is a PDO whey cheese, which according to standards (Greek Code of Foods 2012) is produced in Crete. It has moisture <55% and fat-on-dry matter >45% and is produced from whey that comes from sheep and goat milk cheeses. Initially, a Myzithra type coagulum is produced using whey supplemented by small ruminants' milk in a ratio of 15%. After draining, 1.5 to 2% salt is carefully incorporated into the mass, which then is placed in fabric bags. Pressure is applied onto the bag for about one week. During this period, the curd is acidified and obtains sour flavour. Then, the pressurized curd is placed carefully into barrels avoiding void spaces and kept at 10°C for at least two months before consumption.

Manouri is a soft whey PDO cheese with very rich flavour and particular creamy texture. Its specific characteristics are due to the sheep and goat cheese whey with at least 2.5% fat, in which small ruminants' milk or cream is added. According to standards (Greek Code of Foods 2012), Manouri moisture is <60% and fat-on-dry matter >70% and it is produced exclusively in Thessaly, Central and Western Macedonia. Normally, its shape is cylindrical and it has a rather sweet and aromatic taste. Milk or cream up to 25% is added in the whey along with 1% salt at 70 to 75°C and heating continues up to 88 to 90°C for 15 to 30 minutes. Manouri drains in cylindrical fabric bags for four to five hours. It is consumed fresh or after a limited time of storage at 4 to 5°C. Traditionally, Manouri was produced from whey of Batzos without any supplementation as described by Lioliou et al. (2001). The Batzos coagulum was "beated", i.e., was severely cut and extremely stirred to enrich whey with fat.

11.3 Greek fermented milk products

In Greece all types of fermented milks, i.e., set, stirred and strain yoghurt, drinking yoghurt as well as products with acidified milk and cereals are produced. According to the Greek Code of Foods (2012), Greek yoghurt is classified to two main categories: the natural or plain yoghurt products and the yoghurt desserts. For the production of natural yoghurt products only milk (raw, UHT or frozen) and yoghurt culture are allowed to be used. In the case of yoghurt desserts, additives such as colours, aromatic substances, stabilizers, fruits, nuts and powders of milk origin, i.e., Skimmed Milk Powder (SMP), Whey Protein Concentrates (WPCs) and Sodium Caseinates can be added. Both categories include set type, strain or stirred type products. Therefore, there is a big variety of such products in the Greek market, i.e., traditional yoghurt, natural set type yoghurt, natural stirred yoghurt, set type yoghurt dessert, stirred yoghurt dessert, drinking type (a diluted or low viscous fluid) and strain (concentrated) yoghurt. Among them, the traditional and the strain yoghurt are the most famous.

11.3.1 Traditional yoghurt

Greek traditional yoghurt (Fig. 11.15) is a special yoghurt product, which has a solid fat layer like crust on its surface. Actually it is a set type yoghurt produced from non homogenized cow or sheep or goat milk or mixtures of them, usually cow milk to sheep milk 50:50 and sheep milk to goat milk 90:10, according to the process shown in Fig. 11.16.

Figure 11.15 Greek traditional yoghurt.

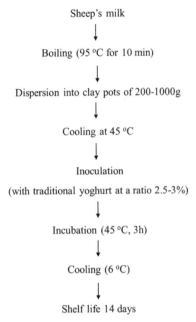

Sheep's milk

↓

Boiling (95 °C for 10 min)

↓

Dispersion into clay pots of 200-1000g

↓

Cooling at 45 °C

↓

Inoculation

(with traditional yoghurt at a ratio 2.5-3%)

↓

Incubation (45 °C, 3h)

↓

Cooling (6 °C)

↓

Shelf life 14 days

Figure 11.16 Manufacturing diagram of Greek traditional yoghurt from sheep milk.

Traditionally, in homemade production, raw milk is filtered through cloth and then, it is boiled for several minutes under continuous stirring. The boiled milk is distributed to small pots from clay and remains until the temperature drops to about 45°C. Meanwhile fat globules conglomerate onto the top and thus natural fat separation occurs forming the characteristic fat layer. At 45°C the milk is inoculated with traditional yoghurt produced on the previous day, so that bacteria are as active as possible. Inoculum is prepared after removing its fat layer and diluting it slightly with the heated yoghurt milk. Inoculation is made carefully using a spoon or a sterilized syringe, avoiding the disruption of the fat layer. Incubation takes place at 40 to 42°C for about three or four hours. After incubation yoghurt pots are kept in cold stores for one week (Zerfiridis 2001b).

The industrial manufacture of Greek traditional yoghurt differs from the household production in several points. At first, milk composition is standardized so that after heat treatment the total solids (TS) content is at least 10% higher than the TS content of the milk that was used. Heat treatment takes place in plate heat exchangers or in milk tanks with double walls and equipped with mechanical agitation and temperature adjustment. Heated milk is dispersed mechanically into plastic retail pots of 220 to 300 g and a dosimeter pump inoculates them. Incubation takes place in closed rooms, i.e., incubators with stable temperature, followed by quick cooling to avoid post-acidification. Finally the pots are sealed with a plastic film and then with the plastic lid. Industrially manufactured Greek traditional yoghurt has longer shelf life, i.e., up to 15 days (Zerfiridis 2001b).

Traditional yoghurt has different organoleptic characteristics from the set type yoghurts. It has a soft body with lower firmness values than those of set type yoghurts because of its lower total solids content due to fat concentration onto the top. Very few studies have been conducted concerning the composition of the traditional type yoghurts. It has been reported that the use of 48 hour refrigerated cow milk or sheep milk does not affect the pH of the produced yoghurt but deteriorates slightly the taste of that made from cow milk (Kehagias 1982). Moreover, Serafeimidou et al. (2013) have shown that CLA content in cow milk traditional yoghurt decreases during 14 days of storage at 5°C, while in sheep milk significantly increases. Finally, it was shown that using specific cultures traditional yoghurt may possess multifunctional health effects such as probiotic properties and bioactive peptides with angiotensin I-converting enzyme (ACE)-inhibitory activity (Papadimitiou et al. 2007). Compositional and nutritional information of traditional Greek yoghurt are shown in Tables 11.13 and 11.14, respectively.

11.3.2 Strained yoghurt

Strained yoghurts are concentrated products having high level of milk solids, i.e., ~20% total solids with ~10% fat. According to the Greek Code of Foods (2012), strained yoghurt from (any kind) of milk is the product, which results from whole milk yoghurt after draining part of the serum (acid whey). This product should contain at least 8% fat except cow milk strained yoghurt, which should contain at least 5% fat and should be natural yoghurt with no additives such as sugar, fruits, stabilizers, etc. Stained yoghurt is consumed either alone or as the base of a famous Greek food dish named tzatziki.

Strained yoghurt was traditionally manufactured in bags made from animal skins; the yoghurt serum was removed while the concentrated product remained in the bag. Nowadays, in household strained yoghurt production, traditional bags were substituted by cloth bags (Fig. 11.17). After reaching acidity of 85°D, i.e., 0.85% lactic acid, the yoghurt is transferred in a cloth bag through which the serum is expelled. Draining takes place at low temperature in order to avoid post acidification (Zerfiridis 2001b). In natural yoghurt the content of the milk total solids is about 14 g/L while in Greek strained products it raises upto 21 to 23 g per litre. This concentration can be achieved by overnight draining of 5 to 10 litres of stirred natural yoghurt using a cloth bag, into a cold room.

In industrial manufacture the serum is removed either by centrifugation or by ultrafiltration. As with gravity drainage, both techniques remove the desired quantity of water together with lactose and mineral salts from

Table 11.13 Average composition of special Greek Yoghurts, expressed as percentages on cheese mass.

Yoghurt	pH	Acidity	Fat	Protein	Carbohydrates	Ash	Total Solids	References
Traditional								
sheep milk [a]	4.77	1.19	6.08	4.71		0.93	20.63	Serafeimidou et al. 2013
sheep milk [b]			6.60	5.17	4.50			Serafeimidou et al. 2013, Vitalioti 2012
cow milk [a]	4.60/4.52	0.71/0.88	5.28/3.90	3.95/3.7	-/3.57	0.93/0.8		
cow milk [b]			3.90	3.70	5.20			
Set type								
goat milk [a,c]	4.15/-	0.93/-	4.92/5.68	4.06/4.2	4.80/-	0.81/-	14.61/15.6	Kaminarides and Anifantakis 2004, Kehagias et al. 1989
goat milk [a,d]	3.97	0.98	3.97/3.69	3.22/3.03	4.81	0.80	3.97/3.69	Kaminarides and Anifantakis 2004, Kehagias et al. 1989
Strained								
cow milk [b]			2	8	4		20	Zerfiridis 2001b
cow milk [b]			10	5.9	3.5			Zerfiridis 2001b

[a]experimental yoghurts; [b]market yoghurts; [c]milk from local breeds; [d]milk from Alpine/Saanen breeds.

Table 11.14 Particular nutritional characteristics of special Greek Yoghurts. Fat expressed as percentage of yoghurt mass; SFA, saturated; MUFA, monounsaturated; PUFA, polyunsaturated fatty acids expressed as percentages of fatty acids; CLA, conjugated linoleic acid mostly C9, t11-C18:2 expressed as percentage of fatty acids; CHOL, cholesterol expressed in mg per 100 g yoghurt; Ca, P, Mg, K, Na expressed in mg per 100 g yoghurt; Se expressed in µg per 100 g yoghurt.

Lipid fraction

Yoghurt	Fat	SFA	MUFA	PUFA	CLA	CHOL	References
Traditional							
Sheep	6.08	73.79	21.61	4.49	0.53		Serafeimidou et al. 2013
Cow	5.28	72.07	24.59	3.26	0.41		Serafeimidou et al. 2013
Cow						12.4	Fletouris et al. 1998
Strained							
Cow						33.6	Fletouris et al. 1998
Sheep						30.1	Fletouris et al. 1998

Inorganic fraction

Yoghurt	Ca	P	Mg	K	Na	Se	References
Set type							
Goat[a]	134/146		13/-		44/-		Kaminarides and Anifantakis 2004, Kehagias et al. 1989
Goat[b]	121/124		13/-		34/-		Kaminarides and Anifantakis 2004, Kehagias et al. 1989
Sheep	191	128	18	117	71		Moschopoulou et al. 2015
Strain							
Cow[c]						20.6	Pappa et al. 2006
Cow[d]						26.9	Pappa et al. 2006
Sheep						21.9	Pappa et al. 2006

[a]milk from local breeds; [b]milk from Alpine/Saanen breeds; [c]yoghurts 0% fat; [d]yoghurts 4% fat.

the initial natural yoghurt to give a smooth concentrated product. A brief description of these two methods, which are applied in the manufacture of strained yoghurt in general, is given by Tamime et al. (2014). Greek strained yoghurt is not very acid and its composition depends on the fat content as shown in Table 11.13. Strained yoghurt is mainly produced with cow milk. When it is made from goat milk the yield is higher but the firmness is lower (Kehagias et al. 1992). Strained yoghurt made from frozen ultrafiltrated sheep milk seems to be of inferior quality compared with those made from evaporated milk, but addition of UF filtrate can restore

Milk

↓

Boiling (95 °C, 10 min)

↓

Transfer of 10-15 L boiled milk into plastic
vessels and cooling at 50 °C

↓

Inoculation at 45-50 °C and incubation until
coagulation

↓

Stirring of the coagulum

↓

Pouring into a cloth bag and draining at ~10 °C to
the desired ratio of milk:product

↓

Mixing of the product, packaging into plastic
containers and storage at 4-6 °C

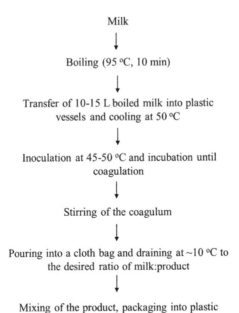

Figure 11.17 Manufacturing diagram of traditional Greek strained yoghurt.

the deficiencies (Kehagias et al. 1980). Compositional and nutritional characteristics of strained yoghurt are shown in the Tables 11.13 and 11.14, respectively.

The Greek strained yoghurt is well known worldwide as Greek yoghurt or Greek style yoghurt. The majority of these products contain high protein content and low fat content, i.e., 0 to 2%. Recently, Tamime et al. (2014) has reviewed commercial strained fermented milks from many countries and showed that there are many differences in the composition among these products. Consequently, they suggested that a typical strained yoghurt or 'Greek strained yoghurt' should have a protein content ≥8 g and carbohydrates content ~5 g per 100 g of product. In contrast, higher carbohydrate content shows that the product has been manufactured by the formulation method, i.e., without removing the acid whey, but fortifying the yoghurt milk with SMP or WPCs, and they propose to be called 'Greek-style yoghurt'.

11.3.3 Set type yoghurt from goat or sheep milk

Set type goat milk yoghurt has been launched in the Greek market rather recently. The procedure for manufacturing this yoghurt type does not differ from that applied in the case of set type cow milk yoghurts. Goat milk is initially standardized for its fat and protein content, is then homogenized at 70 to 72°C, 150 bar, is then heated at 95°C for 5 minutes and is finally cooled at 42°C. The entire quantity of cooled milk is inoculated with an appropriate commercial yoghurt starter and then is dispensed into plastic pots of 200 to 240 g. Sealed pots are incubated up to pH 4.7, usually for four hours. After incubation, yoghurt pots are cooled down to 4 to 6°C, where they are kept until consumption. The shelf life of this product made under strictly controlled hygienic conditions can be 40 days.

According to bibliography, the quality of set type yoghurt made from goat milk of local Greek breeds is better than those of yoghurt produced from goat milk of Alpine or Saanen breeds (Kehagias et al. 1989, Kaminarides and Anifantakis 2004). However, goat milk set type yoghurt is less viscous due to the different content of the casein fractions of goat milk, i.e., lower levels of as_1 casein and higher levels of as_2- and β-caseins than those present in sheep or cow milk (Moatsou et al. 2008, Tamime et al. 2011) and this fact results in softer coagulum (Moschopoulou et al. 2015). Recently, a study was conducted to improve the texture of Greek caprine set type yoghurt by addition of Whey Protein Concentrate (WPC) at percentages 0.5 to 1.5 and results showed that yoghurt fortified with 1.5% WPC exhibited significantly improved rheological properties (Theodorou et al. 2015).

Set type sheep milk yoghurt is also of limited industrial production, but experimental studies have shown that this product is of high quality, even in the case of low fat products, i.e., 0.9%, 2.3% or 3.8% (Kaminarides et al. 2007, Moschopoulou et al. 2015). Moreover, sheep milk mixed with goat milk of improved breeds, e.g., of Alpine breed at a ratio 50:50 can lead to a set type yoghurt product with improved quality (Kaminarides and Anifantakis 2004). The use of frozen sheep milk in yoghurt production is a common practice during summer when sheep and goat milk production is very limited and it has been showed that it does not affect at all the overall quality of the produced yoghurt (Katsiari et al. 2002). Furthermore, it has been reported that whole sheep milk concentrated by reverse osmosis and stored frozen up to eight months produces set type yoghurt of high quality with higher curd consistency and apparent viscosity (Voutsinas et al. 1996). Compositional and nutritional characteristics of set type yoghurt are shown in the Tables 11.13 and 11.14, respectively.

11.3.4 Mixed fermented milk/cereal products

This category includes products that except acidified milk contain cracked wheat or other cereals or flour. Two are the most famous; Xinochondros and Trahanas.

Xinochondros is a traditional organic fermented dry dairy product of Crete Island of Greece, which is consumed fresh as soup or dried as part of traditional Cretan diet. It is usually manufactured towards the end of lactation period using naturally acidified (xino in Greek) ovine milk and coarse wheat called 'chondros' after drying under the sun. Sometimes milled vegetables or extra virgin olive oil and salt are added at very low concentrations. Upto the 1980s, it was exclusively homemade, but today it is also manufactured in small-medium enterprises using mechanical means and drying rooms without, however, to decline from the traditional process.

Five to ten kg acidified and slightly salty milk is boiled and chondros is then added slowly under continuous stirring at a ratio usually 3:1. After heat treatment, Xinochondros paste is transferred onto swallowed dishes in order to cool quickly and increase its viscosity. Next day the mass is cut into coarse pieces by hands or knife, put on wooden tables and exposed to sun for drying, while pieces are overturned periodically. Plastic perforated dishes (sieves) instead of wooden tables are used to allow air circulation for quick and uniform drying, which lasts usually one week. As most of dried products, Xinochondros is affected by molds and also by insects found in storage rooms when stored under conditions with high humidity. *Plodia interpunctella* is the main insect that deteriorates its quality (Anifantakis et al. 2004).

Xinochondros of high quality can be made, using sheep or goat milk and starter cultures as shown in Fig. 11.18.

In the case of Xinochondros from sheep milk, the ratio of milk to chondros is 3.25:1 with a yield product 34.48 ± 0.06%, while in the case of Xinochondros from goat milk the respective values are 2.89:1 and 33.67 ± 0.14% (Anifantakis et al. 2004). Compositional and nutritional characteristics of Xinochondros are shown in Tables 11.15 and 11.16 respectively. The mean composition (g per 100 g) of chondros used in Xinochondros production is fat 1.3, protein 11.32, carbohydrates 72.39 and total solids 89.46.

Trahanas is a cereal-based product produced in many regions of Greece and is classified into four varieties: Trahanas made without milk but with vegetables, called 'Nistisimos Trahanas', Trahanas made with milk and wheat flour, called 'Sweet Trahanas' and two Trahanas varieties made with acidified milk, wheat flour with or without eggs, called 'Sour Trahanas with eggs' or 'Sour Trahanas without eggs' respectively (Georgala 2013). Trahanas that contains milk is manufactured from whole fresh ewe or goat milk or mixtures of them in the summer. Fresh milk is allowed to be acidified either spontaneously or by adding yoghurt culture. Then it is boiled while flour is added gradually under stirring. The ratio of milk to wheat flour is usually 3:1 or 4:1. After cooling, the milk-flour paste is cut in fingers sized pieces, which are subsequently dried under sun (Georgala et al. 2012, Georgala 2013). Composition and nutritional characteristics of Trahanas products containing milk are shown in Tables 11.15 and 11.16 respectively.

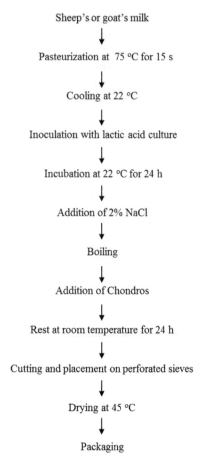

Sheep's or goat's milk

↓

Pasteurization at 75 °C for 15 s

↓

Cooling at 22 °C

↓

Inoculation with lactic acid culture

↓

Incubation at 22 °C for 24 h

↓

Addition of 2% NaCl

↓

Boiling

↓

Addition of Chondros

↓

Rest at room temperature for 24 h

↓

Cutting and placement on perforated sieves

↓

Drying at 45 °C

↓

Packaging

Figure 11.18 Advanced technology of Xinochondros production (Anifantakis et al. 2004).

11.4 Miscellaneous dairy products

There are three special Greek products based on milk fat. These are melted butter (called Tigmeno boutyro), Staka butter (called Stako boutyro) and Staka.

According to the Greek Code of Foods, Tigmeno boutyro or butter for cooking of cow or goat or sheep milk is the milk fat product, which is manufactured after melting fresh butter of the respective milk at low temperature, preferably in water bath. This product should contain moisture and MSNF up to 1% and its acidity should not exceed 10°. Salt addition is permitted up to 1%. While butter is melting, its emulsion is disrupted and thus milk fat (butter oil) and moisture are separated. After moisture removal, the warm milk fat is collected and it is then cooled to form a coarse/granular structure. It is usually kept in glass jars at 4°C.

Table 11.15 Average composition of Greek mixed milk/cereal Products, expressed as percentages on product mass.

Product	pH	Moisture	Fat	Protein	Carbohydrates	NaCl	Ash	Dietary Fiber	References
Xinochondros[a]		7.39	11.79	17.84	57.69	3.09	5.30		Anifantakis et al. 2004
Xinochondros from sheep's milk[b]	5.20	7.34	15.01	19.36	54.95		5.07	2.27	Anifantakis et al. 2004
Xinochondros from goat's milk[b]	5.25	8.09	11.22	15.24	61.76		4.93	2.40	Anifantakis et al. 2004
'Sweet' Trahanas[a]		9.60	4.40	15.30	68.00	1.40	2.90	2.10	Georgala et al. 2012
Sour Trahanas without eggs[a]		10.7	4.30	15.30	65.20	2.6	4.10	1.40	Georgala et al. 2012
Sour Trahanas with eggs[a]		9.2	3.10	15.30	69.40	1.70	2.60	1.00	Georgala et al. 2012

[a]market products; [b]experimental products.

Table 11.16 Particular nutritional characteristics of Greek mixed milk/cereal Products. Fat expressed as percentage of product mass; SFA, saturated; MUFA, monounsaturated; PUFA, polyunsaturated fatty acids expressed as percentages of fatty acids; Ca, P, Mg, K, Na expressed in mg per 100 g product or dry matter occasionally.

Product						References
	Fat	SFA	MUFA	PUFA		
'Sweet' Trahanas	4.40	57.0	23.3	19.8		Georgala et al. 2012
Sour Trahanas[a]	4.30	63.4	22.1	14.5		Georgala et al. 2012
Sour Trahanas[b]	3.10	54.7	24.9	20.4		Georgala et al. 2012
	Ca	Mg	K	Na	Mn	
Xinochndros[c]	320	106		1102		Anifantakis et al. 2004
'Sweet' Trahanas[d]	91	88	348	1154	1.54	Georgala et al. 2012
Sour Trahanas[a,d]	118	60	329	2358	1.24	Georgala et al. 2012
Sour Trahanas[b,d]	78	45	235	1017	0.56	Georgala et al. 2012
	976		40	118	705	

[a]without eggs; [b]with eggs; [c]mg per 100 g product; [d]mg per 100 g dry matter; all market products.

Tigmeno boutyro from mixed goat and sheep milk is very aromatic and is used in cooking to give a special flavour to the prepared food.

Staka and Stako boutyro are produced in Crete Island. Staka is relative to a French product called Roux, while Stako boutyro is similar to Tigmeno boutyro and is used for flavouring pasta or rice pilaf dishes commonly served in weddings or in candies.

Staka is made traditionally from cream of goat milk. Modern Staka is made from unsalted goat butter or from mixture of fresh goat and sheep milk. Cream usually is obtained from the milk of 10 successive milkings. Milk fat is separated onto the top of boiled milk and is naturally acidified. At first, cream is heated under mild and continuous agitation up to boiling temperature, so that the fatty part (butter oil) separates to form Stako boutyro. After removing as much as possible of the butter oil, a small quantity of barley or wheat flour is added to the rest of the boiled cream under continuous heating and agitation until a creamy product, called Staka, is made. Staka is consumed either alone, e.g., served warm after sprinkling with lemon juice, or as part of a food preparation, e.g., added in fried eggs.

References

Aktypis, A., E. Zoidou, E. Manolopoulou, T. Massouras and I. Kandarakis. 2011. Microbiological characteristics of Traditional Greek cheeses «Kefalotyri Kefallonias», Ladotyri Zakynthou» and Arseniko Naxou». 1st Hellenic Food Congress, HVMS, 11–13 November 2011 Thessaloniki, Greece.

Alichanidis, E. 2007. Cheeses ripened in brine. *In*: P.L.H. McSweeney (ed.). Cheese Problems Solved. CRC Press, Cambridge, UK.

Alichanidis, E., E. Anifantakis, A. Polychroniadou and M. Nanou. 1984. Suitability of some microbial coagulants for Feta cheese manufacture. J. Dairy Res. 51: 141–147.

Alichanidis, E. and A. Polychroniadou. 2008. Characteristics of major traditional regional cheese varieties of East Mediterranean countries: a review. Lait. 88: 495–410.

Andrikopoulos, N., N. Kalogeropoulos, A. Zerva, U. Zerva, M. Hassapidou and V.M. Kapoulas. 2003. Evaluation of cholesterol and other nutrient parameters of Greek cheese varieties. J. Food Comp. Anal. 16: 155–167.

Angelidis, A. and A. Govaris. 2012. The behavior of *Listeria monocytogenes* during the manufacture and storage of Greek Protected Designation of Origin (PDO) cheeses. *In*: A. Romano and C.F. Giordano (eds.). Listeria Infections. Epidemiology, Pathogenesis and Treatment. Nova Science Publishers, Inc.

Anifantakis, E.M. 1991. Greek Cheeses, a Tradition of Centuries. National Dairy Committee of Greece, Athens, Greece.

Anifantakis, E., A. Georgala, I. Kandarakis, A. Vamvakaki, E. Moschopoulou and C. Miaris. 2004. [Xinochodros: A traditional organic product of Crete]. Eptalofos publications, Athens, Greece.

Anifantakis, E.M. and G. Moatsou. 2006. Feta and other Balkan cheeses. *In*: A.Y. Tamime (ed.). Brined Cheeses. Blackwell Publishing, Oxford, UK.

Bintsis, T., E. Litopoulou-Tzanetaki, R. Davies and R.K. Robinson. 2000. Microbiology of brines used to mature Feta cheese. Int. J. Dairy Technol. 53: 106–112.

Bontinis, T.G., H. Mallatou, E. Alichanidis, A. Kakouri and J. Samelis. 2008. Physicochemical, microbiological and sensory changes during ripening and storage of Xinotyri, a traditional Greek cheese from raw goat's milk. Int. J. Dairy Technol. 61: 229–236.

European Union (EU), Commission Regulation No. 1107, 1996—The registration of geographical indications and designations of origin under the procedure laid down in article 17 of council regulation (EEC) No. 2081/92. Official Journal of the European Communities L148, 1–10.

European Union (EU), Commission Regulation No. 1829, 2002—Amending to the Annex to Regulation (EC) No. 1107/96 with regard to the name 'Feta'. Official Journal of the European Communities L277, 10–14.

European Union (EU), Publication of an application pursuant to Article 6(2) of Council Regulation (EC) No. 510/2006 on the protection of geographical indications and designations of origin for agricultural products and foodstuffs (2010/C 312/15) 'ΞΥΓΑΛΟΣΗΤΕΙΑΣ' (XYGALO SITEIAS)/'ΞΙΓΑΛΟΣΗΤΕΙΑΣ' (XIGALO SITEIAS). Official Journal of the European Union C312, 25–30 (2010).

European Union (EU), Commission Regulation No. 766/2011 of July 29, 2011 entering a name in the register of protected designations of origin and protected geographical indications [Ξύγαλο Σητείας (Xygalo Siteias)/Ξίγαλο Σητείας (Xigalo Siteias) (PDO)] Official Journal of the European Union L 200/12. August 3, 2011.

European Union (EU), Commission Regulation No. 433/2013 of May 7, 2013 approving non-minor amendments to the specification for a name entered in the register of protected designations of origin and protected geographical indications (Κοπανιστή (Kopanisti) PDO. Official Journal of the European Union L 129/17. May 14, 2013.

Fletouris, D., N.A. Botsoglou, I.E. Psomas and A. Mantis. 1998. Rapid determination of cholesterol in milk and milk products by direct saponification and capillary gas chromatography. J. Dairy Sci. 81: 2833–2840.

Fletouris, D., M. Govari and E. Botsoglou. 2014. The influence of retail display storage on the fatty acid composition of modified atmosphere packaged Graviera Agraphon cheese. Int. J. Dairy Techn. 67: 1–9.

Foundation of Baron Michail Tositsa. 1993. Booklet of the cheese plant [Milk, Cheese and our Cheeses]. Metsovo, Greece.

Georgala, A., E. Moschopoulou, A. Aktypis, T. Massouras, E. Zoidou, I. Kandarakis and E. Anifantakis. 2005. Evolution of lipolysis during the ripening of traditional Feta cheese. Food Chem. 93: 73–80.

Georgala, A., E. Anastasaki, D. Xitos, D. Kapogiannis, T. Malouxos, T. Massouras and I. Kandarakis. 2012. The nutritional value of Trahanas: a fermented milk-cereal nutritional Greek food. In the Special Issue of the International Dairy Federation 1201: IDF International Symposium on Sheep, Goat and other non Cow milk. 151–153.

Georgala, A. 2013. The nutritional value of two fermented milk/cereal foods named 'Greek Trahanas' and 'Turkish Tarhana'. A Review. J. Nutr. Disorders Ther. S11:002. doi: 10.4172/2161-0509.S11-002.

Gerasi, E. Litopoulou-Tzanetaki and N. Tzanetakis. 2003. Microbiological study of Manura, a hard cheese made from raw ovine milk in the Greek island Sifnos. Int. J. Dairy Technol. 56: 117–122.

Greek Code of Foods, Beverages and Objects of Common Use. 2012. Vol. 2, General Chemical State Laboratory, Ministry of Economy and Finance, Hellenic Republic. Greece. www.gcsl.gr.

Kalogridou-Vassiliadou, D., N. Tzanetakis and E. Litopoulou-Tzanetaki. 1994. Microbiological and physicochemical characteristics of 'Anthotyro', a Greek traditional whey cheese. Food Microb. 11: 15–19.

Kaminarides, S., E. Anifantakis and E. Alichanidis. 1990. Ripening changes in Kopanisti cheese. J. Dairy Res. 57: 271–279.

Kaminarides, S., V. Siaravas and I. Potetsianaki. 1995. Comparison of two methods of making kneaded plastic cheese from ewe milk. Lait. 75: 181–189.

Kaminarides, S. and E. Anifantakis. 2004. Characteristics of set type yoghurt made from caprine or ovine milk and mixtures of the two. Int. J. Food Sci. Technol. 39: 319–324.

Kaminarides, S., P. Stamou and T. Massouras. 2007. Comparison of the characteristics of set type yoghurt made from ovine milk of different fat content. Int. J. Food Sci. Technol. 42: 1019–1028.

Kaminarides, S., K. Nestoratos and T. Massouras. 2013. Effect of added milk and cream on the physicochemical, rheological and volatile compounds of Greek whey cheeses. Small Rum. Res. 113: 446–453.

Kandarakis, I. 1981. [Contribution to the study of manufacturing technology of Manouri cheese by the traditional procedure and by means of ultrafiltration]. PhD Thesis. Agricultural University of Athens, Athens, Greece.

Kandarakis, I., E. Moschopoulou, G. Moatsou and E. Anifantakis. 1998. Effect of starters on gross and microbiological composition and organoleptic characteristics of Graviera Kritis cheese. Lait. 78: 557–568.

Kandarakis, I., G. Moatsou, A. Georgala, S. Kaminarides and E. Anifantakis. 2001. Effect of draining temperature on the biochemical characteristics of Feta cheese. Food Chem. 72: 369–378.

Kandarakis, I., T. Massouras, E. Stamati and E. Anifantakis. 2002. Mineral content and trace elements in some Greek dairy products. 26th IDF World Dairy Congress (CONGRILAIT 2002)—Science and Technology Session. September 24–27. Paris, France.

Kandarakis, I., T. Massouras, E. Manolopoulou, D. Karagiorgos, N. Kritikos and E. Anifantakis. 2005a. Study of physicochemical composition and nutrient profile of traditional "Arseniko"cheese of Naxos. In: Proceedings of 2nd International Conference of Traditional Mediterranean Diet. Mediet 2005, Athens, Greece.

Kandarakis, I., T. Massouras, E. Manolopoulou, D. Karagiorgos, N. Kritikos and E. Anifantakis. 2005b. Study of physicochemical composition and nutrient profile of traditional Graviera of Paros. *In*: Proceedings of 2nd International Conference of Traditional Mediterranean Diet. Mediet 2005, Athens, Greece.

Karali, F., A. Georgala, T. Massouras and S. Kaminarides. 2013. Volatile compounds and lipolysis levels of Kopanisti, a traditional Greek raw milk cheese. J. Sci. Food Agric. 93: 1845–1851.

Katsiari, M.C. and L.P. Voutsinas. 1994a. Manufacture of low fat Feta cheese. Food Chem. 49: 53–60.

Katsiari, M.C. and L.P. Voutsinas. 1994b. Manufacture of low fat Kefalograviera cheese. Int. Dairy J. 533–553.

Katsiari, M.C., L.P. Voutsinas, E. Alichanidis and I.G. Roussis. 2000a. Lipolysis in reduced sodium Feta cheese made by partial substitution of NaCl by KCl. Int. Dairy J. 10: 369–373.

Katsiari, M.C., E. Alichanidis, L.P. Voutsinas and I.G. Roussis. 2000b. Proteolysis in reduced sodium Feta cheese made by partial substitution of NaCl by KCl. Int. Dairy J. 10: 635–646.

Katsiari, M., L. Voutsinas and E. Kondyli. 2002. Manufacture of yoghurt from stored frozen sheep milk. Food Chem. 77: 413–420.

Katsiari, M., E. Kondyli and L. Voutsinas. 2009. The quality of Galotyri type cheese made with different starter cultures. Food Control. 20: 113–118.

Kehagias, C., G. Kalanzopoulos, G. Kotouza-Markopoulou and E. Anifantakis. 1980. [Manufacture of yoghurt from concentrated frozen ewe milk]. Agric. Res. 4: 83–93.

Kehagias, C. 1982. [Manufacture of yoghurt from refrigerated ewe or cow milk]. Agric. Res. 6: 281–288.

Kehagias, C., A. Zervoudaki and C. Parlama. 1989. Influence of composition and additives on properties of set type yoghurt from goat milk. Small Rum. Res. 2: 35–45.

Kehagias, C., L. Kalavrithinos and C. Triadopoulou. 1992. Effect of pH on the yield and solids recovery of strained yogurt from goat and cow milk. Cult. Dairy Prod. J. 8: 1014.

Kondyli, E., M. Katsiari and L.P. Voutsinas. 2008. Chemical and sensory characteristics of Galotyri type cheese made using different producers. Food Control. 19: 301–307.

Kondyli, E., E. Pappa and A.M. Vlachou. 2012. Effect of package type on the composition and volatile compounds of Feta cheese. Small Rum. Res. 108: 95–101.

Kondyli, E., T. Massouras, M.C. Katsiari and L.P. Voutsinas. 2013. Lipolysis and volatile compounds of Galotyri type cheese made using different procedures. Small Rum. Res. 113: 432–436.

Laskaridis, K., A. Serafeimidou, S. Zlatanos, E. Gylou, E. Kontorepanidou and A. Sagredos. 2013. Changes in fatty acid profile of Feta cheese including conjugated linoleic acid. J. Sci. Food Agric. 93: 2130–2136.

Lekkas, C., A. Kakouri, E. Paleologos, L. Voutsinas, M. Kontominas and J. Samelis. 2006. Survival of *Escherichia coli* O157:H7 in Galotyri cheese stored at 4°C and 12°C. Food Microb. 23: 268–276.

Lioliou, K., E. Litopoulou-Tzanetaki, N. Tzanetakis and R.K. Robinson. 2001. Changes in the microbiota of Manouri, a traditional Greek whey cheese, during storage. Int. J. Dairy Technol. 54: 100–106.

Litopoulou-Tzanetaki, E., N. Tzanetakis and A. Vafopoulou-Mastrojiannaki. 1993. Effect of lactic starter on microbiological, chemical and sensory characteristics of Feta cheese. Food Microb. 10: 31–41.

Litopoulou-Tzanetaki, E. and N. Tzanetakis. 2011. Microbial characteristics of Greek traditional cheeses. Small Rum. Res. 101: 17–32.

Lopez-Exposito, I., L. Amigo and I. Recio. 2012. A mini review on health and nutritional aspects of cheese with a focus on bioactive peptides. Dairy Sci. Technol. 92: 419–438.

Mallatou, H., C.P. Pappas and L.P. Voutsinas. 1994. Manufacture of Feta cheese from sheep milk, goat milk or mixtures of these milks. Int. Dairy J. 4: 641–664.

Mallatou, H., E.C. Pappa and T. Massouras. 2003. Changes in the free fatty acids during the ripening of Teleme cheese made with ewe, goat, cow milk or a mixture of ewe and goat milk. Int. Dairy J. 13: 211–219.

Mallatou, H., E. Pappa and V. Boumba. 2004. Proteolysis in Telemes cheese made from ewe, goat or a mixture of ewe and goat milk. Int. Dairy J. 14: 977–987.

Mallatou, H. and E.C. Pappa. 2005. Comparison of the characteristics of Teleme cheeses made from ewe, goat and cow milk or a mixture of ewe and goat milk. Int. J. Dairy Technol. 58: 158–163.

Massouras, T., E. Pappa and H. Mallatou. 2006. Headspace analysis of volatile flavour compounds of Teleme cheese made from sheep and goat milk. Int. J. Dairy Technol. 59: 250–256.

Michaelidou, A., E. Alichanidis, H. Urlaub, A. Polychroniadou and G. Zerfiridis. 1998. Isolation and identification of some major water-soluble peptides in Feta cheese. J. Dairy Sci. 81: 3109–3116.

Michaelidou, A., M.C. Katsiari, E. Kondyli, L.P. Voutsinas and E. Alichanidis. 2003. Effect of a commercial adjunct culture on proteolysis in low fat Feta type cheese. Int. Dairy J. 13: 179–189.

Michaelidou, A., E. Alichanidis, A. Polychroniadou and G. Zerfiridis. 2005. Migration of water soluble nitrogenous of Feta cheese from the cheese blocks into the brine. Int. Dairy J. 15: 663–668.

Moatsou, G., I. Kandarakis, A. Georgala, E. Alicahnidis and E. Anifantakis. 1999. Effect of starters on proteolysis of Graviera Kritis cheese. Lait. 79: 303–315.

Moatsou, G., I. Kandarakis, E. Moschopoulou, E. Anifantakis and E. Alichanidis. 2001. Effect of technological parameters on the characteristics of Kasseri cheese made from raw or pasteurized ewe milk. Int. J. Dairy Technol. 54: 69–77.

Moatsou, G., T. Massouras, I. Kandarakis and E. Anifantakis. 2002. Evolution of proteolysis during the ripening of traditional Feta cheese. Lait. 82: 601–611.

Moatsou, G., E. Moschopoulou, A. Georgala, E. Zoidou, I. Kandarakis, S. Kaminarides and E. Anifantakis. 2004a. Effect of artisanal liquid rennet from kid and lamb abomasa on the characteristics of Feta cheese. Food Chem. 88: 517–525.

Moatsou, G., E. Moschopoulou and E. Anifantakis. 2004b. Effect of different manufacture parameters on the characteristics of Graviera Kritis cheese. Int. J. Dairy Tech. 57: 215–220.

Moatsou, G., E. Moschopoulou, D. Mollé, V. Gagnaire, I. Kandarakis and J. Léonil. 2008. Comparative study of the protein fraction of goat milk from indigenous Greek breed and from international breeds. Food Chem. 106: 509–520.

Moatsou, G. and A. Govaris. 2011. White brined cheeses: a diachronic exploitation of small ruminants milk in Greece. Small Rum. Res. 101: 113–121.

Moatsou, G., E. Moschopoulou, A. Beka, P. Tsermoula and D. Pratsis. 2015. Effect of natamycin-containing coating on the evolution of biochemical and microbiological parameters during the ripening and storage of ovine hard Gruyere type of cheese. Int. Dairy J. 50: 1–8.

Moschopoulou, E., I. Kandarakis and E. Anifantakis. 2007. Characteristics of lamb and kid artisanal liquid rennet used for traditional Feta cheese manufacture. Small Rum. Res. 72: 237–241.

Moschopoulou, E., E. Zoidou, L. Sakkas, C. Kalathaki, A. Liarakou, A. Stamos, A. Chatzigeorgiou and G. Moatsou. 2015. Physicochemical, textural and antioxidant properties of set type yoghurt made from sheep or goat milk. Poster presentation in 7th IDF International Symposium on sheep, Goat and other non cow milk. Limassol, Cyprus, 23–25 March 2015.

Nega, A., C. Kehagias and G. Moatsou. 2012. Traditional cheeses: Effect of cheesemaking technology on the physicochemical composition and mineral contents. In the Special Issue of IDF 1201. IDF International Symposium on Sheep, Goat and Other Non-Cow Milk, 129–131.

Nega, A. and G. Moatsou. 2012. Proteolysis and related enzymatic activities in ten Greek cheeses varieties. Dairy Sci. Technol. 92: 57–73.

Nikolaou, E., N. Tzanetakis, E. Litopoulou-Tzanetaki and R.K. Robinson. 2002. Changes in the microbiological and chemical characteristics of an artisanal low fat cheese made from raw ovine milk during ripening. Int. J. Dairy Technol. 55: 12–17.

Papadimitriou, C.G., A. Vafopoulou-Mastrojiannaki, S. Silva, A.-M. Gomes, F.X. Malcata and E. Alichanidis. 2007. Identification of peptides in traditional and probiotic sheep milk yoghurt with angiotensin I-converting enzyme (ACE)-inhibitory activity. Food Chem. 105: 647–656.

Papageorgiou, D.K., A. Abrahim and S. Doundoumakis. 1998. Chemical and bacteriological characteristics of Pichtogalo Chanion cheese and mesophilic starter cultures for its production. J. Food Prot. 61: 688–692.

Papanikolaou, Z., M. Hatzikamari, P. Georgakopoulos, M. Yiangou, E. Litopoulou-Tzanetaki and N. Tzanetakis. 2012. Selection of Dominant NSLAB from a Mature Traditional Cheese According to their Technological Properties and *in vitro* Intestinal Challenges. Food Microb. Saf. 77: 298–306.

Pappa, E., I. Kandarakis, G. Zerfiridis, E. Anifantakis and K. Sotirakoglou. 2006a. Influence of starter cultures on the proteolysis of Teleme cheese made from different types of milk. Lait 86: 273–290.

Pappa, E., I. Kandarakis, E. Anifantakis and G. Zerfiridis. 2006b. Influence of types of milk and culture on the manufacturing practices, composition and sensory characteristics of Teleme during ripening. Food Control 17: 570–581.

Pappa, E., C. Pappas and P.F. Surai. 2006c. Selenium content in selected foods from the Greek market and estimation of the daily intake. Sci. Total Environ. 372: 100–108.

Pappa, E., I. Kandarakis and H. Mallatou. 2007. Effect of different types of milks and cultures on the rheological characteristics of Teleme cheese. J. Food Engin. 79: 143–149.

Prodromou, K., P. Thasitou, E. Haritonidou, N. Tzanetakis and E. Litopoulou Tzanetaki. 2001. Microbiology of 'Orinotyri', a ewe milk cheese from the Greek mountains. Food Microb. 18: 319–328.

Psoni, L., N. Tzanetakis and E. Litopoulou-Tzanetaki. 2003. Microbiological characteristics of Batzos, a traditional Greek cheese from raw goat milk. Food Microb. 20: 575–582.

Psoni, L., N. Tzanetaki and E. Litopoulou-Tzanetaki. 2006. Characteristics of Batzos cheese made from raw, pasteurized and/or pasteurized standardized goat milk and a native culture. Food Control 17: 533–539.

Rogga, K.J., J. Samelis, A. Kakouri, M. Katsiari, I. Savvaidis and M. Kontominas. 2005. Survival of *Listeria monocytogenes* in Galotyri, a traditional Greek soft acid curd cheese, stored aerobically at 4°C and 12°C. Int. Dairy J. 15: 59–67.

Serafeimidou, A., S. Zlatanos, G. Kritikos and A. Tourianis. 2013. Change of fatty acid profile, including conjugated linoleic acid (CLA) content during refrigerated storage of yogurt made of cow and sheep milk. J. Food Compos. Anal. 31: 24–30.

Tamime, A.Y., M. Wszolek, R. Božanić and B. Özer. 2011. Popular ovine and caprine fermented milks. Small Rum. Res. 101: 2–16.

Tamime, A.Y., M. Hickey and D. Muir. 2014. Strained fermented milks: a review of existing legislative provisions, survey of nutritional labelling of commercial products in selected markets and terminology of products in some selected countries. Int. J. Dairy Technol. 67: 305–333.

Theodorou, S., L. Sakkas, E. Zoidou, A. Stamos, A. Chatzigeorgiou, G. Moatsou, O. Gerogianni and E. Moschopoulou. 2015. Effect of fortification with whey protein concentrates on rheology and sensory profile of set type yoghurt made from goat milk. Poster presentation in 7th IDF International Symposium on sheep, goat and other non cow milk. Limassol, Cyprus, Greece. March 23–25, 2015.

Tsorotioti, S., C. Nasopoulou, M. Detopoulou, E. Sioriki, C.A. Demopoulos and I. Zabetakis. 2014. *In vitro* antiatherogenic properties of traditional Greek cheese lipid fractions. Dairy Sci. Technol. 94: 269–281.

Tzanetakis, N., E. Litopoulou-Tzanetaki and K. Manolkidis. 1987. Microbiology of Kopanisti, a traditional Greek cheese. Food Microb. 4: 251–256.

Vafopoulou, A., E. Alichanidis and G. Zerfiridis. 1989. Accelerated ripening of Feta cheese, with heat-shocked cultures or microbial proteinases. J. Dairy Res. 56: 285–296.

Valsamaki, K., A. Michaelidou and A. Polychroniadou. 2000. Biogenic amine production in Feta cheese. Food Chem. 71: 259–266.

Veinoglou, B. and E. Anifantakis. 1981. [Dairy science]. Karaberopoulos Publ., Athens, Greece.

Vitalioti, K. 2012. [Manufacture of yoghurt with microfiltrated milk]. M.S. Thesis. Agricultural University of Athens. Athens, Greece.

Voutsinas, L., M. Katsiari, C.P. Pappas and H. Mallatou. 1996. 1. Production of yoghurt from sheep milk which had been concentrated by reverse osmosis and stored frozen. 2. Compositional, microbiological, sensory and physical characteristics of yoghurt. Food Res. Int. 29: 411–416.

Xanthopoulos, V., A. Polychroniadou, E. Litopoulou-Tzanetaki and N. Tzanetakis. 2000. Characteristics of Anevato cheese made from raw or heat-treated goat milk inoculated with a lactic starter. Lebensm.-Wiss. u.-Technol. 33: 483–488.

Zerfiridis, G. 2001a. [Technology of Dairy Products: Cheesemaking], 2nd ed. pp. 155–157. Yiachoudi Publications, Thessalonica, Greece.

Zerfiridis, G. 2001b. [Technology of Dairy Products: Fermented milks, Ice cream, Cream, Butter]. 2nd ed. Yiachoudi Publications, Thessalonica, Greece.

Zerfiridis, G., A. Vafopoulou-Mastrogiannaki and E. Litopoulou-Tzanetaki. 1984. Changes during ripening of commercial Gruyère type cheese. J. Dairy Sci. 67: 1397–1405.

Zlatanos, S., K. Laskaridis, C. Feist and A. Sagredos. 2002. CLA content and fatty acid composition of Greek Feta and hard cheeses. Food Chem. 78: 471–477.

Zlatanos, S. and K. Laskaridis. 2009. Variation in the conjugated linoleic acid content of three traditional Greek cheeses during a one year period. J. Food Q. 32: 84–95.

Zoidou, E., T. Massouras, A. Aktypis, I. Kandarakis and E. Anifantakis. 2011. Microbiological and physiochemical characteristics of Tinos traditional cheese "Tiraki". 1st Hellenic Food Congress, HVMS, Thessaloniki, Greece. November 11–13, 2011.

Zoidou, E., N. Plakas, D. Giannopoulou, M. Kotoula and G. Moatsou. 2015. Effect of supplementation of brine with calcium on the Feta cheese ripening. Int. Dairy J. 658: 420–426.

Index

A

Acidity 178
Additives 156, 157, 165
Amino Acids 74, 75, 79, 88, 93, 94
Antibacterial 178, 183, 184
Antioxidants 151, 152, 162, 163, 178, 181–184, 227, 229, 232–234, 236
Aroma 40, 41, 44, 46, 50–55
Azeitão 201–205, 207, 218

B

Baladi Bread 103, 104, 110–114, 116, 117
Bean Stew 110, 111, 113, 114
Beef 241, 242, 249, 253, 254, 257, 259
Beekeeping 172–174, 176, 177, 181, 183, 185–187
Bees 171–174, 178, 180, 185–188
Blanching 227, 230, 232–234, 236
Bohinj 121–125
Bone 35, 36, 38–45, 48
Bovec 121, 129–131
Brine 3, 10, 13, 268, 269, 271, 273–277, 280–285, 289, 293, 294
Buffalo 2, 12–14, 18, 23, 25–28

C

Calcium 198, 199, 203, 205, 210–214, 216
Caldo Verde 226
Canned Mackerel 65, 68, 81, 83, 84, 86, 88–90, 92–95
Canning 75, 81–86, 88, 91, 93, 95, 96, 104, 106, 108, 109, 117
Carbohydrates 176–179, 182, 183
Carotenoids 227, 234–236
Casein 2, 6, 12–14, 16–21, 23, 24, 196–200, 209–219
Catshark 67, 68, 70–74
Cemen 249–253
Cereals 102–104, 106, 107, 114, 116, 117
Cheese 1–8, 10–19, 21, 22, 121–135, 137, 138, 267–303, 307
Chlorophylls 227, 232, 234, 236
Chorizo 156, 160, 161, 163–165

Clotting 198, 202, 209–216
Coagulum 2, 8, 9, 21, 27, 28
Colour 232–235, 242–244, 246, 247, 250, 252, 254–256, 259
Cooking 105, 106, 108, 110, 227, 233, 235, 236, 254–256, 258–262
Cow 2, 13, 14, 19, 23, 25–28, 124, 126, 127, 129, 135, 136, 138, 139, 267, 273, 274, 276, 282, 284, 286, 289, 293, 295, 296, 299, 300, 304, 306–310, 312
Curd 196, 198, 202, 203, 206, 207, 209, 212, 214–216, 218–220, 269, 271, 273–276, 280–283, 285, 287, 289–291, 293–296, 298–300, 303, 310

D

Dalmatinska Panceta 56
Dalmatinska Pečenica 56
Dalmatinski Pršut 35–39, 48–50, 53–56
Dalmatinski Šokol 56
Dolenjski 132, 133
Dough 106, 110, 113
Dried Litão 65, 68–71, 74–76
Drniški Pršut 38–41, 48
Dry-Cured Ham 35, 38, 48–60, 141, 142, 144, 145, 148–150, 154
Drying 35, 36, 38, 41, 43–45, 47, 51, 56, 68–71, 73–75, 77–79, 82, 86, 141, 144, 148, 157, 159–162, 165, 233, 234, 242–245, 249, 250, 253

E

Electrical Conductivity 176, 177
Enzymes 199, 206, 209–216

F

Faba Beans 102–106, 108–110, 113, 115–117
Falafel 109–112, 114
Fat 123–127, 129, 130, 132, 133, 135–139, 142–144, 146–149, 151, 156, 157, 159–161, 164, 165, 196–198, 200, 202–204, 213, 218, 219, 241, 242, 244, 247, 249–255, 257, 259, 260

Fatty Acids 49–55, 58, 66, 74, 75, 79, 81, 82,
      87, 88, 90, 146–152, 157, 163, 197–200, 204
Fermentation 104, 106, 107, 109, 155, 156,
      159–161, 241–243, 245, 247
Feta 267–269, 271–274, 296, 300
Fibre 111
Fish 65, 88, 91–93
Flavour 37, 50, 142, 145, 146, 148–150, 153,
      156, 159, 162, 163, 230, 233–235, 240,
      242–244, 250, 259
Folate 114–117
Freezing 104, 106, 109
Fresh Cut 227, 229, 230, 237
Frozen Storage 227, 230–233
Fuet 156, 161, 163, 164

G

Galega Kale 226–237
Germination 104–109, 114–117
Goat 2, 23, 25, 27, 267, 269, 274, 276, 280–284,
      287, 289, 294–300, 302–304, 307–312, 314
Grana Padano 1–5, 8, 11

H

High Intramuscular Fat 142, 143
Honey 171–191
Honeydew 174, 175, 177–181, 186
Hydroxymethylfurfural 174, 183

I

Iberian 142–155, 157
Istarski Pršut 35, 38, 41–45, 48–50, 52–55

K

Karst 127, 131, 132
Kasseri 267, 289–291, 293
Kavurma 244, 253–256
Kefalograviera 278, 283, 284
Kefalotyri 276–280, 282–284, 300, 303
Kislo Mleko 139
Kočevski Gozdni Med 186
Kofte 240, 256–259, 262, 263
Koshari 110, 111, 113, 114
Kraški Med 186, 187
Krčki Pršut 35, 38, 45–48

L

Lactic Acid Bacteria 6, 241–243, 248, 252, 253
Lactose 5, 7, 10, 15, 19, 23, 25
Ladotyri Mytilinis 278, 280, 281
Legumes 102–104, 106–109, 114–116
Lipids 66, 72, 74, 76, 79, 80, 82, 86–88, 90–92
Loins 75–78

M

Maillard 176, 181, 183
Mascarpone 1, 2, 18–22
Mead 187–191
Meat 141–157, 159–162, 164, 165, 240–244,
      246, 248–250, 253–262
Meatballs 240, 256–262
Micelles 199, 209–215
Milk 1–9, 11–19, 21–25, 27, 121, 122, 124–129,
      131–139, 196–220, 267–270, 273–276,
      280–287, 289, 290, 293–214
Minerals 86, 93, 95, 96, 103, 106, 109, 110, 114,
      176, 177, 179
Modified Atmosphere Packaging 233
Mohant 121–124
Moisture 202, 214–220
Mozzarella 1, 2, 11, 12, 15, 17, 18
MUFA 146, 148, 151–153
Multifloral 175, 176, 179, 181, 182, 184, 185,
      187
Muxama 65, 67, 68, 75–81, 96

N

Nabet Soup 106, 110, 114
Nanos 127–129
Nectar 171, 174, 175, 177, 178, 180, 181, 186,
      187
Nutritional 226–230, 234–236

P

Packaging 39, 41, 45, 48, 252, 253, 262
Parmigiano Reggiano 1, 2, 4, 5, 8, 9, 11
Pasta Filata 289
Pastirma 240, 249–253
Pig 142–157, 160, 161
Pollen 175, 176, 180, 181, 185, 187
Pork Meat 42, 46
Pregreta Smetana 136, 137
Pressing 37, 38, 41, 43, 44, 47
Protected Designation of Origin 1
Protein 38, 48, 51, 57–59, 72, 74, 79, 80, 84,
      86–89, 91–93, 96, 122, 123, 125, 127, 130,
      132, 133, 135–139, 143, 147, 149–151, 162,
      164, 240, 247, 250, 252, 259, 260
PUFA 66, 72, 79, 87, 88, 91, 146, 148–152

R

Refrigerated Storage 230, 231
Rennet 3, 5, 6, 9, 122, 124, 126, 128, 129, 137,
      215, 216, 273, 275, 276, 280–283, 287, 289,
      294, 296, 298, 299
Ricotta 1, 22–29

Ripening 35, 38–41, 43–45, 47, 48, 56, 142, 145, 153, 155–157, 159–164, 202, 203, 207, 210, 214, 217, 218, 220, 268–271, 273, 274, 276, 277, 280–282, 284–286, 289–292, 296, 297, 299, 301

S

Salchichon 156, 161, 164
Salt 67, 70, 74–76, 78, 79, 123, 125, 127, 129, 130, 132, 133, 135, 138, 141, 142, 144, 156, 157, 159–161, 199, 200, 209, 220, 241, 242, 247, 250, 251, 254, 255, 257–260, 270, 271, 273–277, 281–286, 289, 291, 292, 295–301, 303, 311, 312
Salting 35, 37–39, 41, 43–48, 53, 55–57, 68–70, 73, 75, 77–79, 82
Sausages 141, 146, 148, 155–165
Serpa 201–207, 217, 218
Serra da Estrela 201–203, 205–208, 217, 218
SFA 146, 148, 150, 151
Sfela 271, 274, 275
Sheep 2, 23, 25, 27, 28, 127, 129, 131, 132, 196–209, 212, 215, 216, 219, 220, 267, 269, 273, 274, 276, 280–284, 287, 289, 294–300, 302–312, 314
Shredded 226, 227, 229, 230, 233, 237
Škrlupec 136
Skuta 121, 137–139
Slovenski Med 185, 186
Smoking 38, 41, 45, 47, 55, 56

Soaking 104–110
Soup 226
Starter 2, 5–13, 15, 17, 18, 241–243, 245, 247, 252
Sterilization 81, 85–88, 91, 93
Sucuk 240–248

T

Telemes 271, 273, 274, 300
Texture 232–235, 241, 244, 250, 252, 259, 260
Thistle Flower 206, 209, 210, 215
Tolminc 121, 125–127
Trnič 133–135
Tuna 65, 67, 68, 75–81, 83, 87, 93, 94, 96

V

Vegetables 226, 227, 229–232, 234–236
Vitamins 74, 79, 86, 87, 91, 96, 103, 106, 107, 109–111, 114–116, 227, 228, 233–235

W

Water Content 177, 186, 187
Whey 1, 2, 5, 7–10, 12–15, 17, 19–28, 202, 203, 209, 213–218, 271, 273, 276, 281, 285, 293, 295, 298–304, 306, 309, 310

Y

Yield 196, 199, 200, 214, 218–220
Yoghurt 267, 269, 283, 304–311

Printed and bound by CPI Group (UK) Ltd, Croydon, CR0 4YY

01/11/2024

01782622-0008